Economic Aspects

of

Atomic Power

Prepared by the Cowles Commission for Research in Economics, The University of Chicago; initiated by the Committee on the Social and Economic Aspects of Atomic Energy of the Social Science Research Council

* * *

Other books on the social aspects of atomic energy, initiated by the Committee:

Coale: The Problem of Reducing Vulnerability to Atomic Bombs

Cottrell and Eberhart: American Opinion on World Affairs in the Atomic Age

ECONOMIC ASPECTS

OF

ATOMIC POWER

*An Exploratory Study
under the Direction of*

SAM H. SCHURR and
JACOB MARSCHAK

PUBLISHED FOR THE COWLES COMMISSION
FOR RESEARCH IN ECONOMICS
BY PRINCETON UNIVERSITY PRESS
1950

PRINTED IN THE UNITED STATES OF AMERICA
BY THE MAPLE PRESS COMPANY, YORK, PA.

CONTRIBUTORS

Research Associates

SAM H. SCHURR

GEORGE PERAZICH

EDWARD BOORSTEIN

Research Consultants

HERBERT A. SIMON

HAROLD H. WEIN

MILTON F. SEARL

PREFACE

THE FUTURE may hold many peaceful economic uses of nuclear processes. The present study is confined to those applications that seem least remote today. Among these applications, we have dealt only with those based on a continually controlled release of energy within a permanent structure (the "reactor") rather than on explosions. The released energy results from the fission (splitting) of heavy atoms such as plutonium. No structural materials are available that could resist the high temperatures obtained when fission energy is released uncontrolled as when a plutonium bomb explodes. This seems to exclude the possibility of constructing reactors for a controlled release of another type of atomic energy, that resulting from the fusion (bringing together) of light atoms such as hydrogen isotopes; for such fusion can start only at very high temperatures and will therefore probably be initiated only by the explosion of a fission bomb. By limiting our study to the controlled release of energy we leave out of consideration the peaceful use of exploding plutonium or hydrogen bombs for, say, the leveling of mountains or (as has been suggested) the melting of the arctic ice-cap. We are also not concerned in this book with the economic consequences of the potential destruction that atomic weapons may bring about or with the economic implications of the defense measures—such as the decentralization of cities—that the existence of such weapons may hasten.

Those applications that today seem likely to become practical first may not, in the long run, be the most important ones. The generation of electricity from the heat created by nuclear fission, while not at present completely worked out, is generally thought feasible; so is the transportation of (relatively) low temperature heat over short distances, as for residential heating. On the other hand, the conversion of fission energy into electric energy without passing through a steam or gas turbine, or the direct use of the high temperature heat of a nuclear reactor (for example, to melt metals) may not be feasible at all. But, if feasible, the economic importance of these applications would overshadow that of merely substituting a nuclear reactor for the coal or oil furnace of an otherwise essentially unchanged power plant. Nor do we know what uses will be made of cheap radioactive elements and compounds, another product of nuclear fission. Perhaps the most important though less immediate applications will be due to the new knowledge of matter, both dead and living, which scientists hope to acquire by using radioactive "tracers." For example: if, helped by these new

vii

research tools, we learn to imitate the action of green leaves in absorbing the sun's energy, both uranium and coal may, at some time and for some countries, acquire a formidable competitor, and the effect on food supplies may be even more important.

Even when limited to applications that seem least remote today, the study is still only exploratory. This must be emphasized. The technology of power-making reactors is still in an experimental stage. But even if all the technological data relevant to the production of electricity from nuclear fuel were available, they would not solve the economist's problem but only secure him a better start. The economist has first to estimate, for different areas of the world, the price relation between electricity produced from conventional fuels and water power and electricity produced from the new fuel, whose main virtue is its enormous energy content per unit of weight and hence its cheap transportability over great distances. This price relation helps to evaluate the role of the new source of power in various countries and various industries. Such evaluation is a vast and laborious task, widely ramified by the diversities of geography and of industrial technology. But even if accomplished, this task would still not complete the economist's query. We are interested in the potential sources of demand for atomic energy mainly because we want ultimately to judge the overall effect of the new invention upon the economy of the nation and of the world. This effect works itself out in a sequence of complicated repercussions of one economic sector upon another.

Thus the study is exploratory in a double sense. First, technological data which include future trends in the techniques of generating and using energy are incomplete. One has, therefore, to use a hypothetical range instead of a single figure. This often results in a set of alternatives. The final choice between them must be made later. A very significant figure in this range is the estimated minimum cost for producing atomic power. Working with this figure, which we know to be the lowest conceivable cost for atomic power when produced by techniques now envisaged, we gain a general picture of the scope of the economic changes which could result, at best, from the use of the new energy source. The various determining factors are thus assigned their approximate relative weights. The extremes of optimism and pessimism are put into their proper places.

Second, the book is exploratory also in the sense that it involves a new attempt to formulate an economic theory of the effects of an invention. Since this subject is complicated, its treatment could only be tentative.

What we said above about the economist's task, starting with the technological data of the new invention and ending with the evaluation of its effects upon the economy as a whole, has determined the outline of the book. It begins with the questions of technical feasibility, the availability of raw materials, and the possible cost and other economic characteristics of atomic power (Chapter I). This cost is then compared (in Chapter II), for various areas of the world, with the cost of electricity from conventional sources. This analysis is followed by the study of the potential applicability of atomic power in several industries which are, or could become, important consumers of electricity or heat (Chapters III–XII): the production of aluminum, chlorine and caustic soda, phosphate fertilizer, cement, brick, flat glass, iron and steel, railroad transportation, and residential heating. (We were not able to complete our preliminary studies of the usability of atomic energy in several industries: ferro-alloys, copper, lead, zinc, and pulp and paper; in the production of nitrogen fertilizers; in some phases of agriculture including irrigation; and in ocean transportation.) With the empirical background provided by the regional and industrial analysis, the study proceeds to sketch a theoretical outline for an estimate of the economic effects of atomic power—first, on the economy of a highly industrialized country like the United States (Chapter XIII); then, on the industrialization of so called backward areas of the world (Chapter XIV). But our limited resources did not permit a close analysis by individual countries (which would correspond to the analysis by industries in Chapters III–XII). Such individual country-by-country analyses would be a very useful next step in studying economic implications of atomic power.

The analysis is supported by four maps of the world. Map 1 gives, for various areas, the cost of electric power generated from conventional sources. For thermal electricity, these are cost estimates for power generated in a modern thermal plant on the hypothesis that the construction cost of such a plant relative to that of an atomic power plant is the same as in the United States. Maps 2 and 3 give the world distribution of water power and fuel resources. As explained in Chapter II, these maps help to judge the degree to which the new source of power may compete in a given area with old sources. Obviously this gives only a partial answer to the question of potential markets for atomic power. Even where there is a comparative cost advantage in favor of atomic power, the demand for atomic or for any other kind of power may be small, depending on the density of population and its purchasing power or else on the presence of particular raw materials or markets that may give rise to a demand for electricity if it becomes sufficiently cheap. Accordingly, Map 4 gives the distribution of people over the globe. Population density is a determinant of demand for electricity (given the cost), not only because of the demand of households to run electric lamps and possibly other domestic appliances but also because of the power needs of local transportation and other public services, of retail business, and of other industries serving the local market. A glance at Map 4 reveals its main shortcoming as a demand indicator: a similarly densely populated area in Europe and in Asia would be given equal weight. It is, of course, not the population per square mile but the purchasing power (existing or potential) per square mile that really matters. Unfortunately, a corresponding refinement of Map 4 would require detailed information on the geographical distribution of real income and the potentialities for its growth, by relatively small areas (since electricity cannot be economically transmitted over a large radius). Such data are not available at present even for the existing levels of income except for a few areas of the world. For the United States, the dependence of the electricity demand of private homes upon both price and real income is studied in Chapter XIII.

Another reason why population density, or even the density of "dollars per square mile," is not a sufficient indicator of demand for electricity is the role of industries that produce goods or services for consumers outside of the area served by the power plant in question. Attracted by cheap energy, such industries may in turn draw more people and more buying power into the area. For the United States, some of the major power-consuming industries are studied in Part Two, and the findings may help us in judging for other areas. But, as already stated, the list of industries studied is very incomplete.

In short, if Maps 1 to 3 indicate for a given area a high price for electricity from coal, oil, or water power, and this area appears dark on Map 4, there is a presumption for a potential demand for atomic power—but this presumption must be reconsidered in the light of other information. A dark area on Map 4 may promise little demand for electricity of any kind if it is inhabited by very poor people. Of course, atomic power may contribute to the development of such an area and thereby increase the income of the people; this question is considered in Chapter XIV. On the other hand, a very light area may become a consumer of electricity if it has certain mineral ores, or is located favorably for an airfield, or has potentialities for irrigation, etc.

This study borders on technology and geography. They provide the main data, however incomplete. But still another and even less complete kind of data would have to be known if we should claim to arrive at definitive conclusions. These unknown data are the political decisions of the future.

Political factors have been indicated and weighed in various places of the book. To begin with, atomic power can (though it need not) be obtained as a by-product in the making of nuclear weapons. In this case it can be a subsidized product, military expenses being in general borne by the taxpayer. If the armaments race is mitigated by an international agreement, other economic consequences might arise, as indicated in Ap-

pendix B of Chapter I. Therefore a straight comparison of the money
cost of energy from new and old sources ceases to be the only guide in determining the extent to which, in a given country, the new source of energy
can compete with the old ones.   Moreover, this is due not only to diplomatic
and military but to other less conspicuous politico-economic decisions as
well. As shown in Chapter II, the relative price and availability of the
various types of energy can depend on the government subsidies for coal
and freight rates (as in Russia) and on the political control of imports and
exchange rates, whether undertaken for reasons of security or of domestic
economic policy.

The basic comparison, of course, is not that of money costs but of "real
costs" in the following sense: To achieve a given level of present and future
national consumption, and a given degree of national security, will it take
more of a country's resources to produce an additional kilowatt-hour of
power from atomic or from conventional sources? To be sure, policy
makers, whether democratic or dictatorial, may not always answer or may
not even ask this question with complete clarity—because of sectional interests or because they are lacking information or competence. Yet it
would be unwise not to press this question of "real cost" when the interest
of our own economy and security is discussed. It would be equally unwise
to assume that other nations ignore this question and are thus bound to
waste their resources foolishly. Now, this question of "real cost" is indeed
answered by money-cost comparison—to the extent that private men, and
even government agencies, compete in markets and try to avoid losses. To
this extent, and this extent only, are money-cost comparisons meaningful.

Many other political variables had to be treated as unknown. In each
case an assumption or a set of alternative assumptions had to be explicitly
stated and the implications explored. For example, the effects of atomic
power upon the national income (in Chapter XIII) were estimated on the
assumption that at no time would the government of an industrial country
tolerate unemployment due to a fall in effective demand that could be offset
by appropriate fiscal and monetary measures. While many political observers have advanced this assumption, they would not be too surprised if
it failed. The political science of today does not tell us what determines
the business-cycle policy of a government. If instead of making our assumption we had said that governments sometimes do and sometimes don't fight
unemployment, we would have given up any possible benchmark for a
reasoned evaluation of the effects of the new invention upon national income.

Intangibles of future political history naturally affect also the discussion
of the effects of atomic power upon the industrialization of backward countries (Chapter XIV), a process that includes not only the physical construc-

tion of mills, roads, harbors, houses, but also its financing by foreign loans or by reducing domestic consumption; it includes a change in birth rates, in sanitary standards, and in the patterns of education; and it presupposes the exercise of policing power. All of this is strongly determined by the domestic policy makers of a country as well as by their partners and opponents in world politics. The economist can but indicate the importance of these variables. He can only ask questions of the anthropologist, the sociologist, and the political scientist.

The fact that the economics of atomic power not only depends on technology but is embedded in general social and political conditions was early recognized by the Social Science Research Council when it appointed a Committee on the Social Aspects of Atomic Energy. This committee represented various social sciences together with physics. Its members were Winfield W. Riefler (Chairman), Bernard Brodie, Rensis Likert, Jacob Marschak, Frank W. Notestein, William F. Ogburn, Isidor I. Rabi, and Henry De W. Smyth. The committee approved the preliminary outline of the present book.

The Rockefeller Foundation undertook to finance the study by awarding a grant to the Cowles Commission for Research in Economics, The University of Chicago. An additional grant by the Life Insurance Association of America helped to complete the study. It is a pleasant duty to thank, on behalf of the study group, the organizations that have initiated the work and provided the necessary funds and facilities.

While the two codirectors of the study share equal responsibility for its shortcomings, Sam H. Schurr had the additional, the major, burden as the author or coauthor of most of Parts One and Two of the book. He wrote Chapters I, III, and IV, and participated in the authorship of Chapter II with Edward Boorstein, of Chapters V–IX and XI with George Perazich, and of Chapter XII with Milton F. Searl, now of Stanolind Oil and Gas Company, Tulsa, Oklahoma. Harold H. Wein, of the U. S. Department of Justice, is the author of Chapter X. The concluding chapters, XIII and XIV, which constitute Part Three, were contributed by Herbert A. Simon, now of the Carnegie Institute of Technology, Pittsburgh. Edward Boorstein and George Perazich, as full-time members of the research staff of the study, contributed in a substantial degree to defining the subject matter and formulating the general approach.

Cartographic advice was given by Robert L. Carmin, now of the Department of Geology and Geography, Michigan State College. He also drew Maps 1 and 3; Map 2 was drawn by Robert E. Stanley. Ruth Frankel Boorstin edited the final copy of the book and, together with Jane Novick, Editorial Secretary of the Cowles Commission, saw the book through the press. Both were assisted by Jean Curtis. William B. Simpson, Assistant Research Director of the Cowles Commission, helped in the final arrangements with the publisher. John R. Menke, now of the Nuclear Development Associates, New York, was attached to the study at its beginning and continued later to help with technical advice, as did, at a later stage, Richard L. Meier, now of The University of Chicago.

Each chapter was repeatedly revised by the authors and the codirectors to coordinate the results into a single whole and to take account of suggestions and comments. Some of the preliminary results of the study were published as it progressed, to invite public comment and criticism.[1] All parts of the study were circulated, at different stages of drafting, to specialists in various fields. We have, in fact, drawn heavily on their advice, both in laying out the study (e.g. in selecting those energy-consuming industries that should be analyzed), and in using their detailed critical comments on individual chapters. It is natural that in a field as new as this, conflicting opinions should be expressed. We have tried to weigh them all carefully and to incorporate in this volume the suggestions of these specialists. But they do not share responsibility for the final text, with which indeed they would not always have agreed. We are particularly indebted to A. B. Kinzel, Union Carbide and Carbon Research Laboratories, because of the general guidance he gave us from the very beginning. Various aspects of this study were also discussed to our benefit with H. J. Barnett of the Program Staff, U. S. Department of the Interior, and with Leo Szilard, Professor of Biophysics, The University of Chicago. Ansley J. Coale, Institute for Advanced Study, Princeton University, Mordecai Ezekiel of the Food and Agricultural Organization of the United Nations, and Ward F. Davidson, Consolidated Edison Company of New York, have read and given extensive suggestions on many parts of the book.

We are indebted to many others who took the pains to read drafts of certain chapters and to offer suggestions as to revision, or who offered advice at various stages during the course of the study.

The Preface and Chapter I (Economic Characteristics of Atomic Power) have benefited from the technical advice or criticism of the following persons: Sir Wallace Akers, Imperial Chemical Industries Ltd., London, England; Bruce K. Brown and G. W. Watts, Standard Oil Company of Indiana; Harrison Brown, Institute for Nuclear Studies, The University of Chicago; W. P. Dryer, Stone and Webster Engineering Corporation; Clark

[1] These were: (1) Cowles Commission Special Paper No. 1: "Nuclear Fission as a Source of Power," by John R. Menke (reprinted from *Econometrica*, Vol. 15, October 1947, pp. 314–334); (2) Cowles Commission Special Paper No. 2: "The Economic Aspects of Atomic Power," reprints of papers by Jacob Marschak (from the *Bulletin of the Atomic Scientists*, Vol. 2, September, 1946); and by Sam H. Schurr with comments by Philip Sporn and Jacob Marschak (from *American Economic Review, Proceedings*, Vol. 37, May 1947, pp. 98–117); (3) "Atomic Power in Selected Industries," by Sam H. Schurr (*Harvard Business Review*, Vol. 27, July 1949, pp. 459–479).

Goodman, Department of Physics, Massachusetts Institute of Technology; Joseph E. Loftus, Teaching Institute of Economics, The American University; C. Rogers McCullough and Charles A. Thomas, Monsanto Chemical Company, St. Louis; E. W. Morehouse, General Public Utilities Corporation, New York; Walton Seymour, Program Staff, U. S. Department of the Interior; John A. Simpson, Institute for Nuclear Studies, The University of Chicago; F. H. Spedding, Institute for Atomic Research, Iowa State College, and his colleagues—D. S. Martin, A. F. Voigt, and H. A. Wilhelm; Philip Sporn, American Gas and Electric Service Company, New York; G. O. Wessenauer, Manager of Power, Tennessee Valley Authority; and Eugene P. Wigner, Professor of Physics, Princeton University.

S. D. Kirkpatrick and his colleagues on the staff of McGraw-Hill Publishing Company have given useful technical comments both on Chapter I and on the analysis of various industries treated in Part Two: P. W. Swain, Editor of *Power*, A. E. Knowlton of the *Electrical World*, Norman Beers and Keith Henney of *Nucleonics*, T. R. Olive, Roger Williams, Jr., and R. F. Warren of *Chemical Engineering*. George Havas, Kaiser Engineers, Inc., Oakland, California, made helpful suggestions concerning aluminum, cement, and iron and steel. L. A. Matheson, Physical Research Laboratory, Dow Chemical Company, commented on aluminum, iron and steel, and chlorine and caustic soda. Professor Cyril Smith, Institute for the Study of Metals, The University of Chicago, gave us comments on the chapters on aluminum and iron and steel. J. H. Walthall, Division of Chemical Engineering, Tennessee Valley Authority, made suggestions on the aluminum and electrochemical industries. In the list that follows we shall gratefully mention others who generously placed their special knowledge at our disposal, together with the particular industries on which they offered suggestions:

Aluminum: Francis C. Frary, Aluminum Research Laboratories, Aluminum Company of America; Ivan Block and Samuel Moment, Bonneville Power Administration, Portland, Oregon; and Irving Lipkowitz, Economic Research Department, Reynolds Metal Company. *Chlorine and Caustic Soda, and Fertilizers:* Roscoe E. Bell, Coordinator Western Phosphate Fertilizer Program, U. S. Department of the Interior; Zola G. Deutsch, Deutsch and Loonam, New York; K. D. Jacob, Bureau of Plant Industry, Soils and Agricultural Engineering, U. S. Department of Agriculture; Glenn E. McLaughlin, National Security Resources Board, Washington, D. C.; and Chaplin Tyler, E. I. DuPont De Nemours and Company. *Flat Glass, Cement, and Brick:* R. W. Allison and J. J. Svec of *Ceramic Age;* C. H. Hahner, Glass Section, National Bureau of Standards; F. G. Schwalbe, Toledo Engineering Company. *Iron and Steel:* E. P. Barrett, Long Beach, California (formerly with the U. S. Bureau of Mines); Isaac Harter, The

Babcock and Wilcox Corporation; J. R. Miller and C. F. Ramseyer, both of H. A. Brassert and Company; and Earle Smith, Chief Metallurgist, Republic Steel Corporation. *Railroad Transportation:* Julian Duncan, Interstate Commerce Commission; Thor Hultgren, National Bureau of Economic Research; W. S. Lacher, Engineering Division, Association of American Railroads; T. M. C. Martin, Bonneville Power Administration, Portland, Oregon; and John I. Yellott, Director of Research, Locomotive Development Committee, Baltimore. For comments on the appendix on the use of nuclear power plants on locomotives we are indebted to Professor E. Wigner, Princeton University. The chapter on *Residential Heating* was read and criticized by J. C. Butler, Illinois Maintenance Company, Chicago; J. E. Koch, Power Plant Supervisor, Chicago Union Station Company; and William H. Ludlow, Committee on Instruction and Research in Planning, The University of Chicago.

Our industrial analyses were also helped by more general comments on the part of Alexander Gourvitch, Division of Economic Stability and Development, United Nations; Carl Kaysen, Department of Economics, Harvard University; Walter Rautenstrauch, Department of Industrial Engineering, Columbia University; R. M. Weidenhammer, U. S. Department of Commerce; and James Zilboorg, General Electric Company, Mexico City.

In the analysis of the costs and resources of power in various parts of the world (Chapter II) and of the effects of atomic power on national or regional economies and on the industrialization of backward areas (the two concluding chapters), we have had the help of the following: Colin Clark, Director, Bureau of Industry of the Queensland Government, Australia; N. B. Guyol, Division of Economic Stability and Development, United Nations; Chauncy D. Harris, Professor of Geography, The University of Chicago; Norman Kaplan, Illinois Institute of Technology; Simon Kuznets, National Bureau of Economic Research; Conrad G. D. Maarschalk, New York City; Professor Kenneth May, Carleton College; Professor Frank W. Notestein, Office of Population Research, Princeton University; Professor Harvey S. Perloff, The University of Chicago; John D. Sumner, Professor of Economics, University of Buffalo. We also drew on helpful comments from S. C. Gilfillan, author of *Sociology of Invention*.

All materials consulted in the course of preparing this volume were unclassified from the standpoint of national security; the expert comment was provided with the understanding that the manuscript was for publication.

JACOB MARSCHAK

Chicago
March, 1950

Contents

ANALYTICAL TABLE OF CONTENTS

PART ONE

TABLES, MAPS, AND GRAPHS

MAPS

GRAPHS

Part One

ECONOMIC COMPARISONS OF
ATOMIC AND CONVENTIONAL POWER

CHAPTER I

Economic Characteristics of Atomic Power

CERTAIN information is available today about atomic power which is extremely useful in defining the boundaries within which economic analysis can proceed. The facts which are known consist, in part, of the basic scientific information which has been revealed in such official documents as the report prepared for the War Department by H. D. Smyth,[1] and the various papers submitted to the United Nations Atomic Energy Commission by the government of the United States.[2] There are also the "informed official judgments" on many critical points found mainly in reports of the United States Atomic Energy Commission, and the "informed unofficial judgments" of many scientists who have been associated with the development of atomic energy in the United States and other countries. Such judgments have been of great importance to us in preparing this chapter.

Most of the unanswered important questions in the economic analysis of atomic power remain unanswered not because the information is being kept secret, but simply because it is not yet available, even to the United States Atomic Energy Commission.

Useful atomic power has not yet been produced, except, perhaps, as a by-product of the operation of research reactors, nor has a commercial atomic power plant been designed. The scientists who are attempting to design such a plant are faced with numerous difficult engineering problems, such as the development of new materials able to withstand the unusual operating conditions of nuclear reactors. Clearly, therefore, the most important item of economic information—data on the cost of producing atomic power—cannot be available at the present time. Still, enough appears to be known about the physical characteristics of the process by which atomic power may be produced to indicate at least the broad limits within which the cost, relative to the cost of conventional thermal power, will eventually fall.

[1] *Atomic Energy for Military Purposes*, Princeton University Press, 1946.

[2] *The International Control of Atomic Energy, Scientific Information Transmitted to the United Nations Atomic Energy Commission*, Department of State Publication 2661, Washington, Government Printing Office, 1946, contains the more important papers.

3

Two known facts have been particularly useful in providing a starting point for our analysis. The first is that while 1 lb. of coal can be transformed into about 1 KWH of electric power, 1 lb. of atomic fuel, fully consumed, would yield about 2½ million KWH of electric power, that is, 1 lb. of nuclear fuel is the equivalent of approximately 1,250 tons of bituminous coal.[3] The economic importance of this fact is clear: by ordinary standards, the cost per unit of energy of transporting the new fuel is negligible. We may say that, in effect, atomic energy will be produced from a weightless fuel, and this in turn suggests that its use could be an important factor in minimizing the wide differences in energy costs throughout the world.

The second important fact is that relatively pure uranium metal, the key mineral in the fission process, cost about $20 per lb. in 1943.[4] As we shall see, there is a reasonable possibility that 1 lb. of uranium can be made into 1 lb. of atomic fuel, which would mean that the energy equivalent of 1,250 tons of coal might cost about $20.[5] This suggests that the new fuel might be available throughout the world at an unusually low cost. This does not mean necessarily, as we shall see, that the conversion to useful power would be cheap; this will depend mainly on the cost of the plant and equipment needed.

These two hypotheses—that this new fuel might be available everywhere in the world at about the same cost, and that this might be a very low cost—indicate at least that this subject warrants further study to fix cost and other economic characteristics more closely. What will it cost to make the energy available in useful form? Will the form in which the energy is made available, and the necessary plant and equipment, limit its use? Will raw materials be available in large enough amounts to constitute a significant addition to the world's stock of energy resources? How great are the developmental problems which must be solved before atomic energy can be used commercially? We will consider these and other questions in this chapter.

We begin in this chapter with a section setting forth certain economic characteristics of atomic power which can largely be inferred from the physical nature of the production process. The second section considers the ore reserves of the source materials of nuclear fuel and inquires whether

[3] This comparison assumes the transformation of heat to electric power at an overall thermal efficiency of about 25% for both coal and nuclear fuel. In our analysis of the cost of ordinary thermal electricity in Chapter II, we will assume a somewhat higher efficiency.
[4] Smyth, *op. cit.*, p. 93, para. 6.14.
[5] The cost of producing metallic uranium may have risen somewhat since; the standard price in New York for metallic uranium, announced in December 1949 by the Atomic Energy Commission, is $50 per lb. We may note, too, that this metal, while highly refined by ordinary standards, would not have the extreme purity required for nuclear reactors (*New York Times,* December 1, 1949, p. 17, col. 2).

these will be sufficient to support a large atomic power industry.   The third section discusses the available estimates of the cost of producing atomic power and derives those cost figures which we will use in subsequent chapters.   Since the analysis is made without regard to the political factors which may strongly affect the economics of producing and using nuclear power, Appendix B, illustrating how the international control of atomic power may produce economic effects, is included.

A. PHYSICAL AND ECONOMIC FEATURES OF USEFUL ATOMIC POWER

Atomic energy in useful form is released in a completely new type of furnace, at first called an "atomic pile," which has come to be known as a "nuclear reactor."   Descriptions setting forth the basic physical characteristics of nuclear reactors and associated facilities have appeared in numerous publications.[6]   We select for discussion here only those characteristics which seem to have an important bearing on the economics of producing and using atomic power.

1. The Uses of Atomic Power

Nuclear fission, which is the source of atomic energy, is also the source of intense radioactivity.   The only known method for protecting personnel from the deadly effects of the radioactivity and neutrons generated by fission is to surround the reactor with a massive shield which, in current practice, consists mainly of concrete.   The necessary shielding at present is reported to be several feet in thickness.   This means a weight of at least 100 tons is needed for comparatively small reactors.[7]   These dimensions need not hold for all time because new types of alloys and ceramic materials may be developed which will permit the shield to be thinner.[8]   But at best the shield will continue to be extremely heavy.

The weight and dimensions of this protective shield, even allowing for improvements, will limit the use of atomic power.   Shielding requirements

[6] Leon Svirsky, "The Atomic Energy Commission," *Scientific American,* Vol. 181, July 1949; and "Atomic Energy 1949," *Business Week,* No. 1026, April 30, 1949, contain useful descriptions for the general reader.   Additional references, of a more technical nature, are listed under "Electric Power" in *An International Bibliography on Atomic Energy,* Vol. 1, Atomic Energy Commission Group, United Nations, Lake Success, 1949.

[7] Sir Wallace Akers, "Metallurgical Problems Involved in the Generation of Useful Power from Atomic Energy," Thirty-Seventh May Lecture to the Institute of Metals, *Journal of the Institute of Metals,* Vol. 73, July 1947, p. 673.   M. C. Leverett, "Some Engineering and Economic Aspects of Nuclear Energy," Declassified Document, United States Atomic Energy Commission, MDDC-1304, December 3, 1948, p. 7.   J. A. Wheeler, "The Future of Nuclear Power," *Mechanical Engineering,* Vol. 68, May 1946, p. 403.

[8] F. H. Spedding, "Chemical Aspects of the Atomic Energy Problem," *Bulletin of the Atomic Scientists,* Vol. 5, February 1949, p. 48.

will be such that small mobile units like automobiles and trucks will not be driven by nuclear reactors.[9]   On the other hand, the use of nuclear reactors in large mobile units such as ocean-going vessels has been considered quite practical.   In fact, the view is often expressed that one of the first applications of nuclear reactors will be in the propulsion of naval vessels.[10]   This use is judged desirable if only for the military advantage to be gained from extending the distance which such vessels could travel without refueling. The practicability of using nuclear reactors in airplanes and locomotives is less certain.   The United States Atomic Energy Commission has actively pushed research on nuclear reactors for aircraft propulsion and there now appears to be some chance that this use, possibly in pilot-less craft, will be feasible.   Here, too, the possible military advantage of increasing the distance a plane can travel on a single fueling is undoubtedly of the utmost importance.   Research on the application of nuclear reactors to railroad locomotive propulsion has, so far as is generally known, not yet been undertaken.

The most feasible use of nuclear reactors for energy production from a technical standpoint (and probably the most promising economic use) is in stationary plants for the production of power.   Such plants, which would produce power for general distribution, or which would be built by a large industrial concern to satisfy its own power requirements, obviously will not be unduly handicapped by shielding requirements.   Although this report is devoted almost wholly to stationary plants, certain general points discussed here will also be applicable to mobile units.   We consider mobile reactors only in connection with our analysis of the possible use of atomic energy by railroads, selected because of their economic importance (Chapter XI). Since, with a limited research program, we could include just one transportation industry, ocean shipping and air transportation are omitted.

Following the practice in most discussions of the economics and engineering of atomic energy, our analysis will be primarily in terms of electric power production from the energy generated in the nuclear reactor.   We will focus on electricity rather than on heat (the form in which energy appears in the nuclear reactor) because electricity is the energy form which has proved most applicable to central generation for distribution over wide areas.   If stationary nuclear power plants are really to permeate modern economic society, they will have to do so either through the production of electricity

[9] Sumner Pike, "Atomic Energy in Relation to Geology," Speech at the Annual Meeting of the American Association of Petroleum Geologists and other organizations, St. Louis, Missouri, March 16, 1949, AEC Press Release.

[10] See, for example, the testimony by Mr. Lilienthal in the Hearings before the Sub-Committee of the Committee on Appropriations, on the Independent Offices Appropriation Bill for 1950, House of Representatives, 81st Congress, 1st Session, Part 1, pp. 1088–1089.

or by some novel method. However, in dealing principally with electricity we do not preclude the possible applications of direct heat, since the production of atomic electricity implies the prior production of heat. Therefore, in discussing the economic and physical characteristics of electric power from atomic energy we will, in effect, be discussing these characteristics (with certain obvious modifications) for nuclear heat as well. As a result, we will be able to apply the information on atomic power presented in this chapter (which relates primarily to electricity) to an analysis of the economic feasibility of using nuclear heat directly as, for example, in the district heating of residential areas (Chapter XII).

As long as nuclear heat is not needed at temperatures above that required for converting heat to electricity (as is true, for instance, of district heating) we will be justified in borrowing data from the analysis made for electricity in this chapter; but when heat is needed at considerably higher temperatures, as in metal smelting or cement manufacture, the use of such data may be unjustified. The problem may be stated in the following simple terms: the higher the temperature required, the smaller the likelihood that the use will be realized because of the extremely difficult engineering problems of operating nuclear reactors at high temperatures. Even so, estimates referring to electric power from atomic energy are the best guide we now have to the probable economic characteristics of nuclear reactors producing high temperature heat, provided we clearly understand this fundamental difference: that economic and engineering judgments which are applicable to atomic electricity may prove much too optimistic where the production of heat at extremely high temperatures is concerned.

2. Nuclear Fuel[11]

Three fuels will be available for use in nuclear reactors: uranium 235, uranium 233, and plutonium. The only one of the three which occurs naturally is uranium 235, which is an isotope of natural uranium constituting about 0.72% of the total (1 part in 140). The other fuels are created artificially: plutonium from uranium 238, the isotope which constitutes almost the whole of natural uranium, and uranium 233 from thorium. Uranium 235 may be considered the primary nuclear fuel in the sense that it is the

[11] Although this section will include an occasional reference to purified U-235, which can be obtained only through isotope separation, the separation process itself is not discussed, because we believe that in the future the production of U-235 probably will be considered a less economic route to the production of useful power than the production of plutonium or uranium 233.

Of course some existing plants for separating the U-235 isotope probably will continue to operate and may even be improved and enlarged. (Akers, *op. cit.*, pp. 670, 674, 675. John W. Irvine, Jr., "Heavy Elements and Nuclear Fuels," Chapter XI in *The Science and Engineering of Nuclear Power*, Vol. I, edited by Clark Goodman, Cambridge, Mass., Addison-Wesley Press, 1947, p. 373; also cf. Clark Goodman, "Future Developments in Nuclear Energy," *Nucleonics*, Vol. 4, February 1949, p. 3.)

only one which exists in nature, and must be available (if only in the form of natural uranium) before a start can be made on the production of the other two fuels.

The large-scale transmutation of uranium 238 to plutonium, and of thorium to uranium 233, can only be accomplished in a nuclear reactor. Our wartime supply of plutonium was produced in the Hanford reactors in Richland, Washington. These reactors also generated a large amount of energy (possibly at a rate as high as 500,000 to 1,500,000 kilowatts in the form of heat[12]), but this energy was allowed to go to waste primarily because it was available at too low a temperature to perform useful work. Reactors of the same general type as those at Hanford, but producing useful power and nuclear fuel simultaneously, are likely to be extremely important in the commercial atomic power industry. Essentially, therefore, while these reactors are using up plutonium—or one of the other two fissionable substances—they will also be producing new plutonium or uranium 233.

a. FUEL PRODUCTION AND CONSUMPTION IN THE NUCLEAR REACTOR. A very clear description of the possible relationship between the consumption and production of fissionable substance in the nuclear reactor of the future has been given by Professor J. A. Wheeler.[13] "In future attempts to accomplish this regeneration process there will be one or [the] other of two possible outcomes. It may be that each day of operation of the pile of the future will burn up, let us say, one kilogram of the original fissionable material and regenerate out of U-238 or similar inert material possibly only 0.9 kilograms of new fissionable material such as plutonium. In this case the plant can continue operation only if we supply from outside each day 0.1 kilogram of new fissionable material. . . . The other possible outcome is more favorable to economy of raw materials. In every day of operation in which we destroy one kilogram of fissionable material we synthesize from an inert substance more than one kilogram of new fissionable material, say, for example, 1.1 kilograms. In this case we leave one kilogram of the new product in the plant to make up for the losses of the day and remove the other 0.1 kilogram to help start up a new pile. . . . In case we can achieve this outcome, there is no need for us to supply new fissionable material to our plant from the outside, except to get it started. . . . Evidently we have only to design a plant with sufficiently good regeneration characteristics in order to use for power purposes all of the uranium, not merely the rare constituent U-235."

In an informative article, Sir Wallace Akers adds to this description by indicating that thorium may prove particularly important in the breeding of fissionable substance. He points out that "there seems to be a possibility,

---

[12] Smyth, *op. cit.*, p. 104, para. 6.41.
[13] Wheeler, *op. cit.*, p. 402.

and quite a good one, that the number of neutrons ejected when U-233 [which is transmuted from thorium] undergoes fission, together with the neutron capture cross-section of thorium, may make it possible to build a pile containing U-233 and thorium in which more nuclei of U-233 are formed than are 'burnt up,' so that the supply of U-233 will be continually increased."[14]   It is conceivable, then, but by no means certain, that breeding might be carried through largely on the basis of thorium.   It appears likely that at least it will be an important supplement to uranium as a raw material for producing new fissionable materials in nuclear reactors.

Despite the possibilities indicated in these quotations, we have the word of the United States Atomic Energy Commission that the "engineering difficulties associated with breeding are enormous."[15]   The Commission indicates that one of the major problems is that the conditions which are essential for the efficient breeding of new fissionable substance may not be the same as those needed for economy in operating nuclear power reactors. In addition, efficiency in breeding will require frequent, quite expensive chemical reprocessing of the uranium or thorium in the pile in order to separate the wastes (fission products) which accumulate in the reactor during its operation from the unchanged uranium or thorium, and to sep- arate these from the plutonium or uranium 233.[16]

An experimental reactor which would breed fissionable materials has already been designed at the Argonne National Laboratory, and the Atomic Energy Commission has let contracts for its construction.   Although it is not certain that this reactor will work, the scientists who produced the design are confident of its success.[17]   However, this reactor is not designed to produce useful power; whatever power it does produce will be an inci- dental by-product.

The possible difficulty of reconciling efficient breeding with economic power operation suggests that at least two types of reactors might be used in the atomic power industry: one type which would specialize in breeding fissionable materials, but which would also produce some power, and an- other type which, while specializing in producing power, might also produce some new fissionable substance.   Presumably the former class of reactors would have lower fuel costs but higher operating costs for producing power than the latter class, and vice versa.   Even if this should happen, it need not influence our cost analysis if we view the atomic power industry as being

[14] Akers, op. cit., p. 677.
[15] United States Atomic Energy Commission, Fourth Semiannual Report, Washington, G.P.O., 1948, p. 45.
[16] Ibid., p. 45.
[17] United States Atomic Energy Commission, "AEC Selects Contractor to Build First Nuclear Reactor at Testing Station in Idaho," AEC Press Release, November 28, 1949; also excerpts from news conference transcript, New York Times, November 29, 1949, p. 9, col. 3.

composed of groups of reactors each of which includes a combination of both reactor types; then particular economies of operation anywhere in the group could be reflected equally in lower costs for each reactor in the group. It is not unlikely that, if it is necessary to have the two types of reactors, simplified accounting procedure along these general lines might be adopted in the commercial atomic power industry.

However, it is by no means certain that efficient breeding and power production cannot be combined in a single reactor.[18]   At the Knolls Atomic Power Laboratory in Schenectady design is under way on a reactor which would produce significant amounts of electric power "without depleting— and perhaps even increasing—the national supply of fissionable material."[19]

We shall base our analysis on the assumption that nuclear reactors will make at least enough new fissionable substance to replace the fuel consumed in producing their energy output.   This assumption implies one important operating characteristic: that the continuing fuel supply for the operation of nuclear power plants will be in the form of natural uranium or thorium. Since, as we shall see, these materials are very cheap per unit of energy, this means that the cost of the day-to-day fuel supply of the average nuclear reactor will probably be extremely low.   If two types of reactors are required to accomplish both efficient breeding and economic power production, this assumption would apply to certain reactors only if, as suggested above, they are considered part of a "reactor group."

b. THE INITIAL INVESTMENT OF FUEL IN THE NUCLEAR REACTOR. The amount and kind of fissionable material which a nuclear reactor needs in order to begin operating must be distinguished from its continuing fuel supply.   Although the continuing fuel supply of the reactor (viewing reactor operations perhaps as a group) may well be in the form of natural uranium or thorium, the initial investment in fuel probably will have to include some relatively pure fissionable material in the form of uranium 235, plutonium, or uranium 233.   Since these materials (which are also used as the explosive charge for the atomic bomb) are relatively costly and are available only in limited amounts, at least at the present time, great efforts probably will be made to economize their use.   For example, Lyle Borst, Chairman of the Reactor Science and Engineering Department, Brookhaven National Laboratory, has stated: "The efficiency of a power plant will be judged to a large extent by the power generation rate for a given inventory of nuclear fuel."[20]

From present indications, the initial fuel supply of nuclear reactors usu-

[18] AEC Press Release, November 28, 1949, *op. cit.*

[19] Svirsky, *op. cit.,* p. 39.

[20] Lyle B. Borst, "Industrial Application of Nuclear Energy," *The Commercial and Financial Chronicle,* Vol. 167, March 4, 1948, p. 32.

ally will consist both of natural uranium (1 part fissionable out of 140) and pure fissionable substance, say, plutonium, which will have the effect of enriching the fissionable content of the natural uranium.   Since unenriched natural uranium can itself sustain a nuclear chain reaction, as it has done in the Hanford reactors, the question arises why enrichment should be required.   The reason seems to be that the flexibility in design important in the development of practical power reactors can probably be obtained only at the cost of including these scarce fuels in the reactor's fuel supply.[21] Increasing the proportion of fissionable substance in the reactor's fuel elements strengthens the likelihood that neutrons necessary to propagate the chain reaction will be captured by fissionable materials rather than by inert materials.   This results in greater freedom to include pipes for the circulation of the heat transfer medium and other materials needed in the production of useful power within the reactor, since there will now be a greater margin of neutrons to lose by capture by these materials without causing the chain reaction to die out.

Our analysis will be based on the assumption that pure fissionable substance will have to be added to natural uranium in the reactor's initial fuel investment.[22]   In this event the economics of producing nuclear power will be affected in two chief ways.   First, the cost per kilowatt of constructing nuclear power plants probably will include a not insignificant charge for fuel in the form of pure fissionable materials.   Fuel cost will therefore enter into both the overhead cost, i.e. fixed charges on the investment (as fuel in an expensive form), and the direct operating expenses (as fuel in a very inexpensive form) of the nuclear power plant.   Second, the speed with which the nuclear power industry can get under way, as well as the rate at which it can grow, will be determined partly by our ability to provide the pure

[21] Akers, *op. cit.*, p. 674; David Lilienthal, "Atomic Energy and American Industry," Speech before the Economic Club of Detroit, October 6, 1947, AEC Press Release; R. F. Bacher and R. P. Feynman, "Introduction to Atomic Energy" in *The International Control of Atomic Energy,* Department of State Publication 2661, *op. cit.,* pp. 21–22.

[22] Although it seems reasonable to assume that pure fissionable substance will be required in the initial fuel investment in nuclear reactors, this is not the only possible assumption.   Dr. R. F. Bacher has stated in describing the American reactor development program "The main part of the reactor development program is built around the construction of units which utilize fissionable material in purified or enriched form as the fuel elements. . . . One of the questions now under consideration is whether a practical reactor can be constructed using natural uranium for fuel and designed to operate primarily for power production with minimum replacement and reprocessing of fuel elements." (Speech on The Development of Nuclear Reactors before the American Academy of Arts and Sciences, Boston, Massachusetts, February 9, 1949, AEC Press Release.   Also see the introductory essay on Nuclear Reactors by J. D. Cockcroft in *An International Bibliography on Atomic Energy,* Vol. II, United Nations Atomic Energy Commission, Preliminary Edition, February 14, 1949.)   However, there is no indication that as much work is being done yet on the design of such a reactor as on reactors using enriched fuel.

fissionable substance required for the initial fuel investment.   Since we do
not now know how much pure fissionable material a reactor will require to
begin operations, and at what rate breeding of new fissionable materials to
meet this need will take place, conjecture—both official and unofficial—
must be substituted for real information.

The United States Atomic Energy Commission mentions the building
up of fuel inventories among the problems which "introduce a time factor
measured in years into any discussion as to when nuclear energy can make a
significant contribution to the supply of power now available from other
sources."[23]   This extremely vague estimate tells us at least that a large-scale
atomic power industry could not spring into existence in one or two years, if
only because of the need for producing pure fissionable substance for the
initial fuel investment.

How many years will be required to build up fissionable stocks to the
point where atomic power can make a significant contribution to the total
energy supply?   Breeding increases stocks in much the same way that interest is compounded in a savings account; although the rate of increase in
stocks per year is fixed, the absolute amount of increase will grow from year
to year as the size of the existing stock of fissionable materials increases.
Obviously, therefore, the early years of the building-up process will present
the greatest difficulty in providing fuel for reactors.

It has been estimated that it might require several decades to breed
enough fissionable material to support nuclear reactor capacity equal to the
present-day thermal requirements of the entire world.[24]   However, long
before stocks would be built up to a level where they could all at once support an atomic power industry capable of satisfying the world's energy
requirements, responsible officials would begin to divert materials to start
the atomic power industry on what they would want to be a process of
orderly growth.

It is difficult to say when the first allocations of fuel stocks for power
reactors could be made, or how much capacity could be supplied at the
beginning.   But we may note that Leo Szilard, a pioneer in the American
development of atomic energy, estimated in 1945 that sometime between
1949 and 1958 the United States might be able to divert enough fissionable
material to generate about 15 million kilowatts of electricity.[25]   This figure
may be compared with the gross additions to electric power capacity in the
United States in 1947 of about 2 million kilowatts.   Hence, the stocks which

[23] United States Atomic Energy Commission, *Third Semiannual Report,* Washington,
G.P.O., 1948, p. 13.
[24] M. H. L. Pryce, "Atomic Power: What are the Prospects?" *Bulletin of the Atomic
Scientists,* Vol. 4, August 1948, pp. 245–248.   Thermal requirements include all heat
and power, not just electric power.
[25] Hearings before the Special Committee on Atomic Energy, United States Senate,
79th Congress, 2nd Session, Part 2, Washington, G.P.O., 1946, p. 269.

might be diverted to power uses by 1958 (according to this estimate) could take care of all gross additions to capacity at the 1947 rate for about 7 consecutive years.

If we shift our attention from when fuel stocks will be large enough to allow atomic power to account for a significant part of total energy production to when stocks will be large enough to enable atomic power installations to constitute a significant part of current additions to power plant capacity, we move from a time scale measured in decades to one measured by a 5 or 10 year period (other limiting factors, which are discussed later, not considered).[26]

The important question then becomes: Will fissionable material supplies be great enough to support the growth of the atomic power industry over time to the point where it can really become a major energy supplier? Pryce, in his estimate, assumes that the breeding of nuclear fuel might be at a rate of 10% per year.   The actual rate might be somewhat different depending on how successfully the technical problems associated with breeding are handled.   But let us take the 10% rate as a reasonably good estimate and compare it with the rate at which the output of electric utilities has grown in the United States.   Between 1902 (the first year of the statistical record) and 1942, electric utility production grew at an annual rate of 10.7%.   The estimated breeding rate would, therefore, just about keep pace with the average annual rate at which electric power production grew. This breeding rate falls short of the rate at which production increased during the industry's early years—averaging from 10% to 18% during the first two decades of the twentieth century—and is greater than the average rate, during the thirties, of between 5% and 8%.[27]   Although these comparisons should not be carried too far, they suggest that breeding will permit the nuclear power industry to expand at a rate which compares well with the history of electric power production in this country.

3. Plant and Equipment

In keeping with our general approach in this chapter, we are not devoting this section to a systematic enumeration and description of the facilities and equipment required to produce atomic power.   Instead, we try to describe those production factors that have important economic implications. Three general characteristics of plant and equipment stand out: (1) con-

[26] Cf., for example, Clark Goodman's estimate that: "Were it not for the urgent military demands on the technical personnel and atomic resources, industrial atomic energy would be supplying electricity and comfort heating in many localities within the United States ten years from now." ("Future Developments in Nuclear Energy," *op. cit.*, p. 9.)

[27] J. M. Gould, *Output and Productivity in the Electric and Gas Utilities, 1899–1942*, New York, National Bureau of Economic Research, 1946, p. 39.

ventional facilities will be required for converting nuclear heat to electricity; (2) the operation of nuclear reactors will be possible only in conjunction with the operation of associated chemical processing facilities, either nearby or distant; and (3) the perfection of nuclear reactors must wait on the solution of numerous engineering problems of great complexity.

a. The Need for Conventional Facilities to Produce Electricity. Most experts hold that the atomic plant for producing electricity will be a thermal generating station in which a nuclear reactor has replaced the conventional furnace. Except for the nuclear reactor (furnace) and its associated facilities, the costs of which are unknown, all items of power plant equipment would be approximately the same as in an ordinary electric station burning coal, oil, or natural gas. Sumner Pike of the United States Atomic Energy Commission has indicated some of the added difficulties of the nuclear plant: " . . . in the conventional boiler we have to take the heat out in the form of steam and put it into a turbine . . . in this case [the nuclear reactor] we will have to take out the heat in the form of something very hot, probably not steam, possibly some inert gas, or possibly a liquid metal, and probably will have to do another set of heat exchanging outside of the pile [to eliminate the remaining radioactivity], so that we can get something that will run either a conventional steam turbine or some form of gas turbine."[28]

In a conventional thermal electric plant, the costs associated with those facilities which will also be required in the atomic electric station are by no means negligible. So we can obtain a firm estimate of the minimum below which the cost of atomic power cannot possibly fall so long as the atomic power plant continues to require these facilities. This goes a long way toward providing us with one of the "boundaries within which economic analysis can proceed" to which we referred earlier.

This approach to estimating minimum atomic costs is reliable only so long as the basic technological assumption on which it rests holds true. But the technology of atomic power is brand new. Is it not possible that a method will be found for obtaining electricity without going through the conventional process for passing via heat to electricity? As our answer we have the statement of one expert, who occupies a key position among those engaged in nuclear reactor development in the United States, that despite having "looked long and hard for a trick method of getting electrical energy directly from the chain reaction," none has been found.[29] Another authority expresses his opinion with even greater finality: "No practical means

[28] Sumner Pike, "The Work of the United States Atomic Energy Commission," Speech before Cooper Union Forum, New York City, January 13, 1948, AEC Press Release, p. 11.
[29] Borst, op. cit., p. 4.

has yet been devised, or is likely to be devised, of converting the fission energy directly into electric or mechanical energy."[30]   This problem, then, is not in the same category as the many other difficult problems which must be solved before useful atomic power can be produced, for while there is a promising approach to these problems, no even vaguely promising method has been put forward for by-passing the heat engine in the production of electricity from nuclear fission.

b. THE NEED FOR CHEMICAL PROCESSING FACILITIES.   If the need for conventional facilities for producing electricity introduces a familiar element into an otherwise novel process, the need for chemical processing facilities emphasizes that part of the process of producing electricity from nuclear fission has no counterpart in conventional power plants.   Why chemical processing will be needed, and what the difficulties involved are has been well stated by David Lilienthal: "When fission takes place energy is released. . . . But something is left behind, the result of splitting the nucleus.   This mixture of the isotopes of many elements is very radioactive. Inside the reactor, where all this takes place, the effect of these 'hot' elements on the operation of the reactor is bad.   Some of the new elements absorb neutrons readily, so that their effect is to quench the nuclear reaction.   The fission products—the nuclear ashes—must be removed and the remaining nuclear fuel recovered to be used again.   This is a chemical separation job that presents some highly complex and difficult problems of chemistry, chemical engineering, and metallurgy—and it must be carried out by remote control to protect the workers from lethal radiations.   I need not stress the troubles of this kind of operation by mechanical proxy, nor need I stress the economic importance of efficient recovery of the unspent nuclear fuel [i.e. pure fissionable substances], for this is material that may not under any circumstances be wasted."[31]

The chemical process, then, is designed to separate unspent fissionable material, some of which will have been created in the operations of the reactor, and unused raw material (i.e. uranium 238 or thorium) for the manufacture of fissionable material from the fission products which would otherwise lower the efficiency of the reactor operation.   The process is made difficult not only by the lethal characteristics of the materials, but also by the fact that the extraction of the fissionable material involves the separation of very low concentrations of one element from high concentrations of other elements.[32]   Moreover, economy in the use of raw materials probably will require that a single charge of uranium or thorium be recycled through

[30] Pryce, *op. cit.,* p. 245.
[31] Lilienthal, "Atomic Energy and American Industry," *op. cit.,* p. 17.
[32] Irvine, *op. cit.,* p. 365.

the chemical process several times.   Despite the difficulties, the process has been described as being "parallel to the familiar operations of the inorganic chemical industry."[33]

The most important economic implication of the need for chemical processing appears to arise from the possibility that there will be decided economies in large-scale chemical facilities.[34]   The effect of this might be to make the optimum size of atomic power plants considerably greater than the capacity required in most consuming regions, with unit costs rising sharply for plants smaller than optimum size.   An attempt will undoubtedly be made to circumvent this problem through the centralization of chemical processing facilities, so that a single chemical plant will be used by numerous atomic power stations.   The success of this attempt will hinge on whether the radioactive materials that these plants will process can be transported safely over extended distances at a reasonable cost.

Whether the transportation of radioactive materials will be feasible cannot now be settled definitely, but there is some evidence that it can be done.   Thus, Arthur V. Peterson of the United States Atomic Energy Commission's Division of Production has stated that slugs containing fission products are transferred in large quantities at Hanford;[35] we know from another source that the transportation from piles to the Hanford chemical plant a good many miles away is by railroad.[36]   (There were three reactors at Hanford during the war—more are being, or have been, built—widely separated from one another in the Hanford reservation, which covers an area more than half as large as Rhode Island.[37])   Peterson also points out that in its current operations the Atomic Energy Commission has used special convoys to ship amounts of radioactive materials judged to be too great for safe shipment by ordinary commercial facilities.   Additional evidence that methods for transporting fission products might be successfully worked out is found in some of the technical information submitted by the United States to the United Nations.   It is explained that the American plan for international control of atomic power would probably require the centralization of chemical facilities under international control and the

[33] "Technological Control of Atomic Energy Activities" in *The International Control of Atomic Energy*, Department of State Publication 2661, *op. cit.*, p. 161.

[34] "Atomic Energy, Its Future in Power Production," *Chemical Engineering*, Vol. 53, October 1946, pp. 131, 133; Leverett, *op. cit.*, pp. 8, 10.

[35] "Digest of Proceedings," Seminar on the Disposal of Radioactive Wastes, Sponsored by the United States Atomic Energy Commission, January 24–25, 1949, AEC Press Release, January 30, 1949, p. 19.

[36] John A. Wheeler, "Inspection of Manufacturing Processes: Possibility of Detection" in *Reports and Discussion on Problems of War and Peace in the Atomic Age*, The Committee on Atomic Energy of the Carnegie Endowment for International Peace, New York, 1946, p. 56.

[37] *Atomic Energy Development, 1947–1948*, U. S. Atomic Energy Commission, Fifth Semiannual Report, Washington, G.P.O., 1949, pp. 24–27.

shipment of "partly consumed materials" from nationally-controlled power plants to these chemical plants for reworking.[38]

c. ENGINEERING PROBLEMS IN REACTOR DESIGN.   Although this study is not the place to catalogue the engineering problems which must be solved before atomic power is a practical reality, it is helpful to consider briefly the general nature of these problems, particularly in connection with the time scale of atomic power development in the United States and other countries.

Perhaps the best introduction to the general nature of the problems is found in the following two quotations from experts who have worked on reactor designs.   Borst indicates the difficulty of finding suitable structural materials for the reactor: "It is one of the ironies of nature that common engineering materials, such as ferrous metals and most nonferrous metals, absorb neutrons and may not be used inside thermal neutron reactors. Those which do not absorb neutrons lack mechanical strength, and undergo excessive corrosion at the necessary temperatures."[39]   E. R. Gilliland voices a more general complaint: "In removing heat from a reactor, there are a number of considerations, but to an engineer, the chief one appears to be that the physicist prefers that he keep his equipment out of the reactor.   Apparently, nearly any material used in the reactor is objectionable.   If a gas like helium is used, while not objectionable from its nuclear properties, it is not a good moderator [i.e. is relatively inefficient in slowing the speed of neutrons] and hence increases the size of the reactor.   Many of the liquids require structural materials for the passages through which they flow that are objectionable in thermal reactors.   The main objective is to remove the heat without disturbing the physical characteristics of the reactor any more than is absolutely necessary."[40]

The crucial need is for new materials to perform ordinary functions: to provide structural support, to carry off heat, to supply channels for the passage of the heat transfer medium, etc.   The problem of structural materials is particularly critical and becomes more so the higher the desired temperature: the objective here is to discover substances which can withstand high temperatures and intense radiation and yet not absorb too many neutrons as, for example, steel does.   While the challenge is a formidable one, we are assured by an official of the Atomic Energy Commission that there is "every reason to expect a satisfactory solution over a period of time."[41]   It is worth noting that the solution probably will depend on

---

[38] "Technological Control of Atomic Energy Activities," op. cit., pp. 179, 183.

[39] Borst, op. cit., p. 4.

[40] E. R. Gilliland, "Heat Transfer," Chapter 10 in The Science and Engineering of Nuclear Power, op. cit., p. 323.

[41] Pike, "The Work of the United States Atomic Energy Commission," op. cit., p. 11. The reactor development program of the Atomic Energy Commission includes a reactor for the primary purpose of testing materials which may be used in future reactor con-

developing essentially new materials with peculiar nuclear or chemical or physical properties—for example, discovering how to produce pure zirconium in the form of a structural material,[42] or refining familiar materials to new standards of purity.[43]

Engineering problems resemble the problem of building up fuel inventories, which we considered earlier, in the sense that they also introduce a time factor measured in years into any consideration of when atomic power will provide an important part of current additions to power supply. The Atomic Energy Commission's General Advisory Committee has reported that under the most favorable assumption on technical developments "we do not see how it would be possible . . . to have any considerable portion of the present power supply of the world replaced by nuclear fuel before the expiration of 20 years."[44]   Although this statement is negatively phrased, it is in one sense a highly favorable forecast.   For if a considerable portion of the world's power supply *is* accounted for by nuclear power 20 years from now, this would imply that a sizable percentage of current additions to power capacity would be accounted for by atomic power many years earlier.   We may note, by way of example, that diesel locomotives already are well advanced technologically as the statistics on locomotive additions by American railways in recent years attest.   But even though diesels have in recent years constituted about 90% of the new locomotives ordered by our railroads, they account for only about 30% of the gross ton miles of freight service by all locomotives.[45]

Although the Advisory Committee referred to world power supply, we may suppose that its members were thinking primarily of the United States when considering the time scale required for overcoming these complex engineering problems. If the development of atomic power proceeds wholly on an individual national basis, the time scale for many other countries would probably be considerably slower (even assuming the same absolute effort is put forth). Other countries, even those which are highly industrialized, might be faced with greater problems, as the experience of the French suggests, when they tried to draw on industry for the highly purified materials needed to develop their first experimental atomic pile.[46]

---

struction.   Construction of this reactor is scheduled to begin in the Spring of 1950 (AEC Press Release, November 28, 1949, *op. cit.*).

[42] *Atomic Energy Development, 1947–1948, op. cit.*, p. 64.

[43] Spedding, *op. cit.*, p. 49.

[44] Atomic Energy Commission, *Fourth Semiannual Report, op. cit.*, p. 46.

[45] "Monthly Comment on Transportation Statistics," Interstate Commerce Commission, Bureau of Transport Economics and Statistics, Washington, October 11, 1949. The percentage of freight by diesels refers to the first seven months of 1949; in a similar period in 1948, freight service by diesels was 20% of the total.

[46] L. Kowarski, "Atomic Energy Developments in France," *Bulletin of the Atomic Scientists*, Vol. 4, May 1948.

B. URANIUM AND THORIUM RESOURCES

There would obviously be little point in investigating the commercial possibilities of atomic power if uranium and thorium resources were not sufficient to support a high level of energy production for a long time. Almost no quantitative information on the subject of resources of uranium and thorium raw materials has been officially released since it became known that these minerals could be used as a source of fissionable material. The information on resources dating from before that time is inadequate, mainly because discovered reserves of minerals are in large degree dependent on demand, and the demand for uranium and thorium was formerly quite small.

However, highly significant, although non-quantitative, information has been released by American officials.   Mr. Lilienthal has said: [47] "Statements have been made and widely publicized, that there is only enough uranium ore to last a relatively brief period, and so the prospects of benefit from atomic power must go glimmering.   This is simply not so.   It appears clear that atomic energy is on a sound basis for an indefinite period in the future. . . .   As explicitly as national security permits, the Atomic Energy Commission wishes to state that it finds no basis in fact for these statements about extremely limited uranium ore supplies. . . .   Instead, the contrary is true about uranium supplies.   Estimated known reserves have already increased substantially since usefulness of uranium as a source of fissionable material became firmly established."

What is the nature of the resources of thorium and uranium which will supply the fuel for the production of atomic energy?   The highest grade deposits of uranium are in the form of pitchblende, the highest grade thorium ores in the form of monazite sands.   The largest known concentrations of both these materials occur outside the United States: pitchblende mainly in Canada and the Belgian Congo, and monazite mainly in India and Brazil.   It is likely that in his statement Lilienthal was thinking not of these foreign resources, but primarily of other ores occurring in the United States.   These other ores undoubtedly include both the carnotite deposits of Colorado, which have been mined in the past primarily for their vanadium content, and also certain oil shales and other marine sediments which exist in this country in great abundance but which are not now being mined at all.[48]

[47] Press Conference on "Uranium Supplies," Denver, Colorado, December 17, 1948, AEC Press Release, pp. 1, 2.

[48] Also see the testimony by Carroll Wilson, General Manager of the Atomic Energy Commission, before the Sub-Committee of the Committee on Appropriations of the House: "We believe that by the development of low-grade sources, which we are investigating extensively today, we will be able to get uranium out of materials which contain very, very small amounts of uranium.   Thus far we have used high-grade deposits."

The possibility that uranium will be obtained from oil shales has received much attention in discussions of the availability of sources of fissionable material. In an official statement, "Major Classes of Uranium Deposits," the Atomic Energy Commission states:[49] "It has long been known that certain oil shales and other marine sediments, including phosphatic beds, contain very small quantities of uranium. Sweden, for example, has announced that she is building a small atomic pile and intends to derive uranium from her oil shales to feed this pile. According to published statements Swedish shale deposits containing many millions of tons of 'ore' run around 0.02% uranium oxide. . . . By-product uranium from oil shale or phosphate industries may play a part in the development of atomic energy in different parts of the world. The AEC expects to exhaust every possibility of this character in the United States."

Also along these lines, Dr. Gustav Egloff, research director of The Universal Oil Company, has spoken of the possibility that shales may in the future be exploited for both their oil and uranium content, thereby reducing the cost of producing each material.[50] Another student of the subject expresses the view that shales may prove to be our major reserves of uranium.[51]

The emphasis on oil shales serves to highlight this important possibility for the United States as well as for other countries: that eventually supplies of uranium and thorium may come mainly from mineral ores not formerly mined for these materials, primarily because they contain amounts which used to be considered too low for commercial exploitation. Supporting this view is the information that in the Union of South Africa the by-product production of uranium by the Witwatersrand gold mining industry holds such great promise that the Minister of Industrial Production has said: "We believe that we are able to say that the Union of South Africa may produce more uranium than any other country of the world."[52]

(Hearings on the Independent Offices Appropriation Bill for 1950, 81st Congress, *op. cit.*, p. 1109.)

[49] AEC Press Release, December 9, 1948, p. 2.

[50] *New York Times,* September 18, 1947, p. 27, col. 1. Based on a speech before the American Chemical Society.

[51] Clark Goodman, "Distribution of Uranium, Thorium and Beryllium Ores and Estimated Costs," Lecture notes (No. 26) at Clinton National Laboratories, 1947, unclassified.

[52] Quoted in "Uranium Resources" speech by J. K. Gustafson, Manager of Raw Materials Operations, United States Atomic Energy Commission, before a Metallurgic Colloquium, Massachusetts Institute of Technology, Cambridge, Massachusetts, March 9, 1949, AEC Press Release, p. 9. We may note that the Atomic Energy Commission announced, on November 10, 1949, that representatives of the American and British governments were then discussing with the South African Government "problems relating to the production of uranium in the Union of South Africa." (AEC Press Release, November 10, 1949, p. 1.)

The price of uranium and thorium ore may rise if deposits of a considerably lower grade than those used today are brought into production.   How could a change in the price of uranium or thorium affect the cost of producing atomic power?   We presented earlier the assumption that an individual reactor or a "reactor group" would make at least enough new fissionable substance (plutonium or uranium 233) from natural uranium or thorium to replace the fissionable material which had been used up in generating power.   This means that the entire weight of either uranium or thorium, rather than just a part, could be made into nuclear fuel, with the result that the "burning up" of one pound of either metal would generate about 2½ million kilowatt hours of electricity.   With this relationship in mind, we can see immediately that even very large increases in the cost of obtaining uranium ore could not have a significant effect on the cost of atomic power.   For example, with uranium at the wartime price of $20 per lb. its cost per kilowatt hour of electricity (assuming 100% conversion to nuclear fuel) would be 0.008 mills.   Now, if the price of uranium should increase tenfold to $200 per lb., its cost per kilowatt hour of electricity would be only 0.08 mills, and even if it were to multiply in price one hundredfold to $2,000 per lb. its kilowatt hour cost would still be only 0.8 mills.   At $2,000 per lb., uranium would be no more costly in terms of its energy content than coal at about $1.60 per ton (at the same thermal efficiency).   We can evaluate these figures better when we note that electric utilities paid on the average close to $5 per ton for their coal in 1946.

This is indeed an amazing fuel, which could absorb enormous increases in the cost of prying it from the earth and refining it and still remain cheaper than the cheapest coal.   Of course, to produce electricity from uranium will require facilities not needed in an ordinary thermal electric station, as we have already seen, which will affect the cost of energy production.   But at any rate this much is true: the cost of atomic power, whatever it may be, probably will not be seriously affected by possible changes in the cost of uranium and thorium.

So far, then, we have decided on quite general grounds, first, that uranium and thorium resources will be available for a large atomic power industry, and second, that possible changes in the cost of getting these metals from their ores will not seriously affect the cost of producing atomic power. Attempts have been made to deal with both these questions in more definite terms by estimating the minable quantities of different grades of resources of the two materials and the costs of exploiting them.   We will summarize one of these estimates, which represents the joint efforts of Professor Clark Goodman of The Massachusetts Institute of Technology (who estimated the quantity and grade of reserves) and John R. Menke, formerly associated with Clinton National Laboratory at Oak Ridge (who estimated the cost of mining and refining).[53]

According to the Goodman-Menke estimates, reserves of uranium and thorium in the United States which are "definitely minable" equal about 100 million lbs. These consist mainly of carnotite. They estimate that the cost of producing the metal from these reserves will be about $5 per lb., once mining and recovery techniques are brought to a high level of commercial efficiency.[54] In addition to the "definitely minable" reserves, they estimate that the United States possesses reserves of 1 billion lbs. of uranium and thorium which are "probably minable." These consist mainly of oil shales, a type of ore which, as we mentioned, is already being mined in Sweden. They estimate that it should cost about $50 per lb. to produce metal from shales.[55] They estimate that there also exist vast reserves of uranium and thorium in ores ranging from "probably worthless" to "possibly minable," which we do not consider here.

While these are, of course, extremely rough estimates, as Goodman and Menke have been careful to point out, they are consistent with the qualitative statements quoted above. It may be helpful to summarize the Goodman-Menke estimates as follows (assuming complete conversion of uranium and thorium to fissionable material):

	Approximate number of years reserves would be sufficient to produce electric power at the 1946 level of output in the U. S.	The cost of uranium or thorium in the cost of producing electricity (mills per KWH)
"Definitely minable" reserves	1,000	0.002
"Probably minable" reserves	10,000	0.020

[53] C. Goodman, Lecture notes at Clinton Laboratories, *op. cit.;* J. R. Menke, "Nuclear Fission as a Source of Power," *Econometrica,* Vol. 15, October 1947, pp. 321–324.

[54] The cost estimate derived by Menke is particularly interesting. For numerous minerals produced from underground mines he collected data on the price of the refined mineral and the percentage of the mineral contained in marginal ore bodies. By multiplying the price and the percentage of mineral content he derived a synthetic value for the cost of mining and refining the amount of mineral contained in a given unit, say, 1 lb. of ore. This led to the discovery that the cost of mining and refining *per lb. of ore* of widely differing minerals is remarkably alike, a result which Menke attributes to the similarity of many of the operations performed, no matter what the mineral. In other words, then, the cost of minerals won by similar mining methods is determined in good part by *the number of lbs. of ore mined per lb. of pure mineral.* This principle, with appropriate figures derived from the calculations, was applied in estimating the cost of uranium and thorium from different grades of ores.

[55] This may be compared with an estimate by Josef Eklund of the Swedish Geological Survey, that uranium can be extracted from Swedish shale at a cost of about $23 a pound (U. S. Department of the Interior, Bureau of Mines, *Minerals Yearbook, 1947,* Washington, G.P.O., preprint, p. 1209).

These calculations obviously should not be taken as a precise measure of the number of years our reserves of atomic energy raw materials will last. Not only are the quantities of reserves extremely rough estimates, but in addition electricity consumption will rise greatly over the 1946 level; furthermore, atomic energy may be used as more than just a source of electricity. The figures should be interpreted as meaning that probably uranium and thorium, at a cost which will be a negligible item in the total cost of electricity (0.002 mills or 0.020 mills per KWH may be considered almost zero cost for all practical purposes) will be available in the United States in large enough amounts to support a high level of atomic energy production for many hundreds of years. And since there are indications that oil shales (and other marine sediments) which form our major reserves according to this estimate are available in many different parts of the world,[56] in addition to higher grade reserves in a few locations, there is reason to believe that a similar generalization might also be valid for a large number of other countries.

C. THE COST OF ATOMIC POWER

1. The Conceptual Basis of the Cost Estimates

As a starting point for estimating the cost of atomic power, we have chosen the cost of producing electric power in a conventional power plant based on ordinary fuel. Two interrelated reasons determined our choice. First, there are important points of similarity between such a plant and an atomic power plant; costs in the conventional plant will, therefore, provide some guidance in estimating atomic costs. Second, in determining the cost of producing electricity in an ordinary thermal power plant certain accounting and engineering values are applied. These values are only to some extent standardized, and therefore an element of arbitrariness enters into their selection. Since our interest is in comparing the costs of atomic and conventional power, we shall obviously want to use the *particular* engineering and accounting values which we have applied in deriving our thermal power cost figures as benchmarks for estimating the cost of atomic power.

In obtaining our figures for the cost of ordinary thermal electricity we want to show how costs will vary among regions depending principally upon their distance from coal mines (or other energy resources) and the cost of transporting the fuel there. The reason for highlighting variations in the cost of conventional power resulting from the transportation of fuel is, of course, that atomic energy will, by contrast, be based on a fuel whose energy yield per unit of weight is so great that its transportation will be practically cost-free. The thermal electricity costs shown throughout this report

[56] See Atomic Energy Commission statement quoted on p. 20.

(mainly in Chapter II) are, therefore, derived as follows: all costs other than fuel in producing thermal electricity are given a constant value and differences in the cost of power are made to depend solely on variations in the cost of fuel. In determining what the constant cost should be, and in translating fuel cost into its equivalent electricity cost, we have selected factors which are in accord with the best practices in large modern power plants.

Our figures on conventional power costs, therefore, are not real values but estimates. They differ from the atomic cost estimates in that there are engineering and commercial data from which to select the assumed values. But the specific cost figures on conventional thermal power in this report come out as they do because of the assumptions which have entered into their derivation; the specific atomic cost figures will come out as they do because they are being constructed primarily for comparison with the estimates of the cost of conventional power. There is, in reality, an inbreeding of the two sets of estimates: the conventional costs are made to reveal the variations of only one cost factor which we judge to be of critical importance to the comparison; the atomic cost figures borrow (in part) certain assumed values from the conventional cost estimates to make them as truly comparable with those figures as the basic data permit.

2. The Cost Estimates

a. FACTORS USED IN DERIVING COSTS OF ORDINARY THERMAL ELECTRICITY.[57] In estimating the cost of ordinary electricity we assume that the power is being produced in a plant of the most modern design with a capacity of 75,000–100,000 kilowatts operating at an average rate of 50% of capacity (i.e. it produces 4,380 KWH annually per kilowatt of capacity, a reasonable rate of capacity utilization for central stations in general).[58]

The costs which we hold constant for this plant are (in 1946 prices):

(1) Fixed charges at 3.3 mills per KWH. This is based on:
    a. Investment in plant and equipment at $130 per kilowatt of capacity,
    b. Fixed charges on investment at the rate of 11% per annum.
(2) Operating costs other than fuel at 0.7 mills per KWH.

b. ESTIMATED MINIMUM COST OF ATOMIC POWER. From these values we can immediately estimate the costs below which atomic costs probably

[57] Summarized from the discussion in Chapter II.
[58] Plant capacity must, of course, be built to satisfy peak electricity requirements, while average demand is generally below this peak. Hence, the almost universal (and unavoidable, considering the nature of demand) tendency among central stations toward relatively low utilization of capacity.

cannot ever fall, so long as the heat of nuclear fission is converted to electric power by conventional means. As we have seen, the cost of uranium and thorium will constitute a negligible item in the cost of producing nuclear power. Fixed charges and operating costs other than fuel will, therefore, make up almost the entire cost. About ⅔ of the investment shown above (about $90 per kilowatt of capacity) represents the cost of power-generating equipment which would have to be duplicated in the atomic power plant. The fixed charges on this investment, as well as such operating costs as repair and maintenance on these facilities, should be the same in the atomic plant as in the conventional plant. If, in order to estimate the *lowest conceivable* cost of atomic power, we assume that the remaining costs shown above for the ordinary thermal plant would also be roughly the same in the atomic plant (an optimistic assumption, if only because the unique facilities required in the atomic plant appear to be more costly than the facilities they replace in the ordinary thermal station), we have 4.0 mills per KWH as the minimum cost of atomic power (at a 50% rate of capacity use). This figure says, in effect, that atomic power at best would be the cost equivalent of ordinary thermal power based on a costless fuel.

c. Cost Estimates Derived from Published Studies of Atomic Power. In the cost estimate just made, there was no attempt to determine how expensive the unique facilities and operations required to produce atomic power would be compared to those they replace in the conventional thermal plant. However, these costs have been separately estimated in several studies of atomic power already published, from which we can derive additional estimates of atomic costs. These may be considered less speculative than the preceding estimate of minimum cost since they explicitly consider the unique features of atomic power production.

The figures we will use have appeared in several different reports. The most authoritative of these is a report submitted to the United Nations Atomic Energy Commission by the representative of the United States and prepared by members of the staff of the Clinton Laboratories under the direction of Dr. C. A. Thomas (subsequently referred to here as the Thomas report).[59] Another extremely useful, although completely unofficial, report was prepared by a committee of social and physical scientists at the University of California under the direction of Professor J. B. Condliffe (subsequently referred to as the California report).[60] While these two reports constitute our chief source of cost information, they have been supplemented by four other reports. These include a paper by M. C. Leverett, who participated in the preparation of the Thomas report, which presents more detailed information on some of the calculations underlying the pub-

[59] "Nuclear Power" in *The International Control of Atomic Energy,* Department of State Publication 2661, *op. cit.,* pp. 121–127.
[60] "Atomic Energy, Its Future in Power Production," *op. cit.*

lished Thomas figures;[61] a paper by Ward F. Davidson, research engineer, Consolidated Edison Company of New York, Inc., which contains an unofficial critique of the Thomas estimate;[62] and two independent cost studies, one by J. R. Menke, now with Nuclear Development Associates, Inc., New York,[63] and the other by C. F. Wagner and J. A. Hutcheson of the Westinghouse Electric Corporation.[64] Menke's study contributed in a considerable degree to the formulation of the general approach followed in this cost analysis.

Although some of the reports contain their own estimates of the cost of producing nuclear power, we do not use these cost figures directly. Instead we attempt, in the remainder of this section, to fit the figures underlying their estimates into the conceptual framework governing our cost calculations. The resulting estimates give the cost of nuclear power as it might have been estimated in, for instance, the Thomas or California report, if the estimates were being made for comparison with the costs of conventional thermal power as estimated in this report.

(1) *Investment in Plant and Equipment.* The investment in the plant and equipment required for producing atomic power may be considered in two parts: those items of plant and equipment which will be the same as in the ordinary thermal power plant, and those items which will be unique for the nuclear power plant. We estimate the cost of the repeated facilities to be $90 per kilowatt of capacity. As we noted earlier, this is approximately $2/3$ of the investment we assume for ordinary thermal power stations, a proportion which represents a compromise among those suggested in some of the studies mentioned above.[65]

In dealing with the estimates which have been made of the necessary investment in novel nuclear facilities, we begin with the California and Thomas figures. Although the estimates in both of these reports appear to be based on a fairly detailed consideration of the special facilities required by the atomic power plant, the costs derived are widely different. They show the following values for the specialized nuclear plant facilities (in 1946 prices):

$226 per kilowatt of capacity   (Thomas)[66]
$48   ”   ”   ”   ”   (California)

[61] Leverett, *op. cit.*

[62] "Nuclear Energy for Power Production," by W. F. Davidson, Paper delivered at the Fuel Economy Conference of the World Power Conference, The Hague, 1947 (conference proof).

[63] Menke, *op. cit.*

[64] "Nuclear-Energy Potentialities," by C. F. Wagner and J. A. Hutcheson, *Westinghouse Engineer*, Vol. 6, July 1946.

[65] Menke suggests ¾ as the proper proportion, Davidson suggests ½, and Wagner and Hutcheson suggest ⅔.

[66] The figure is taken from Leverett's paper. The Thomas report itself does not show a breakdown of costs.

Both of these estimates are obviously surrounded by a considerable area of uncertainty, since no plant for the commercial production of useful atomic power has yet been built or, so far as we know, even designed. But the wide gap which separates them may result from factors in addition to uncertainty in estimating the costs.

Part of the difference in the two estimates appears to result from the fact that the Thomas estimate is for a plant of 75,000 kilowatts, and the California estimate for a plant of 500,000 kilowatts. The California study is purposely based on a plant of 500,000 kilowatts, an exceptionally large size for power stations, because the authors of that study felt that a plant in this size range would approach the optimum for the chemical separation facilities. This may provide a useful clue to the difference in the two estimates, for if the Thomas estimate for a plant of 75,000 kilowatts includes the full investment in associated processing facilities, it is likely that part of the difference in unit investment would be accounted for by this factor. Although it cannot be definitely determined that the Thomas estimate does provide for individual chemical processing facilities for its atomic power plant, Leverett's comments on the estimate imply that this is so. It is possible, therefore, that if chemical processing facilities could be centralized and made to serve numerous nuclear power plants, the California estimate of unit investment might apply equally to a smaller size plant, and, on the other hand, the Thomas investment for the 75,000 kilowatt power station might be somewhat lower.

However, even if the cost of the chemical plant were wholly omitted from the Thomas estimate, the investment in the remaining specialized nuclear facilities (primarily the nuclear reactor itself) would still amount to $160 per kilowatt (according to Leverett's figures), which is well in excess of the California total investment in the specialized nuclear facilities, including the chemical plant, of $48 per kilowatt. Since there is no information to support the view that the unit cost of these facilities will decline much with increases in plant capacity above 75,000 kilowatts, the most reasonable assumption we can make is that this difference in unit cost must be attributed chiefly to a real disagreement in the evaluation of the cost of these items irrespective of differences in the size of the power plant.

The Thomas estimate appears to be based to some extent on the experience gained in constructing and operating the pioneer Hanford piles and in conducting experimental work at Oak Ridge, while the California report bases its cost estimate on engineering data applicable to analogous industrial processes in which many of the problems of design have already been worked out. It is perhaps proper to assume that the California estimate represents the level to which nuclear power costs might fall only after the commercial atomic power industry has had many years of experience behind it and has carried its technology to a very high degree of efficiency. On the other hand, the Thomas estimate represents the level of costs which will be found in the earliest commercial plants producing nuclear power.

In these circumstances, a figure falling somewhere between the two estimates probably would be a more appropriate estimate than either of them of the level of costs which will prevail in those atomic power plants which are built in, say, the 5 to 10 years following the first commercial installations. For these plants will incorporate the improvements resulting from the lessons learned in constructing and operating the commercial plants of the earliest design, even though they will themselves fall short of the highly efficient designs which will result from the lessons of a good many years of commercial operation. In this connection the following comment in the Thomas report on the estimates made there is significant: "It seems reasonable to expect that the future development of nuclear power will result in the standardization of design and construction, and a material reduction in the investment and operating costs."[67]

The other studies mentioned above all yield investment costs for the specialized nuclear facilities falling between the costs estimated in the Thomas and California reports. Menke states that while present costs are higher, a "reasonably achievable goal" for the cost of these facilities is about $100 to $120 per kilowatt.[68] In their study, Wagner and Hutcheson assume that the specialized atomic facilities might cost from 2 to 4 times the coal plant facilities they replace,[69] while Davidson remarks that a cost 4 times the replaced facilities "would seem to be not unreasonable."[70] Starting with $40 per kilowatt as the cost of the coal plant facilities which would be replaced by specialized nuclear facilities, we have a range of costs of from $80 to $160 per kilowatt based on the factors given by Wagner, Hutcheson, and Davidson.

These figures are extremely rough approximations. It is clear from their derivation that they are useful only as guides to estimating the *general level of costs* of the unique nuclear facilities. Since very little would be gained in our analysis by applying all of the separate atomic power cost figures we could derive from these studies, we select a single value to represent all of them. This figure, a rough average of the separate estimates, is $125 per

[67] "Nuclear Power," *op. cit.*, p. 126.
[68] Menke, *op. cit.*, p. 328.
[69] Wagner and Hutcheson, *op. cit.*, p. 126.
[70] Davidson, *op. cit.*, p. 6. In a study published after this chapter was written, Davidson estimates that the unique nuclear facilities would cost at least 3 times those they replace in the coal-burning station ("Atomic Power and Fuel Supply," *Atomics*, Vol. 5, October 1949, p. 13). Although this factor agrees with that implied in the average value we derive below, our total investment figure derived on this basis is substantially different from the figure estimated by Davidson in this article. The difference is due mainly to the fact that our estimates are based on 1946 prices, while Davidson's estimate refers to 1949 prices.

kilowatt for the unique nuclear facilities.[71]   We assume that this level of costs would apply to a period falling between the early developmental stage of atomic power we believe is implied in the Thomas estimate and the technologically mature atomic power industry we believe is implied in the California estimate.

The estimates of total investment per kilowatt of nuclear power which will be used in this study, therefore, are (in 1946 prices and for a 75,000 kilowatt power plant):

	For the repeated facilities	For the specialized facilities	Total
Maximum estimate	$90	$226	(about) $315
Intermediate "	90	125	" 215
Minimum "	90	48	" 140

The maximum estimate of investment in specialized facilities is taken from the Thomas report, the minimum from the California report, and the intermediate is derived from other studies.   Although we believe that each of these cost estimates probably refers to a different stage in the development of atomic power, we will also use them strictly in the later analysis as alternative estimates of the possible investment in atomic power, any of which might materialize as the "real" stable cost depending on the way in which the many unknown factors shape up in the tough process of development.[72]

(2) *The Rate of Fixed Charges on Investment.*   Fixed charges are generally defined to include cost items like interest, allowance for depreciation and obsolescence, property taxes and franchise costs, insurance, certain administrative expenses, etc.   The characteristic these costs hold in common is that they continue at pretty much the same absolute level no matter what the rate at which the plant is operating.   The shorthand method which is used in approximating fixed charges in estimates such as those we are making here is to figure them on an annual basis at a fixed percentage of investment in plant and equipment, even though not all of the charges are actually incurred on this basis.   For ordinary thermal electric plants we have figured fixed charges annually at 11% of the investment in plant and equipment.   We shall assume the same rate for the nuclear power plant, for reasons explained in Appendix A of this chapter.

(3) *The Total Cost of Producing Atomic Power.*   Table 1 summarizes

[71] It is worth noting that this cost is quite close to the cost which could be derived by averaging the Thomas and California estimates of the cost of the specialized nuclear facilities—$137 per kilowatt.

[72] It is, of course, possible that the cost of building nuclear power plants will be even above our highest figure.   However, with our highest cost figure, which some may consider too low, our later analysis reveals that atomic power would be of negligible economic importance.   It would, therefore, serve no analytical purpose to use any cost figure higher than that yielded by the Thomas report.   Nevertheless, it should be recognized that in the present state of our information it could not be argued that our assumed range of costs includes all the possibilities.

the cost estimates which will be used in the analysis made in subsequent chapters.   Fixed charges and the cost of fuel are based on figures presented in earlier sections.   Operating costs other than fuel have been derived either directly from the California study (minimum cost) or from the Thomas report (maximum cost) or from a range of values based on these and other studies (intermediate cost).   The table also includes the estimate of minimum cost computed by us in an earlier section directly from costs in the ordinary thermal plant excluding fuel; no breakdown of costs is shown for this estimate but its closeness to the estimate derived from the figures in the California report would suggest a similar breakdown.

The designation of the estimates as "minimum," "intermediate," and "maximum" is in accord with the usage explained in deriving the investment figures, i.e. partly an attempt to trace a path of atomic power costs through time, partly a reflection of the basic uncertainty now existing about most of the cost items.   For simplicity, we shall refer to the cost classes in round numbers as 4.0–4.5 (minimum); 6.5–7.0 (intermediate); and 10.0 (maximum).[73]

(4) *Effects of Changing Some of the Basic Assumptions.*   The cost estimates we have presented, as we have stressed, are derived for comparison with figures on the costs of producing ordinary thermal electricity estimated according to certain assumptions.   Wherever warranted, these assumptions have been carried over, to the extent that the data on nuclear energy permit, to our cost estimates on atomic power.   We say, therefore, that our two sets of estimates are comparable.   But it is important to understand that departing from these basic assumptions would not necessarily affect both sets of estimates in the same way.

To attempt to enumerate the many different ways in which changing one assumption or another would alter our cost figures would carry us away from the purpose of this analysis, which is to estimate cost on that basis which has the most general applicability.   We therefore use "general purpose" assumptions, although they will certainly be inapplicable in

---

[73] These estimates cover generating costs only; there will also be costs for transmitting the electric power to consumers.   We neglect transmission in the cost comparisons in Chapter II, on the assumption that atomic power plants would be as well located in an area with respect to power consumers as an ordinary thermal plant.   However, as Ward Davidson points out: "Until years of successful operation have proved their reliability, no prudent engineer would place a large nuclear power station in the heart of a large city. . . . The ever present chance, however small, that something might go wrong and release large amounts of intensely radioactive material would dictate sites in areas where a failure would be less disastrous." ("Atomic Power and Fuel Supply," *op. cit.*, p. 13.)   Extensive research on reactor safeguards has been done by the Atomic Energy Commission, which states that: "Evaluation of the potential hazards of reactors is one of the primary considerations in the location and design of reactors. . . . " ("AEC Reactor Safeguard Committee to Visit United Kingdom," Press Release, August 31, 1949, p. 1.)   By the time atomic power is commercially feasible, reactor safeguards may have reached a high degree of perfection.

TABLE 1. *Estimated total costs of producing nuclear power in a 75,000 kilowatt plant operating at 50% of capacity.*[a]

	Minimum cost	Intermediate cost	Maximum cost
	(mills per KWH, 1946 prices)		
Fixed charges[b]	3.5	5.4	7.9
Fuel[c]	0.002–0.02	0.002–0.02	0.002–0.02
Operating costs (including labor, maintenance, repair and supplies)	0.8[d]	1.2–1.6[e]	2.3[f]
TOTAL	4.0[g]–4.3	6.6–7.0	10.2

[a] This table presents, in simplified fashion, a digest of the information on nuclear costs in the several studies mentioned in the text. In no case do we reproduce the cost estimates made in these studies; instead we have attempted to fit their cost figures into the conceptual framework governing the cost comparisons which will be made in this report. Since in adjusting the figures to meet our needs we have often departed in some respects from the presentation made by the authors of the estimates, we submitted this chapter to them for criticism; their suggested improvements have been taken into account.

[b] Minimum, intermediate, and maximum investment per kilowatt shown earlier times 11% rate of fixed charges, divided by 4,380 (50% of total hrs. per year). In applying the California investment figure to a 75,000 kilowatt plant, we assume the centralization of chemical processing facilities, i.e. that the 75,000 kilowatt plant would be assigned only a portion of the total investment in an optimum-sized chemical plant. On the other hand, the Thomas investment figure probably includes the total investment in chemical facilities; assuming centralized chemical facilities might, therefore, have lowered the maximum estimate. As for the intermediate cost, we know that Menke assumes centralized facilities; the other cost studies do not specify their assumptions on this point. (It is possible that they consider the estimates to be so rough that one or another assumption about chemical facilities would have negligible effects.)

[c] From fuel costs shown earlier.

[d] As assumed in the California report.

[e] The higher figure is the average of the California and Thomas estimates for this item. The lower figure is derived on the assumption that these costs will bear the same relationship to this category of costs in the conventional thermal plant as the assumed investment per kilowatt of capacity in the nuclear plant bears to the investment in the thermal plant. (This assumption, used in place of any firm basis for estimating these costs, rests on the fact that expenditures for maintaining plant and equipment are a significant part of total costs in this category. We assume, then, that the more costly nuclear facilities will have proportionately higher maintenance costs.)

[f] Derived from Thomas report.

[g] Minimum as estimated earlier directly from ordinary thermal plant costs except fuel.

many individual cases. Any investigator interested in a particular locality would have to ask whether all of our assumptions suited the conditions found there.

Two assumptions in particular might prove inapplicable in many localities. These are, first, the assumption that the power plant would have a capacity of 75,000 kilowatts and, second, that it would operate at an average rate of 50% of capacity. How much would atomic power cost compared to conventional thermal power in localities which would require power plants of less than 75,000 kilowatts, and in those places where the average rate of operation might either substantially exceed 50% or fall well below this figure? If we compare the breakdown of total costs between fixed charges and operating costs for nuclear power and ordinary thermal power, we find that fixed charges are of considerably greater relative importance in the nuclear plant.   Thus, for example, comparing the intermediate cost range for nuclear power with the same cost range for ordinary thermal power, we have (in mills per KWH at 50% capacity operation):

	Nuclear power	Ordinary thermal power
Fixed charges	5.4	3.3
Operating costs	1.2–1.6	3.3–3.7
TOTAL	6.6–7.0	6.6–7.0

We see from the comparison that at this particular level of total costs, fixed charges account for about 50% of the total cost of ordinary power, and for about 80% of the cost of nuclear power.   Consequently any change in the assumed conditions which would increase fixed charges per unit of output by a greater percentage than it increased unit operating costs would bring about a larger increase in nuclear costs than in ordinary thermal costs; conversely, a change which would decrease fixed charges per unit of output by a greater percentage than it decreased unit operating costs would bring about a larger decrease in nuclear costs than in ordinary thermal costs.

The direction of change in the cost comparison which would be brought about by assuming a different rate of capacity utilization can be easily evaluated.   If the costs of nuclear power and ordinary thermal power are estimated to be the same at 50% of capacity operations, nuclear costs would almost certainly be higher than thermal costs at, for instance, 30% of capacity operations, because such a change would increase unit fixed charges by a greater percentage than it increased unit operating costs.   They would almost certainly be lower than ordinary thermal costs when operations were at 80% of capacity, since this change would decrease unit fixed charges by a greater percentage than it decreased unit operating costs.   The rise and fall in unit fixed costs are, of course, simply the result of spreading a constant fixed cost bill over a smaller or greater number of kilowatt hours of electricity output.[74]

[74] In the analysis of the possible industrial effects of atomic power (Chapters III–XII), we almost always assume a rate of capacity use of above 50%.   This reflects the fact that most of these industries have an almost continuous and unvarying demand for power, so that they will use a kilowatt of capacity closer to 100% of the time. (See Chapter III, Section A, 5.)

Our power cost comparisons in Chapter II always assume the 50% rate of capacity use, because this is a reasonable average for electric power systems. However, new plant capacity in existing power systems is often an exception.   Such plants may be operated at a high rate of capacity utilization to take full advantage of their greater efficiency, while older plants are used to satisfy peak requirements.   Since it would be particularly advantageous to use atomic power plants in this way (because of the importance of fixed charges in their costs), several of the experts who read a preliminary draft of this chapter urged that a higher rate be used in our comparisons.   This would, of course,

The possible effects on comparative costs of assuming a plant of smaller size than 75,000 kilowatts will depend on what happens to investment per kilowatt and to operating costs as plant size declines. The only information we have on this subject, which is for ordinary thermal plants and is quite unsatisfactory because of the dissimilarity of the plants in many respects, suggests that as size declines from 75,000 kilowatts to about 20,000 kilowatts there is a tendency for investment in equipment per kilowatt to increase by at least 20% over the entire range. Below 20,000 kilowatts, and especially below 10,000 kilowatts, unit investment rises more sharply. Unit operating costs appear to rise throughout by a considerably smaller percentage. Although these data are hardly conclusive, even for ordinary thermal plants, they provide some support for the belief that a comparison in which nuclear costs and ordinary thermal costs were equal at 75,000 kilowatts might at, for instance, 25,000 kilowatts change to one in which nuclear power would be more costly. In view of the uncertain data, we state this not as a conclusion but merely to illustrate how our comparisons might be invalidated if the power plant considered were of considerably smaller size.

APPENDIX A: THE RATE OF FIXED CHARGES

Our cost comparisons between atomic power and conventional thermal power assume the same rate of fixed charges on both. Some cost comparisons made by others allow for a higher rate of fixed charges on the atomic plant than on the conventional thermal plant on the assumption that the unique facilities in the atomic plant will undergo depreciation and obsolescence at a faster rate than the ordinary thermal plant.[75]

For simplicity, depreciation and obsolescence are commonly lumped into a single charge in electric utility cost accounting; in estimating the service life of a new plant provision is made for the effects of both. However, each provides for a decline in plant value in the future arising from different causes: depreciation provides for the wear-and-tear on property; obsolescence for changes in technology resulting in more economical methods of production.[76] We considered the two separately in deciding whether to charge different rates for the atomic and conventional thermal plant.

---

have favored atomic power. On the other hand, a higher than 50% average rate, even in new plants, probably could not be realized in many of the less industrialized regions covered in our worldwide cost comparisons. Therefore, we decided that our "general purpose" cost comparisons were more safely made with the 50% rate.

[75] Walter Isard and John B. Lansing, "Comparisons of Power Cost for Atomic and Conventional Steam Stations," *The Review of Economics and Statistics*, Vol. 31, August 1949.

[76] See, for example, Emery Troxel, *Economics of Public Utilities*, New York, Rinehart and Co., 1947, pp. 353–356.

The first atomic power plants will represent an early design in a new field.   Should we allow for a higher obsolescence charge on the assumption that the value of the first plants will decline very rapidly with future improvements in design?   A higher rate for this reason would be justified, if at all, only for the high end of the range of cost estimates we use.   This implies that the cost of producing atomic power will fall materially below the cost represented in the estimates based on early design (i.e. the Thomas report).   However, if substantial cost reductions are expected soon after the early high-cost plants come into operation, cost estimates for these early plants may have very limited economic significance.

The problem of obsolescence may be stated in more general terms: Will there be a long-term persistent downward trend in atomic costs which will result in a higher obsolescence rate for atomic power than is now normal for conventional power plants?   In this case both our maximum and intermediate cost estimates would be affected.   Suppose that a nuclear power plant and a conventional thermal plant produce electricity at the same total cost, figuring obsolescence on both at the same rate.   If there is reason to expect that improvements in techniques will so reduce atomic costs in the future as to require a faster obsolescence rate than in the past, both types of power plants would be equally affected; no matter whether one or the other plant is constructed, its cost accounting would need to allow for the anticipated declines in nuclear power costs.   With respect to obsolescence, it is therefore incorrect to use a different rate for atomic and conventional power plants when making cost comparisons between them.

With the advent of atomic power a higher rate of obsolescence than is customary for ordinary thermal plants might be applied to *both* types of plants.   How this would affect our cost comparisons cannot be clearly foreseen.   On the one hand, the nuclear plant might be expected to suffer in the comparison because the costs of investment in plant and equipment are relatively more important than for ordinary thermal power plants.   On the other hand, nuclear plants might anticipate being able to keep abreast of improvements in production techniques by replacing only that part of their unique facilities which would become outmoded.   The ordinary thermal plant, not being able to incorporate improvements in nuclear facilities, might have to anticipate replacing its entire furnace and boiler with nuclear facilities.   In this circumstance it is impossible to say to what part of the investment in each type of plant the higher obsolescence rate should be applied.

Should a different depreciation rate be assumed for the two types of plants?   Is it not possible that the physical life of the unique nuclear facilities, at least the nuclear reactor itself, will be shorter than that of the facilities they replace because of the conditions of intense radiation under which they will operate? Two important considerations in reaching a decision on how to handle depreciation are: (1) To a considerable extent the unusual operating conditions for the nuclear reactor are reflected in increased maintenance charges. In our intermediate and maximum cost estimates we allow for higher maintenance costs, which might include the periodic rebuilding of those portions of the plant which deteriorate most rapidly. (2) Investment cost in the specialized nuclear facilities apparently will include a sizable item for the initial inventory of fissionable material. This material probably will, as we have seen, have a perpetual life in the sense that it will continually reproduce itself. It is questionable, then, if any depreciation allowance should be made for this item of investment cost.

We have decided that within the framework of our cost comparisons the available information does not justify a different depreciation and obsolescence rate for the nuclear plant as compared with the ordinary thermal plant. Since for other items in fixed charges (taxes, interest, etc.) there is no reason to expect that the rate will be different in the two types of plant, we use the same overall rate in computing fixed charges for both.

APPENDIX B: SOME ECONOMIC IMPLICATIONS OF THE CONTROL OF ATOMIC ENERGY

The generation of significant quantities of atomic power will involve the production and use of large amounts of exactly the same kind of fissionable materials that are used in the atomic bomb. This gives rise to political considerations of critical importance in any program for the production of useful atomic power. With either international control of the production and use of atomic power or governmental control individually exercised by each nation, there are likely to be consequences of economic importance. We will try to suggest what some of these may be, without exploring the subject in detail.

If there is no international agreement and each nation develops atomic energy independently, as many are now doing, several kinds of economic effects could follow:

(a) Since fissionable materials and energy are a joint product of reactor operations, the energy might be converted to useful form and made available for commercial purposes. This might be at a lower price than otherwise owing to heavy military subsidization of the reactor's operations (say, in the form of a very high price for fissionable material).

(b) On the other hand, useful power might not be produced at all, if the conditions required for the generation of useful power were found to be inconsistent with the maximum rate for breeding fissionable materials for use in the bomb. That the two operations may not be wholly compatible has been suggested by some of the specialists who have written on this subject.[77]

(c) Or useful power might be produced, but in much smaller amounts than otherwise, owing to the fact that fissionable materials would be persistently diverted to bomb manufacture. There would be fewer reactors (or smaller ones) because of an unwillingness to tie up too much material in reactors, in order to have more on hand in the form of bomb components. (It is worth noting that there is a problem here even with respect to military production alone: how much of a given stock of fissionable materials should be processed into bomb components, and how much should be invested in new reactors which would in time produce more fissionable material than is invested in them?)

(d) Useful power might be produced, but in regions which were selected because of their relative invulnerability to attack. Only by accident would locations based on this criterion coincide with the location which might have been chosen on purely economic grounds.

(e) For most nations the time scale of atomic power development probably would be much slower than if they could draw upon the pooled technical and scientific knowledge and financial assistance of all nations, and particularly upon what the United States could contribute to an international development authority.

An agreement on the international control of atomic energy might also have serious economic implications. The control scheme which has been given the most detailed elaboration is that formulated by the Board of Consultants to the Secretary of State's Committee on Atomic Energy (subsequently referred to as the Acheson-Lilienthal plan),[78] which formed the basis for the atomic energy proposals submitted to the United Nations by the American representative, Bernard Baruch, in June 1946. It does not seem likely that this plan will be accepted. Yet it is useful to consider the possible economic consequences of the Acheson-Lilienthal plan as a basis for the discussion of other plans for international control.

[77] See, for example, the discussion in Chapter I, A.2.a. Also Dr. Szilard's statement that: " .·. . the fastest ways of producing fissionable material might not necessarily be those which permit the utilization of heat for steam production." (Hearings before Atomic Energy Committee, United States Senate, 79th Congress, *op. cit.*, p. 270.)

[78] *A Report on the International Control of Atomic Energy*, House Document No. 709, 79th Congress, 2nd Session, Department of State Publication 2498, Washington, G.P.O., March 16, 1946.

A key provision of the Acheson-Lilienthal plan was that commercial atomic power should be produced in two types of installations: (a) those which produce fissionable substances; and (b) those which cannot produce fissionable substances. The former were considered dangerous because they make materials which can be used in manufacturing bombs, and would for that reason be internationally managed; the latter were considered relatively "safe" and their management would be left in national hands.[79] The "safe" installations would rely on the internationally-managed installations for all the fissionable materials they required in their operations.

Let us note first that under this arrangement the volume of dangerous (or primary) capacity would set the size of the world atomic energy industry, since dangerous reactors would produce the fuel required by the safe reactors. If, therefore, there were any disposition to limit the volume of primary capacity in order to minimize the dangers of diversion of materials or seizure of production plants by the country in which they are located (dangers which will exist even under the control scheme), the production of useful atomic power might be smaller than it would otherwise be.

Whether the size of the world atomic power industry would be limited by these considerations would depend on the manner in which such a control scheme, if ever adopted, were administered. But the plan is also characterized by what might be called a "built-in" restriction on the size of the atomic power industry. That is to say, reactors which might be able to produce fissionable materials would be prevented from doing so as a security measure. Since fissionable material would be needed as part of the initial fuel investment in new reactors, the ability of the atomic power industry to grow might be severely limited. But this is not necessarily a net disadvantage of the control scheme, for if there were no international control a similar limitation might arise, as we have seen, from the diversion of fissionable material to bomb manufacture.

The importance of the restriction implied by such a control scheme would be determined in large measure by the comparative size of the safe and the dangerous nuclear reactor capacity. If there were comparatively few safe reactors (i.e. those not producing fissionable materials) the drag on the growth of the atomic power industry might not be serious. But to have very little safe capacity compared to dangerous capacity would defeat one of the main purposes of the plan. The authors of the Acheson-Lilienthal plan actually expected to have an almost even distribution of

[79] The relative safety of the latter type of plant is enhanced under the Acheson-Lilienthal plan by the provision that they use as fuel "denatured" fissionable substances, i.e. fissionable material rendered relatively ineffective for the production of atomic bombs, through mixture with an inert isotope of the same substance.

nuclear power capacity as between safe and unsafe reactors.[80]   If the two types of nuclear reactor capacity were about equal, the problem of adequate fissionable material supplies for power reactors could be quite severe; for there might be difficulties not only in producing enough fissionable materials for building new reactors, but simply in producing enough to take care of the needs of existing reactors.

As we explained earlier, we have assumed in our analysis that reactors will make at least enough new fissionable materials in the course of their operations to replace the material consumed in the production of power. In fact we assume that they can make somewhat more than the amount needed to replenish their own stocks of fissionable material—possibly 10% more, according to one estimate—which could be used to build new reactor capacity.   This basic assumption in turn supported another assumption which was of critical importance in the cost calculations: that the continuing fuel supply of the reactor group would consist of natural uranium or thorium, which are extremely cheap per unit of energy produced.

But let us assume now that the safe and dangerous reactor capacities are equal, and that the dangerous capacity produces only 1.1 units of fissionable material each time it burns 1 unit.   In this case, the safe reactors would burn up as much fissionable material as the dangerous ones, since they would be producing an equal amount of power.   Therefore each time 2 lbs., for instance, of fissionable material were used up (1 lb. in safe and 1 lb. in dangerous reactors) only 1.1 lbs. would be produced (in the dangerous reactor).   The existing nuclear reactors would, therefore, continuously need additional fissionable material from outside to keep going—at the rate of 0.9 lbs. for each 2 lbs. burned.   A balance could be achieved only if there were facilities for producing fissionable material outside the nuclear power industry. While plants for separating the U-235 isotope or for producing plutonium from natural uranium (such as those at Oak Ridge or Hanford) might make good the deficit, the costs of producing nuclear power might then be considerably above those we estimate.

Of course, this example is based on uncertain assumptions; it is not intended as a forecast of what would happen if a plan like that of Acheson-Lilienthal were adopted.   Thus we have based our example on an assumed breeding rate of 10% in the dangerous reactors.   This rate argued against the possibility that only as much safe capacity would be built as could be satisfied by the fissionable materials produced in dangerous reactors in excess of their own needs; under these circumstances, safe capacity would constitute only about 10% of dangerous capacity and such uneven proportions would almost nullify the objectives of the control plan.   But it is by

[80] *Ibid.*, pp. 28–29.

no means certain that the 10% breeding rate will actually obtain. If the rate were higher—say, about 30%—safe capacity based on the excess fissionable materials produced in dangerous reactors could constitute about 30% of dangerous capacity. It is unlikely that equality between the safe and dangerous capacities could be sustained on this basis alone, as this would require a breeding rate of 100% (i.e. for each lb. of plutonium used up, 2 lbs. are produced), a higher rate than is implied in the published literature.[81] But a breeding rate of about 30%—taken together with the possibility that natural uranium may be used alone in some reactors— might permit the volume of safe capacity to constitute a sufficiently high percentage of dangerous capacity (although not necessarily equalling dangerous capacity) to satisfy the objectives of the control plan, without significantly increasing operating costs through the need for drawing on costly fissionable material supplies produced outside the nuclear power industry.

[81] The critical factor here is the number of neutrons which are generated during fission in excess of the neutrons required to propagate the chain reaction. Irvine states (*op. cit.*, p. 364): "The number of fission neutrons in excess of the one necessary to maintain the chain reaction is approximately 1.3 per fission." This is in accord with the calculations made by L. Szilard and W. H. Zinn, in "Emission of Neutrons by Uranium," *Physical Review*, Vol. 56, October 1, 1939, p. 619.

CHAPTER II

The Cost of Electricity from Conventional Energy Sources

THE main purpose of this chapter is to present information about the cost of generating electric power from conventional energy sources throughout the world. Costs are known to vary greatly among different regions, depending on the local energy resources, if any, and on the distance over which fuels must be transported when such resources are not locally available. Our purpose is to show systematically the region-to-region differences in the cost of electricity produced from conventional sources, relative to the cost of electricity generated from the heat of nuclear fission.

These estimates are confined to electric power, but electricity accounts for only a small part of total energy use. In the United States, for example, the generation of electricity accounted for only about 15% of all the energy consumed in 1947. Therefore, should our estimates indicate that atomic electricity might be cheaper than electricity from conventional sources in a particular region, this would mean that only a part of that region's total energy requirements could be provided at a lower cost by atomic power. Thus the question arises whether atomic power might not be cheap enough to be used for those purposes, which account for a very large part of total energy consumption, where electricity is not economical at the present time. This question is considered in some of our later chapters. There we analyze the possibility of using energy generated from nuclear fission for such operations as iron ore reduction, railroad transportation, residential heating, glass melting, and so forth, where electricity is, by and large, not competitive with other energy forms according to present-day technology and with existing price relations among energy sources.

The estimates we have made of the cost of electricity from conventional sources are shown on Map 1. Our understanding of these cost estimates will be aided by examining the underlying world pattern of energy resource availability. We have, therefore, also included a world map of the location of coal and oil resources (Map 3), and another world map of the location of water power resources (Map 2). In using these maps, it is well to refer to the map showing the world distribution of population (Map 4).[1]

The chapter is in two main sections. The first part discusses the data, concepts, purpose, and methods underlying the derivation of the cost estimates, and summarizes the information contained in Map 1. The second part considers certain economic factors important to the comparison between atomic power and conventional power, not brought out in the preceding discussion of costs.

A. THE WORLD MAP OF ELECTRICITY COSTS

Map 1 has been constructed only for the purpose of comparing the costs of electricity produced from mineral fuels and water power with the estimated costs of atomic power. The figures shown on this map cannot be interpreted properly unless the purpose and the concepts underlying their derivation are understood.

1. The Nature of the Cost Figures

a. THE HYPOTHETICAL THERMAL ELECTRIC POWER PLANT. All of the costs shown for thermal electricity, i.e. power generated from conventional fuels, are estimates rather than actual costs. These estimates are of what the cost of producing electricity in the particular region would be if it were generated in a 75,000 to 100,000 kilowatt capacity plant of the most modern design, operating at an average rate of 50% of capacity. We hold constant in all our estimates the cost of building such a plant, its operating characteristics, and all operating costs other than the cost of fuel. Therefore, variations in the estimated costs of thermal electricity shown on the map arise only from differences in the cost of fuel among regions. Our purpose is to compare atomic power not with old and possibly inefficient existing power plants, but with the best that modern technology can provide.

The costs which we hold constant for the hypothetical plant assumed in the thermal cost estimates are (in 1946 prices, for comparability with the atomic cost estimates) :[2]

---

[1] This map was reproduced by kind permission of Clarence B. Odell, Division of Geography and Cartography, U. S. Department of State. Since cities are represented on the map by spheres which are proportional in volume to the metropolitan population, the shading of the rural and town areas is somewhat exaggerated in relation to the metropolitan areas.

[2] The cost and operating characteristics of the hypothetical thermal power plant were derived from a number of sources, including consultation with specialists in power plant engineering.

The investment cost of $130 per kilowatt of capacity is a reasonable engineering rule-of-thumb figure for costs prevailing in 1946. It must be realized, however, that actual costs per kilowatt in power stations constructed in that year would vary quite widely depending on conditions of construction, the climate, the type and cost of fuel, the amount of coal storage needed, etc. Although a single value on investment per kilowatt abstracts from all these variations and is, therefore, not to be taken literally in

(1) Fixed charges at 3.3 mills per KWH. This is based on:
   a. Investment in plant and equipment at $130 per kilowatt of capacity.
   b. Fixed charges on investment at a rate of 11% per annum.
(2) Operating costs other than fuel at 0.7 mills per KWH.

Every one of the thermal cost figures shown on the map, therefore, contains a charge of 4 mills per KWH for all costs of production other than fuel. To get the total figure, we add to this the cost of fuel, on the basis of an assumed thermal efficiency of 34% in converting the energy contained in the fuel to electric energy (i.e. 10,000 BTU of heat input per KWH of electric energy produced). Thus, for example, where standard grade American bituminous coal containing 26,200,000 BTU per short ton is used for power generation, the estimating procedure would, for different coal costs, yield electricity costs as shown in Table 2. This table could be expanded to include other fuels, as well as coal at higher prices.

We believe that the cost factors just set forth are in accord with the most advanced American engineering practices. But it must be understood that we have used these figures to make cost estimates for widely differing regions throughout the world. It is conceivable that for some

---

respect to costs at a specific site, it is useful in a broad survey such as ours, which is mainly concerned with measuring the effects of variations in fuel costs. (We may note that a cost of around $130 agrees roughly with the investment in thermal power quoted in other studies. See, for example, "Atomic Energy, Its Future in Power Production," *op. cit.;* Thomas, "Nuclear Power," *op. cit.;* Davidson, "Nuclear Energy for Power Production," *op. cit.,* p. 6; Philip Sporn, Atomic Energy Discussion, *American Economic Review, Papers and Proceedings,* Vol. 37, May 1947, p. 113; and Isard and Lansing, *op. cit.,* footnote 12.)

The proper rate of fixed charges in thermal power plants has been a subject of much dispute. The overall rate for privately financed thermal plants is generally placed at between 12% and 15%. However, this allows for a higher rate of interest than prevails on government borrowing. Since it is reasonable to assume government borrowing at 3% for the atomic power plant, we decided, wholly for reasons of comparability, to use the same interest rate for the thermal power plant. The other elements in the total have been assumed at 4% for depreciation and obsolescence, and 4% for taxes, insurance and miscellaneous overhead and administrative expenses. (See "Cost of Energy Generation, Second Symposium on Power Costs" in the *Proceedings of the American Society of Civil Engineers,* Vol. 64, April 1938, Part 1, particularly the separate papers contributed by W. F. Uhl, John C. Page, C. F. Hirshfeld and R. M. Van Duzer, Jr., and Ezra B. Whitman. For an overall rate roughly comparable with ours see Isard and Lansing, *op. cit.* [after substituting the lower interest rate for government borrowing], and Menke, *op. cit.*)

Operating costs other than fuel are based mainly on data in "Fourth Steam Station Cost Survey," by A. E. Knowlton, *Electrical World,* Vol. 112, December 2, 1939, adjusted for price change between the prewar period and 1946.

A thermal efficiency of 10,000 BTU per KWH is equivalent to about ¾ lb. of standard grade bituminous coal per KWH of power produced. Average practice in thermal stations in the United States in 1947 required 1.31 lbs. of coal per KWH. Our assumed thermal efficiency is about equal to that of the most efficient straight steam-electric plant in the United States (Philip Sporn, *op. cit.,* p. 113).

of these regions the resulting cost estimates, which purport to be the lowest possible costs for thermal power in the area, would, in fact, be somewhat higher than actual costs in a modern plant serving the area. This kind of error might arise, especially since investment per kilowatt might have been somewhat less if we had assumed a power plant of lower thermal efficiency. The objective in any specific case of power plant construction is, of course, to minimize total costs, so that where fuel is very cheap it might be economical to sacrifice increased thermal efficiency in order to save on fixed charges. We have chosen to maximize thermal efficiency in our assumed plant because the fuel costs in most parts of the world—and certainly in those regions where atomic power might be competitive— are such that total production costs probably are minimized in a plant of high thermal efficiency.

TABLE 2.   *Electricity cost corresponding to varying coal costs according to estimating procedure used for Map 1.*

Cost of coal ($ per short ton containing 26,200,000 BTU)	Generating costs other than fuel (mills per KWH at 50% of capacity use)	Cost of 10,000 BTU of fuel (in mills)	Total generating costs (mills per KWH at 50% of capacity use)
2	4	about 0.8	4.8
4	4	” 1.5	5.5
6	4	” 2.3	6.3
8	4	” 3.0	7.0
10	4	” 3.8	7.8

The dependence of thermal efficiency upon investment in plant is one of the limitations of our estimates. (Another limitation, which is inherent in the conversion of foreign currencies into American dollars, will be taken up later.) These estimates are merely approximations of the real situation which might prevail in any locality at some future time with respect to the competitive position of atomic power and conventional power.

b. THE COST OF FUEL. Because our estimates of the cost of atomic power are in terms of 1946 prices, we have also tried, wherever possible, to show our estimates of the cost of power from conventional sources in 1946 prices. Thus, the assumed investment per kilowatt and the operating costs other than fuel refer to prices prevailing in 1946. In principle, therefore, the fuel costs which enter into our electricity cost estimates also should be for 1946.

However, at the time our research on costs was being done, 1946 fuel prices were difficult to find for many of the regions we wanted to include in our survey; in addition, the 1946 price figures were difficult to interpret properly, even when they were available. Many coal mining regions throughout the world had been seriously hurt by the war and its aftermath. Normal channels of coal trade were disrupted as nations like Germany and England, which had previously exported large amounts of coal, were barely able to meet their own requirements. In these and other European countries the required number of mine laborers proved difficult to recruit, and mining machinery and other mining supplies were not available. Consequently, the price of coal attained unusual heights in most parts of the world. The price of oil rose also in response to the general fuel shortage.

The inflated coal prices of 1946 may represent to some extent deep-seated difficulties of the coal mining industry which began to manifest themselves in the postwar period. There is evidence of this in England. But, by and large, it appeared to us that the world-wide fuel shortages would ease, and that 1946 fuel prices in most parts of the world were probably higher than they would normally be in relation to the general level of prices. We decided, therefore, not to use 1946 prices, even where they could be found, but to use prices prevailing in the prewar decade. Usually these costs were for 1937, a year of fairly normal conditions in international fuel production and trade.

In deciding to use these prewar fuel costs we have perhaps made the best possible selection of a uniform set of world-wide fuel prices for our estimates, but we still are faced with the problem of comparability between our estimates of atomic costs and conventional power costs. There was a general rise in prices between the prewar period and 1946, and our figures for the cost of electricity from atomic sources (and for the non-fuel part of the electricity cost from conventional sources) are based on the higher prices of 1946. The main components of these figures are power plant equipment, general construction, and chemical and other specialized facilities for the atomic plant. The failure to provide correspondingly for any increase in fuel prices above those prevailing in 1937 biases our comparisons in favor of conventional sources of electric power. We will make allowances for this later on the basis of data for the United States where we believe that fuel prices in 1946 were, in general, not inflated in comparison with the overall price level, and especially with the cost of construction.

c. COSTS OF HYDROELECTRIC POWER. As we have seen, the thermal electricity costs shown on Map 1 are all derived according to a uniform estimating procedure which, in general, gives assurance that these are the lowest possible costs for thermal electricity in the particular region.

The figures for hydroelectric costs shown on the map are, on the other hand, actual generating costs at existing hydroelectric power stations, no matter when the particular installation was built, or what the special conditions are under which it might operate (e.g. the rate of capacity utilization and the rate of fixed charges are generally those actually prevailing at the particular plant; furthermore, no allowance is made for the fact that hydroelectric power must often be transmitted over longer distances than thermal power, because there is less flexibility in the choice of hydroelectric plant sites).

There are relatively few hydroelectric cost figures on the map, far fewer, as we shall see, than would be justified by the relative amounts of electricity generated from water power in the world.   The reason is simply that actual cost figures for existing hydroelectric plants are not readily available.   In the case of thermal electric plants a uniform, highly efficient power plant was assumed no matter what the region; the replacement of any obsolete plant by such a modern plant was considered possible. Thus the geographical differences in electricity cost were reduced to differences in fuel cost.   But in the case of hydroelectric power the cost of constructing the installation depends on conditions of physical geography and is the major element in cost differences from region to region; so that actual costs of existing installations or special estimates for projected installations are the only useful figures.

There is another important difference: while our cost estimates for thermal electricity purport to represent the costs which might prevail in a new thermal power station as compared to an atomic power plant in any region, the hydroelectricity figures, on the other hand, do not necessarily represent costs in an installation which might be built in the future.   The hydroelectric costs always represent sites which have already been developed; there is the question of whether undeveloped sites with as good natural conditions still exist.   Furthermore, costs in existing hydroelectric installations, which consist almost wholly of fixed charges, reflect construction costs of some other time.   Thus, even if we were to assume that the natural conditions of undeveloped sites in a particular region are no worse than those of existing sites, the old site costs would not be properly comparable with our atomic estimates which are based on 1946 construction costs.   The latter problem is analogous to the problem of fuel costs mentioned above.

d. FOREIGN CURRENCY CONVERSIONS. Since the cost of atomic power has been estimated only for the United States and not for other countries, how is it possible to make international comparisons of atomic and conventional electricity costs?   For any foreign country, for example France, our cost comparisons are set up as follows:

Atomic power costs	Thermal-electric power costs (Map 1)	Hydroelectric power costs (Map 1)
(Consisting mainly of fixed charges on investment): Estimated for the U. S., in American prices.	(a) Fixed charges on investment: Estimated for the U. S., in American prices. (b) Fuel prices: Given in French prices, converted to American prices at a prewar rate of exchange.[3]	(Consisting mainly of fixed charges on investment): Given in French prices, converted to American prices at a prewar rate of exchange.

Although this is the form in which the comparisons are actually shown in this report, they could just as well have been made in the following reversed form, without changing the indicated relations among the costs of different kinds of electricity (provided, of course, the same rate of exchange were used):

Atomic power costs	Thermal-electric power costs	Hydroelectric power costs
Given in American prices, converted to French prices.	(a) Fixed charges on investment: Given in American prices, converted to French prices. (b) Fuel prices: Given in French prices.	Given in French prices.

We would have liked to show all of our comparisons in terms of the second formulation, but this would have caused serious presentation difficulties since it would have involved showing prices in different currency units throughout the map. The relationship has, therefore, been expressed in the reversed form, and everything is shown in American prices, as in the first formulation. But the conditions which should be satisfied by a proper currency conversion rate can be seen more clearly by examining the second formulation of the comparison. Let us, then, examine the required price conversions and their meaning on this basis.

Thus the costs of fuel and hydroelectric power are already expressed in prices actually prevailing in France for these items. The problem, then, is to convert fixed charges on the investment in thermal power stations, and the estimated costs of atomic power, into French prices which would reflect their relative cost in France compared with the costs of those items already shown in French prices.

We apply the same conversion rate to the fixed charges on thermal power and the total costs of atomic power (which consist mainly of fixed charges on investment), both of which have been estimated only for the United States. We are assuming in effect that the *relation* between the investment required in atomic power and the investment required in conventional power which has been estimated for the United States will apply also in France and all other countries. Since we do not know how much an atomic power plant

[3] For almost all countries, the average 1937 rate of exchange was used.

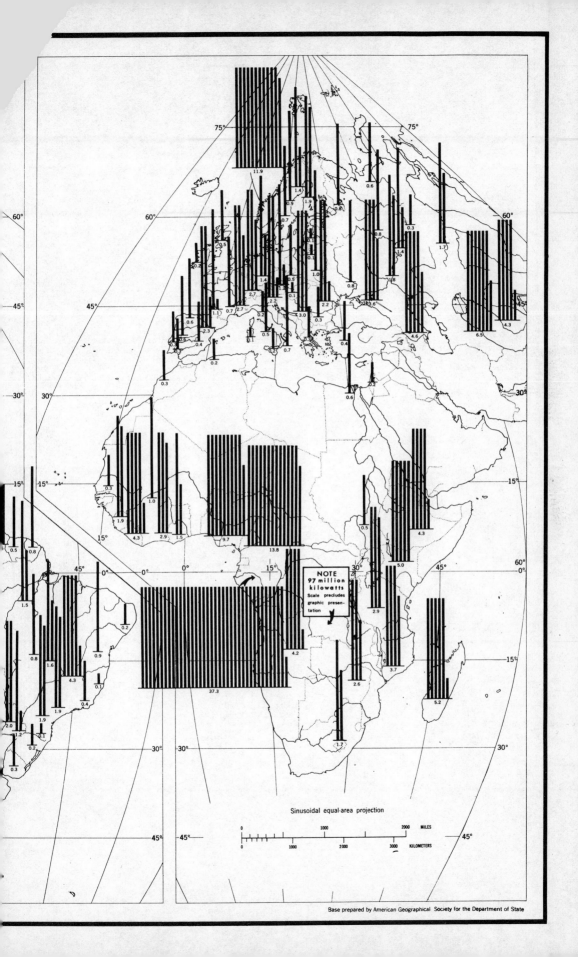

11.9

1.4
0.9 1.9
0.7
0.5
0.2 0.1
0.1

1.4
0.6 2.7 2.2 0.5 1.0
1.1 2.2 0.1
0.7 0.2 3.0 0.3
0.6 2.3 0.1 0.5
0.4 0.7
0.6

0.6
0.8
0.3
1.7

0.8
0.8 0.3
1.4
3.6
4.3
6.5
4.6
0.4
2.2
3.6

0.3
0.2
0.2

0.6
0.2

0.3

0.3
1.0
1.9
4.3 2.9 1.5 9.7 13.8

0.5
4.3
5.0

0.5 0.8
1.5

1.6
4.3
2.0
1.2
0.2
0.3
0.2

0.9
0.1
0.4
0.1
1.9
0.1

37.3

4.2

2.6

1.7

2.9
3.7

5.2

NOTE
97 million
kilowatts
Scale precludes
graphic presen-
tation

Sinusoidal equal-area projection

0   1000   2000 MILES

0   1000   2000   3000 KILOMETERS

Base prepared by American Geographical Society for the Department of State

will cost per kilowatt of capacity in France or any other foreign country, we are using the cost per kilowatt of capacity of a thermal power plant there as our guide, adjusting it to represent the cost of an atomic power plant according to what is now known (or estimated) about the relationship between the required investment in atomic facilities and in conventional power facilities in the United States.   This is justified by the consideration that in a given country, by and large, almost any factor affecting the cost of constructing a modern thermal power plant would also affect the cost of constructing atomic power facilities since the main cost components are the same: construction labor, planning and supervision, construction materials, electrical generators and accessories.   To be sure, the remaining cost factors, viz., the apparatus for generating heat and for pre-processing of fuel and the utilization or disposal of waste, are vastly different technologically in the two cases.   It is unlikely, however, that the economic causes that will determine the geographical variations in the construction cost of these facilities (primarily the cost of skilled labor, manual and otherwise) would be sufficiently different to affect markedly the relation between the total costs of construction of the two types of power plants, per kilowatt of capacity.[4]

There remain two questions: whether, on the one hand, the exchange rate can justifiably be used to translate the construction cost of the assumed modern thermal power plant from American into foreign prices, to be added to the foreign price of fuel, after reducing both to a per-KWH basis; and, on the other hand, how to translate the American estimate of atomic power cost into its foreign counterpart.   For this purpose, the exchange rate should reflect the ratio between the cost per kilowatt of capacity of building and equipping a thermal power station in France (or any other country) and the United States.   Clearly, therefore, in those countries which bought or sold power plant equipment abroad the exchange rate would serve our needs quite well.   Moreover, in a few cases—France, Great Britain, and the Soviet Union (to be discussed presently)—where we had information on the cost of power plant construction, we found that an exchange rate did reflect the relevant cost ratio.[5]   We lacked such material for other countries.   Hence

[4] This discussion is wholly in terms of the investment cost per kilowatt of capacity. In reducing this cost to a per-kilowatt-hour basis, we apply a rate of fixed charges based on American data.   Whether this rate of fixed charges can justifiably be applied in estimating costs in all foreign countries is considered in another section of this chapter (see pp. 71–72).

[5] To know that the exchange rate reflects the relevant cost ratio requires information on the cost per kilowatt of capacity of building and equipping identical thermal power plants in different countries.   As we have seen (footnote 2), construction costs vary among power stations according to the specific circumstances in particular locations; it is, therefore, difficult to find cost figures for different countries referring to exactly the same conditions.   The most we can do then is to get an engineering "rule of thumb" cost for another country corresponding to similar estimates used by American engineers, or an average cost of power station construction, and to compare the resulting cost ratios be-

the conversion via exchange rates may introduce a bias in the case of countries (other than the three just named) that did not import or export power plant equipment.

However, for such countries this bias would affect both sides of the comparison between the costs of atomic power and of conventional thermal power in the same direction although (because of the fuel component of the costs of conventional power) not in the same proportion.

It is in the light of these considerations that the likelihood of competition between atomic and conventional power on the basis of the estimates in Map 1 has to be judged for each country.

A close study of the currency conversion problem had to be made in the case of the U.S.S.R. Although that country did import some foreign electro-generating equipment, the rouble equivalent of the price paid for it,

---

tween the United States and other countries with the exchange rates. This is what we did for Great Britain, France, and the U.S.S.R.

The prewar costs of thermal power stations in the United States generally fell within the range of from $75 to $115 per kilowatt of capacity; the usual engineering estimate of cost during this period was about $100 per kilowatt (see, e.g., *The Electric Power Industry*, by John Bauer and Nathaniel Gold, New York, Harper and Bros., 1939, p. 65; "Progress in the Generation of Energy by Heat Engines," by George A. Orrok in *Proceedings of the American Society of Civil Engineers*, Vol. 63, December 1937, pp. 1887–1888; and Hirshfeld and Van Duzer, *op. cit.*, p. 658).

The costs quoted for France and Great Britain which indicate that the exchange rates reflect the proper cost ratios between their investment costs and the American cost figures just cited are found in two papers presented at the Third World Power Conference: "National Power and Resources Policies," by Ernest Mercier (*Transactions, Vol. IX*, Washington, G.P.O., 1938, p. 98) and "Integration of Electric Utilities in Great Britain," by Harold Hobson (*Transactions*, Vol. VII, p. 626). The French figure quoted—1,600 francs per kilowatt—is the estimated cost of erecting a then modern power plant per kilowatt of installed power; the English figure is not given directly, but was derived from the fixed charges component shown in a breakdown of the average costs of production at all public utility generating stations in Great Britain in 1934–1935. The British figure, although indirectly derived and based, in part, on construction costs in earlier periods, supports the exchange rate as adequately reflecting the ratio of the level of investment per kilowatt of capacity in Great Britain as compared with the United States. We may note that in France the required cost ratio corresponded better to the pre-devaluation exchange rate of 1936 than to the post-devaluation rate of 1937.

Here is how Professor Metzler dealt with the general problem of using prewar exchange rates: "For most countries the period selected as a prewar base was the nine-month period October 1936 to June 1937. . . . The base period is one of relative exchange stability which followed the Tri-Partite Agreement between the United States, the United Kingdom, and France. The exchange rates prevailing in this base period are assumed . . . to be more or less normal prewar rates, but it is probable that in the unstable international economic situation which prevailed before the war no single period can be regarded as a completely normal or balanced period." (Lloyd A. Metzler, "Exchange Rates and the International Monetary Fund," in *International Monetary Policies*, Board of Governors of the Federal Reserve System, Postwar Economic Studies No. 7, Washington, September 1947, p. 5.) On the whole, we have followed Metzler in selecting the period. But note that our problem differs from his in that we are concerned, not with the question of whether an exchange rate between a pair of countries approximates some average of the price-ratios of all commodities, but merely with the question of whether it approximates the ratio between plant construction costs.

calculated on the basis of an official exchange rate, may not reflect the Russian construction costs; the state monopoly of foreign trade and of production appears to make the official exchange rate arbitrary. In addition, the exchange rate was abruptly changed during the prewar years that interest us. In choosing the appropriate conversion rate we applied the criterion just described, i.e. we selected that rate which converted our assumed American cost per kilowatt of capacity in a modern thermal power plant into a Russian price which corresponded most closely to the cost of building such a power plant in Russia. This rate was 1 rouble = $0.19, and all the costs shown for the U.S.S.R. on Map 1 were derived on this basis.

We can illustrate the problem of choosing a proper currency conversion rate for Russia by using rough figures for 1937 of the average costs of erecting a power plant, and the average price of coal at the mine mouth in the United States and the U.S.S.R., as follows:

	U. S.	U.S.S.R.[6]
Power plant, per kilowatt of capacity	$100	500 roubles
Coal, per ton	$ 2	10 roubles

Although we have no way of knowing from these figures how coal costs in Russia and in the United States compare, we do learn from them that relative to the cost of building a power plant, coal is equally costly in Russia and in the United States. Assume now that atomic power costs have been estimated for the United States and are found to consist wholly of the cost of providing the necessary plant and equipment. Let us say further that the cost of plant and equipment is estimated to be twice as much as for ordinary thermal electricity, or $200 per kilowatt of capacity. On the assumption that this relationship would apply also in Russia, the investment in atomic power there would be 1,000 roubles per kilowatt of capacity, compared with 10 roubles per ton of coal. Now, we want to convert the Russian price of coal into American prices for comparison with atomic costs which had been stated in American prices. Suppose we have to choose between two exchange rates, one valuing the rouble at 50¢ and the other at 20¢. The 50¢ rate would give a price of $5 per ton for Russian coal in American prices, the 20¢ rate a price of $2 per ton. Since we have to convert the Russian cost of coal into American prices in such a way that the

[6] The Russian figures are derived from *Electric Power Development in the U.S.S.R.*, prepared by a Soviet scientific committee for the Third World Power Conference (Moscow, INRA Publishing Society, 1936). The power plant investment figure is not given directly in that source, but was estimated from data on the investment in thermal stations per thousand annual KWH, including the investment in mining and transporting fuel for the power station. (The investment figures including mining and transportation are given on p. 486; other data used in deriving a cost for the power stations alone are given on pp. 390–392 and 415–420. For a further discussion of these figures see Chapter XIV, D.1.)

same relationship be maintained between the cost of coal and the investment in power plants as prevailed in Russia, we have to choose the 20¢ rate. A price conversion on this basis would say this: If the investment in ordinary power plants is valued at $100 per kilowatt, and the investment in atomic power is valued at $200 per kilowatt, the comparable figures for Russian coal prices at the mine is $2 per ton. We still do not know that in an absolute sense Russian and American coal costs at the mine are equal; we know only that in terms of the relationship between coal cost and the cost of investment in power plants in Russia, $2 is the proper coal cost to compare with a $200 cost per kilowatt of atomic power, just as $2 is the comparable cost in the United States. This in essence (with approximate values) is our solution to the Russian currency conversion problem.[7]

To sum up: Map 1 does not purport to answer the question: How do conventional electricity costs compare in an absolute sense among the countries of the world? Instead the figures on this map give an approximate answer to this question: Compared with the cost of providing atomic power in any country, as we have estimated this cost in American prices, what is the cost of electricity from conventional energy sources?

2. *The Map Data Summarized*

a. THE TYPES OF INFORMATION CONTAINED.[8] Map 1 contains information on the cost of electricity and the energy sources used in its generation

[7] In applying a prewar exchange rate in 1946 price comparisons we are assuming the same trend in power plant costs in the United States and other countries. We were not able to check this assumption. As indicated below, our adjustment of prewar fuel prices to 1946 prices is also on the basis of price trends in the United States. In principle we do this: We approximate the prewar price relationships between coal and the investment in power plants prevailing in a foreign country. We then adjust this relationship to 1946 prices on the basis of the separate price trends in the United States between the prewar period and 1946 for fuel and power plant construction.

[8] The most valuable general reference on the energy sources used in producing power in different countries was *Energy Resources of the World* (U. S. Department of State, Publication 3428, Washington, G.P.O., 1949), which had been made available to us in preliminary draft through the kindness of Mr. Nathaniel B. Guyol, who directed that study, and of Mr. Wilfred Malenbaum. Other valuable references on energy sources were the articles on various countries contained in the *Transactions* of the First and Third World Power Conferences; *Coal Mining in Europe* (U. S. Bureau of Mines, Bulletin 414, Washington, G.P.O., 1939); and the "Foreign Minerals Surveys" prepared by the U. S. Bureau of Mines.

Our sources of cost information were varied. The *Foreign Commerce Yearbook* (published by the U. S. Department of Commerce) was an important source of fuel export and import prices by countries. Papers presented at the First and Third World Power Conferences provided much cost information on specific countries. Official Statistical Yearbooks of the different countries (or of smaller units) were frequently used. For the Soviet Union, *Electric Power Development in the U.S.S.R., op. cit.,* was our major source of cost information. This was supplemented by more recent texts on the geography of the Soviet Union (in particular, *The U.S.S.R., A Geographical Survey*, by J. S. Gregory and D. W. Shave, London, George G. Hareb, 1944; and *Land of the Soviets,* by N. Mikhailov, New York, Lee Furman and Co., 1939).

For thermal costs in the United States, we relied mainly on a map of fuel costs in "Power and Fuels," by Lincoln Gordon (in *Industrial Location and National Resources,* National Resources Planning Board, Chapter 7, Washington, G.P.O., 1943).

at selected points throughout the world. In general, these points are the major population centers of the different regions. Obviously, the characteristics of such centers vary widely throughout the world depending on the region concerned. The figures shown for each location represent costs there as derived by the methods described in the preceding sections. These costs have been grouped in four classes, designated by the four colors on the map.

Information is shown for regions in widely differing stages of economic development. Some of the regions, like the United States and parts of Europe, have a well-developed electric power industry. At the other extreme, there are regions like central Australia and the Amazon valley in Brazil which are probably almost completely without electric power facilities. Some attempt has been made to picture the difference in electric power development among regions by varying the density of observations included on the map, but this provides merely a very rough approximation of these differences. At best, the varying density on the map shows only that the United States and western Europe have a more highly developed electric power industry than South America or Asia, for example, but it does not purport to tell us anything of the differences in development between countries like India, China, the Union of South Africa, Australia, Brazil, etc.

Where electricity is now being produced the map tries to give a representative picture of the energy sources being used, and the costs of electricity based on these sources, if produced in a modern (thermal) plant. Where electricity is not now being produced, we have tried to determine what the most likely source would be and have made estimates of its costs (if produced in a modern plant) where this could reasonably be done.

The map necessarily incorporates many simplifications regarding the specific information shown. Often choices had to be made on the basis of judgment. It is certain, therefore, that other researchers setting out to do the same job would not have produced a map agreeing with ours at every point even if it were based on the same overall approach.

The following outline sums up the kinds of information contained on the map, and the methods followed in obtaining this information:

(1) *For Thermal Electricity*
    1.1 In those cases where thermal electric facilities exist.
        1.1.1. Shows the major fuels used in a region.
        1.1.2. Where appropriate fuel prices were available for a particular

---

For the costs of hydroelectric power in the United States we relied mainly on H. K. Barrows, "Hydro-Generated Energy," *Proceedings of the American Society of Civil Engineers,* Vol. 64, April 1938, Part 1. Information on very low-cost hydroelectric power for special industrial users came from "Aluminum Plants and Facilities," *Report of the Surplus Property Board to Congress, September 21, 1945,* Washington, G.P.O., 1945.

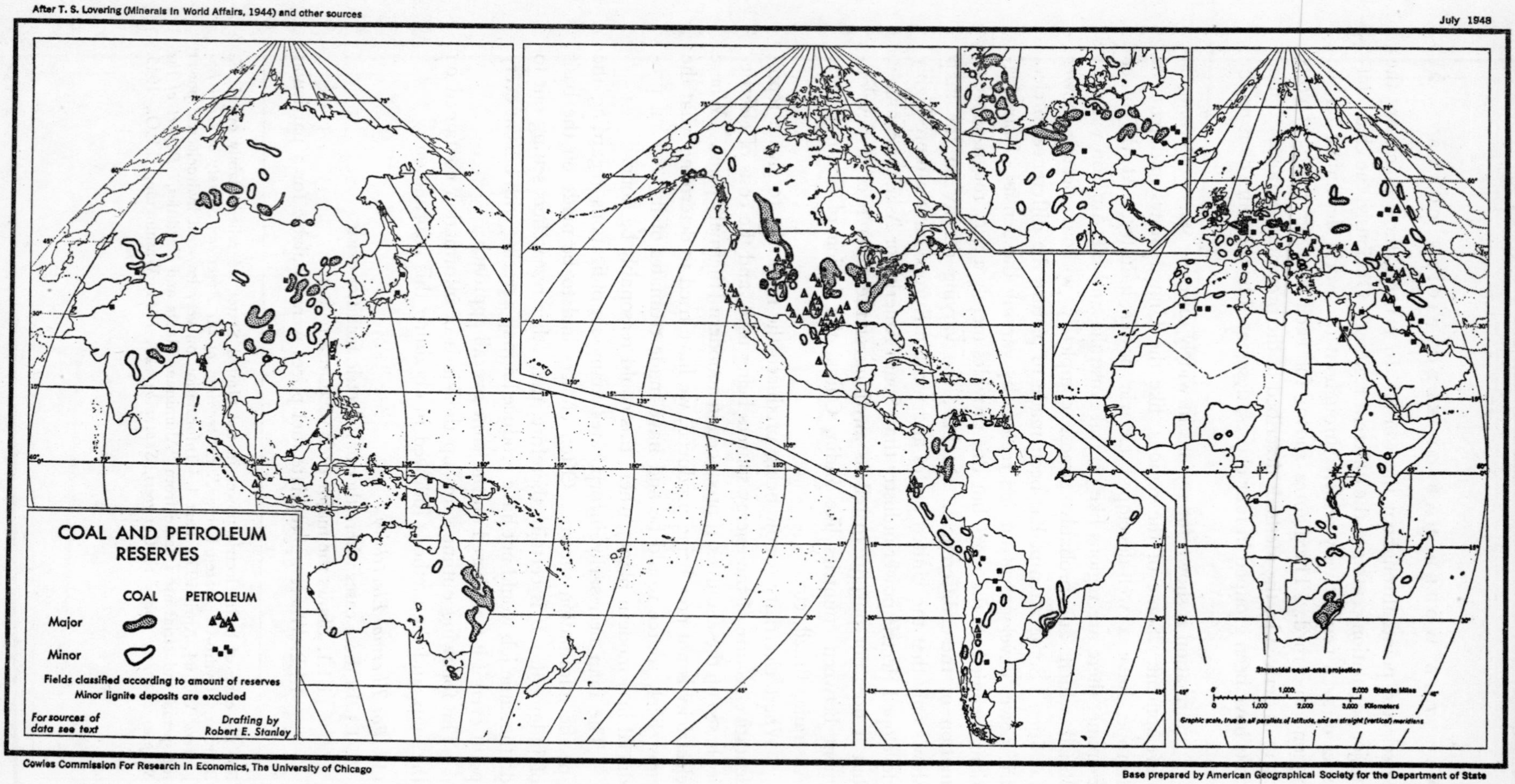
After T. S. Lovering (Minerals in World Affairs, 1944) and other sources
July 1948
COAL AND PETROLEUM RESERVES
COAL
PETROLEUM
Major
Minor
Fields classified according to amount of reserves
Minor lignite deposits are excluded
For sources of data see text
Drafting by Robert E. Stanley
Sinusoidal equal-area projection
Statute Miles
Kilometers
Graphic scale, true on all parallels of latitude, and on straight (vertical) meridians
Cowles Commission For Research in Economics, The University of Chicago
Base prepared by American Geographical Society for the Department of State

Map 3

DISTRIBUTION OF POPULATION

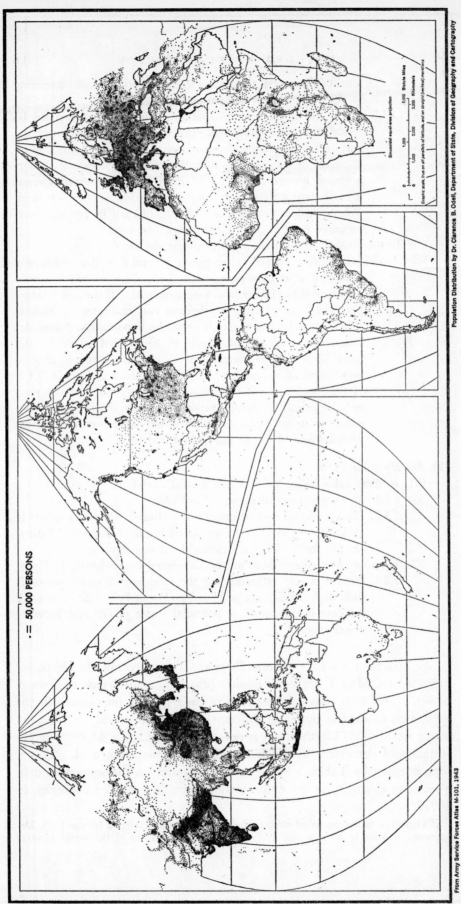

. = 50,000 PERSONS

Sinusoidal equal-area projection

Graphic scale, true on all parallels of latitude, and on straight (vertical) meridians

Statute Miles
0   1,000   2,000
Kilometers
0   1,000   2,000   3,000

From Army Service Forces Atlas M-101, 1943

Population Distribution by Dr. Clarence B. Odell, Department of State, Division of Geography and Cartography

MAP 4

locality they were used to estimate electricity costs according to the method described earlier.

1.1.3. Where no quotations on fuel price in the locality could be found, the cost of fuel was estimated by (a) finding out the cost of the fuel at the closest source of supply, and (b) adding the costs of transporting it to the particular locality. Where a choice between different methods of transportation existed it was assumed that the method with the lowest costs was used. Where actual transportation rates could not be found, representative rates in the United States were used.

1.2 In those cases where there are no existing electric facilities.

1.2.1. Information on the water power potential of the region was examined to determine whether thermal power plants would be important in future power development. If the water power potential is very large, no thermal cost estimates are shown.

1.2.2. Where thermal power plants would be important in future development, estimates are made of which fuel would be the most likely energy source, and of its cost in the locality in prewar prices, according to the procedure described in 1.1.3. If the locality is not on a navigable waterway and is served only by primitive transportation such as beasts of burden, an exact cost is not estimated; instead we indicate only that costs probably would fall in the highest cost class.

(2) *For Hydroelectricity*

2.1. In those cases where hydroelectric facilities exist.

2.1.1. Shows representative costs, if available.

2.1.2. If information on cost could not be found, but water power is an important local source of electricity, the existence of hydroelectric facilities is indicated without cost figures.

2.2. In those cases where there are no developed hydroelectric facilities.

2.2.1. Neither a cost estimate, nor the existence of water power is indicated. Reference to Map 2 on Water Power Resources will indicate if the area has potential water power and in what abundance.

b. THE SOURCES OF ELECTRIC POWER. The sources of electric power indicated on Map 1 are coal, lignite, peat, oil, natural gas, and water power. This map conveys some idea of their relative importance in the generation of electric power in different parts of the world, and for the whole world taken together, but it does not purport to be an exact representation of the extent to which the different sources are used. This is supplemented by Tables 3 and 4, showing figures on the sources used in generating electricity in the entire world, and a few selected countries, in 1937.[9]

[9] These two tables are based almost wholly on estimates developed by the U. S. Department of State in an extensive study of the energy resources of the world (*Energy*

Table 3 shows that the world's output of electricity in 1937 was accounted for mainly by a handful of countries. The seven countries separately shown produced about 75% of the world's total; the United States alone accounted for about 30% of the world's production. Almost 60% of the world's electricity was generated from fuels and somewhat more than 40% was based on water power. However, the comparative importance of fuels and of water power varied greatly among the countries of the world. For example, in the United States about ⅔ of the electricity was generated from thermal sources and ⅓ from water power, thus roughly paralleling the world average, while in the United Kingdom the total consisted almost entirely of thermal electricity, and in Canada the total was almost wholly hydroelectricity. The relative importance of the different fuels used in generating thermal electricity is indicated in Table 4. Almost ¾ of the world's thermal electricity was derived from coal, an amount roughly equal to the world's production of hydroelectricity. The amount of electricity generated from fuels other than coal accounted for less than 15% of the world's total quantity of electricity in 1937.

Although fuels other than coal were a relatively unimportant part of world totals, they were often quite important in individual countries. Lignite, although not widely used, was an extremely important source of electricity in Germany, and a few other countries (including Czechoslovakia) not shown separately in Table 4. Peat was quite important in the U.S.S.R., and natural gas in the United States, particularly in Texas and Oklahoma. Oil was used as a source of electricity in a large number of countries, as indicated in the bottom line of Table 4. It was the most important source used in many of these countries which were, however, small electricity producers. The total amount of electricity generated from oil accounted for only about 3% of the world's total production of electricity.

  c. THE COSTS OF ELECTRIC POWER. As shown in Map 1, the range of generating costs for conventional power sources, calculated for comparison with atomic power costs, is from 0.5 mills to 10.5 mills per KWH, or from considerably less than the estimated minimum cost of atomic power to about the estimated maximum cost. This range includes widely differing costs for different regions of the world, and often for different sections of a single country. These variations can be seen at a glance in terms of the colors representing different cost classes on the map, or, in greater detail, by

---

Resources of the World, 1949 ed., *op. cit.*).
   There are a few instances where certain information shown on Map 1 disagrees with the estimates shown in Tables 3 and 4. For example, our research indicated that a small amount of natural gas was used in electricity generation in the U.S.S.R. This is indicated on the map. Table 3, on the other hand, shows no electricity production from natural gas in the U.S.S.R. Again, we show electric power production from lignite in the United States on Map 1 while in Table 4 this is included under bituminous coal. This reflects a difference in the classification of coals. However, these differences are of little importance in terms of the general pattern of energy use in electricity production in these countries.

TABLE 3. *Production of thermal electricity and hydroelectricity: the world and selected countries, 1937[a].*

Country	Total electricity produced	Per cent of world production	Production of thermal electricity	Per cent of country's total electricity production	Production of hydroelectricity	Per cent of country's total electricity production
	(Production in billions of kilowatt hours)					
United States[b]	118.5	29.8	74.5	64.0	44.0	36.0
Germany	49.0	12.1	42.1	85.9	6.9	14.1
U.S.S.R.	36.4	9.0	29.5	87.0	6.9	18.9
Canada	27.7	6.8	0.5	1.8	27.2	98.2
Japan	27.4	6.7	7.2	26.4	20.1	73.5
United Kingdom	23.0	5.7	22.4	97.3	0.6	2.7
France	18.2	4.4	8.3	45.6	9.9	54.4
TOTAL, above countries	300.2	74.5	184.5	61.9	115.6	38.1
Remainder of the world	103.0	25.4	45.7	44.4	57.3	55.6
TOTAL WORLD	403.2	100.0	230.2	57.4	172.9	42.5

Source: Figures other than for the United States adapted from estimates shown in *Energy Resources of the World*, U. S. Department of State (preliminary edition, 1945).

[a] The figures for the United States are known to refer only to electricity for public use, not to electricity generated by industries for their own use (see Chapter XIII).

[b] Figures from Federal Power Commission, "Consumption of Fuel for Production of Electric Energy, 1946." Excludes a small amount produced from wood and wood wastes.

examining the individual cost estimates. The four color classes, in particular, provide a convenient summary of geographical variations in cost. In this section we explore the factors accounting for these geographical differences, and assess their significance for comparisons with atomic power.

The data presented in the preceding section indicate that on a world scale, only coal and water power are important sources of electricity. To clear the way for the discussion of coal and water power, we first inquire briefly into the minor sources of power.

Two of the minor fuels, lignite and peat, share an important characteristic. They are generally used in power generation only in the immediate vicinity of the places where they are mined, since their calorific value is so low compared with coal that the costs of transporting them, per unit of energy, are generally found to be excessive. Therefore, in evaluating the possible economic importance of atomic power, the competitive position of lignite and peat need be considered in only a very few regions. In these regions, the range of electricity costs shown on Map 1 is from 5 to 7.5 mills per KWH, with most costs falling between 5 and 6 mills. In general, then, atomic power would be competitive in regions using peat and lignite (based on the prewar prices of these fuels) only if it could be produced at less than the estimated intermediate cost.

TABLE 4. *Thermal electricity production according to fuels used: the world and selected countries, 1937.*[a]

(Production in billions of kilowatt hours)

Country	Total production of thermal electricity	Amount produced from coal	Per cent of country's total produced from coal	Amount produced from lignite	Per cent of country's total produced from lignite	Amount produced from oil	Per cent of country's total produced from oil	Amount produced from natural gas	Per cent of country's total produced from natural gas	Amount produced from manufactured gas[b]	Per cent of country's total produced from mfd. gas	Amount produced from peat	Per cent of country's total produced from peat
United States[c]	74.5	59.7	80.6			4.9	6.4	9.8	12.8	0.1	0.2		
Germany	42.1	18.0	43.2	20.0	47.2	0.2	0.4			3.9	9.2		
U.S.S.R.	29.5	17.3	58.6			2.7	9.2			1.4	4.7	8.1[d]	27.5
Canada	0.5	0.5	100.0										
Japan	7.2	7.2	99.8			*	0.2			*	0.2		
United Kingdom	22.4	22.4	99.4							*	0.3		
France	8.3	8.3	100.0										
TOTAL, above countries	184.5	133.4	73.2	20.0	10.2	7.8	4.2	9.8	5.5	5.4	2.8	8.1	4.1
Remainder of world	45.7	34.3	76.7	6.9	13.0	4.0	8.7	0.4	1.0	0.2	0.4	*	0.1
TOTAL WORLD	230.2	167.7	73.8	26.9	10.7	11.8	5.1	10.2	4.6	5.6	2.3	8.1	3.4
Number of countries using each fuel	59	59		8		61		2		7		8	

Source: Figures other than for the United States adapted from estimates shown in *Energy Resources of the World*, (preliminary edition, 1945, *op. cit.*).

* Amount produced was less than 100 million KWH.

[a] See footnote *a* to Table 3.

[b] At least a part of the manufactured gas used in electricity production is byproduct gas generated in coke plants and blast furnaces.

[c] Figures are from Federal Power Commission, "Consumption of Fuel for Production of Electric Energy," *op. cit.* Excludes a small amount produced from wood and wood wastes.

[d] Includes a small amount produced from wood.

Two other minor fuels, natural gas and oil, are little used in power generation primarily because of the pressure of competing demands for the limited supply available.   Power plants will in general buy oil or gas only if the price per unit of energy is no higher than a competing fuel such as coal.   Oil and gas are, however, especially adapted to certain uses other than the generation of electricity.   For example, oil is convenient for transportation uses; and both oil and gas have advantages in household and certain special industrial uses.   Consequently these fuels are used for electric power generation only in special circumstances.   Natural gas is most frequently used for power production in gas-bearing regions where more gas is released (by both oil and gas wells) than is consumed in other uses in the region or can be carried away by available pipelines.   Under these circumstances electric power will be quite cheap, sometimes as low as 4.5 mills per KWH in gas-bearing localities in the United States, the U.S.S.R., and Rumania.   Therefore, although gas use is generally unimportant in power production, nuclear power would not be competitive in those places where it is used, unless generated at a cost not far above the estimated minimum.

Although oil and natural gas account for about equal amounts of power in the world as a whole, the geographical pattern of power use of the two fuels is quite different.   While the use of gas in power production is confined to gas-bearing regions, oil is used to produce power in regions quite remote from oil resources.   This is due to an obvious difference between the two.   While gas can be moved economically only by pipe-line, oil can be shipped overland by tank car, although not so cheaply as by pipe-line, and, more important, it can be shipped overseas by oil tanker at a very low cost. In certain instances it is found economical, in widely separated parts of the world, to use oil to generate electricity.   In this sense oil might be considered a possible universal competitor to atomic power as a source of electricity.   Actually, however, the negligible amount of power involved in the total indicates that other energy sources have generally been found more economical than oil.   These other power sources are water power and coal.   The future role of atomic power as a source of electricity will be determined primarily in competition with them.[10]

(1) *Water Power*.   The lowest generating costs for electric power exist at hydroelectric developments in different parts of the world.   At selected points in several countries, including Norway, Sweden, Yugoslavia,

[10] If the price of coal continues to increase relative to the price of oil and gas, the latter two fuels may become more important in electric power generation in the United States.   Indeed, there is some evidence that they are already gaining in relative importance.   ("Survey Shows Coal Losing to Gas, Oil," *New York Times*, March 8, 1950.)   Only oil could rival coal on a world scale, but this would happen only with fuel prices above those in 1937 and 1946 (the years used in our cost analysis).   Note also that the analysis made below for coal would apply as well, in its essentials, to oil.

Japan, Italy, Switzerland, the United States, and the Soviet Union, hydro-electric power is generated at a cost of 4 mills per KWH or less. Electricity at such low costs is cheaper than the lowest cost now conceivable for atomic power. However, the map also shows cases where hydroelectric power is generated at much higher costs. In fact, on Map 1 hydroelectric costs show a wider variation than electricity generated from any other energy source; it is the only source of energy yielding power-generating costs which fall in each of the four cost classes. This variation in costs results from the wide differences in the natural conditions of water power occurrence from place to place, even within an abundantly endowed country such as Norway. This fact has important implications for cost comparisons among energy sources. Thus, where Map 1 indicates the existence of de-veloped water power but shows no cost, it is incorrect in the absence of any other information to assume that costs at this installation will neces-sarily fall into a low cost class.

As we have indicated, Map 1 includes only developed water power sites. Map 2 shows the distribution throughout the world of potential water power including the amounts already developed.[11] It can be seen that the

---

[11] The data shown on Map 2 refer to the total water power potential, including the amounts already developed; they represent the total power that could be obtained at "ordinary minimum flow." "Ordinary minimum flow" generally refers to that flow of water available from 90% to 95% of the time, i.e. only at 5% to 10% of the time would the flow of water be lower than this amount. The "ordinary minimum flow" is some-times estimated from the average flow during the month of lowest flow. The installed capacity at fully developed water power sites usually will exceed the potential power available under minimum flow conditions, possibly by a factor of 2 or 3.

Our primary source on potential water power throughout the world is the com-pilation prepared by the U. S. Geological Survey, appearing (among other places) in the *Foreign Commerce Yearbook, 1937* (U. S. Department of Commerce, Washington, G.P.O., 1938). However, for some countries, other estimates seemed preferable to those of the Geological Survey. For the Soviet Union, our estimates are calculated from data contained in *Electric Power Development in the U.S.S.R., op. cit.;* for certain other countries, including Brazil, China, and India, our estimates are based on papers sub-mitted to the First and Third World Power Conferences.

We have made regional breakdowns of a country's total water power potential wherever the country was large enough to justify doing so, and when the necessary data could be found. The regional breakdown is ordinarily by a country's major political subdivisions. In Canada these are the separate provinces, in the United States and Brazil the separate states, etc. For some countries the available data only provide a breakdown by other types of regions; for example, in Asiatic U.S.S.R. it is by river basins, in France, Italy, and Sweden by arbitrary geographical divisions. The size of the subdivisions obviously varies from country to country.

The following cautionary note is necessary: "The point must be emphasized that [no compilation of water power statistics now available] can be regarded as final. The countries regarding which exact information is available are limited in number, and it is doubtful whether, even in the most perfectly surveyed territory, a final figure can be realized. . . . We can classify as countries where a first approximation has been made towards the assessment of water power resources the following: U.S.A., Canada, New Zealand, Tasmania, Great Britain, Norway, Switzerland, Austria, Japan, Dutch East Indies, Czechoslovakia and Germany." (*Power Resources of the World*, London, World Power Conference, 1929, p. 54.) Although these words were written more than twenty years ago, they are hardly less valid today.

distribution of potential water power generally follows rainfall and relief. Dry regions have a small water power potential; see, for example, most of Africa from above the 15th parallel to the northern coast, central Australia, most of Arabia, Iran, Turkestan, and Mongolia. Again, flat regions have little water power potential; as, for example, the north European plain from northwest France through Belgium, Holland, northern Germany, Poland, and Estonia; much of the central United States; and Great Britain.

According to Map 2, there are numerous areas which have an abundance of water power. However, in the absence of engineering surveys of specific sites it is not possible to say what part of the potential shown at any place is of economic significance, or, more precisely, at what cost the potential power could be harnessed for use.

(2) *Coal.* Coal is by far the most important source of thermal electricity in the world as a whole. Naturally, electricity based on coal is generated most cheaply in coal-bearing regions. In some cases coal at the mine costs no more than about $1.50 per ton; burning coal at this price our hypothetical thermal plant could generate power at about 4.5 mills per KWH. Coal could be obtained this cheaply in the prewar decade in some sections of the United States, the United Kingdom, the Soviet Union, Manchuria, China, India, and South Africa. However, even in the vicinity of the coal mines, this was an unusually low price for coal. The average mine value of coal in the United States was about $2.00 per ton, which would yield a generating cost of between 4½ and 5 mills per KWH in the hypothetical power plant. Around the world, our estimated costs for electricity generated near coal mines commonly fall between 5 and 5½ mills per KWH.

The range of estimated electricity costs based on coal is second only to the range of costs based on water power. Despite the greater mobility of coal compared with water power, peat, lignite, and gas, the transportation cost still constitutes a significant part of the delivered cost of coal in most places.

For example, in the United States in 1937, the average value of coal f.o.b. mines was $1.95 while the average freight charge per net ton was $2.17. A representative railroad freight charge for a haul of 500 miles was $4.13. Ocean freight charges, although much less per ton-mile than railroad charges, are still considerable because of the long distances that are frequently involved. The average selling value per metric ton of coal raised at the pit in Great Britain in 1937 was $3.75; the average declared value per ton of coal exported was $4.61; freight charges to various ports in Brazil and Argentina ranged from $3.20 to $3.60. The price at these ports, therefore, ranged from $7.80 to $8.20 per metric ton, and at many cities far from the coast the price was $10.50 per ton, or more.

Thus, in the above example, if the average generating cost in a modern power plant at the mine mouth in the United States is estimated at about 4.8 mills per KWH (at a 50% plant factor), a similar thermal plant 500 miles away from the mine by railroad would have generating costs of about 6.5 mills per KWH.   Again, if the average generating cost in a modern plant at the mine mouth in Great Britain is estimated at 5.5 mills per KWH, a similar thermal plant in a port city of Brazil or Argentina, using coal from the same British source, would generate electricity at a cost of about 7.0 mills per KWH; and in inland cities the cost would exceed 8 mills per KWH. Let us note that at the mine mouth the average British and American coal valued at 1937 prices would yield electricity at such low costs that atomic energy would be competitive only if available at a cost near the minimum estimate.   As transportation charges are added to the price of coal, the comparison shifts.   At 500 rail miles from the mine in the United States, the thermal electricity cost would be about the same as the estimated intermediate cost of atomic power; in Brazil and Argentina electricity costs based on British coal would fall between the intermediate and highest cost estimates of atomic power.

These examples have a general significance.   For thermal power generally, coal being by far the major source, the wide regional variations in electricity costs pictured on Map 1 are primarily the result of the different distances involved in transporting fuel from the mine to the power plant. They are, therefore, due to the extremely uneven distribution of mineral fuel resources in the world.   This is shown on Map 3 for coal and oil, the two most widely transported fuels.[12]

The most important coal fields lie in the northern hemisphere.   With the exception of a few regions such as Eastern Australia and the Union of South Africa, the southern hemisphere is deficient in coal.   Even in the northern hemisphere there are many countries which do not possess coal, and many regions that lie far from coal fields.

The distribution of coal resources is reflected to a considerable extent in the statistics of coal production, shown in Table 5.   The ten countries specified produced 90% of all the world's coal before the last war.   Four of them, the United States, Great Britain, Germany, and the Soviet Union, produced about 75% of the world's total.   Within these countries, production was concentrated in a few main mining areas, such as the Appalachian and central fields in the United States, the Ruhr and Silesia in Germany, and the Donetz and Kuznetzk basins in the Soviet Union.

---

[12] Our primary source of information for Map 3 is T. S. Lovering, *Minerals in World Affairs*, New York, Prentice-Hall, Inc., 1944. We used as supplementary sources: *Coal Resources of the World*, XII International Geological Congress, Toronto, Morang and Co., 1913; *Energy Resources and National Policy*, U. S. National Resources Committee, Washington, G.P.O., 1939; and *Coal Mining in Europe, op. cit.*

The production of oil is even more highly concentrated in a few regions than the production of coal. Table 6 indicates that four countries, the United States, the Soviet Union, Venezuela, and the Middle East, account for about 90% of the world's output. (They also contained 90% of the world's proved reserves.) Within the producing countries, output is further concentrated in a few localities; for example, southwestern United States, and the Baku and Emba regions of the Soviet Union.

TABLE 5. *World production of coal by leading countries, 1937.*

Country	Production[a] (in millions of metric tons)	Per cent of world production
United States	449.4	32.5
United Kingdom	244.3	17.6
Germany	234.7	17.0
U.S.S.R. (1936 data)	115.2	8.3
France	44.7	3.2
Japan (1936 data)	41.8	3.0
Poland	36.2	2.6
Belgium and Luxembourg	29.9	2.2
Czechoslovakia	28.9	2.1
TOTAL, above countries	1,225.1	88.5
WORLD TOTAL	1,384.2	100.0

Source: *Energy Resources of the World.*, 1949 ed., *op. cit.*, p. 44.
[a] Includes lignite, expressed in terms of its equivalent coal tonnage in calorific value.

TABLE 6. *World production of crude oil by leading producing regions, 1937 and 1945.*

Region	1937 Production (millions of barrels)	1937 Per cent of world total	1945 Production (millions of barrels)	1945 Per cent of world total
United States	1,279.2	62.7	1,711.1	61.5
U.S.S.R.	193.2	9.5	300.0	10.8
Venezuela	186.2	9.1	321.9	11.6
Near East	117.5	5.8	201.0	7.2
TOTAL, above countries	1,776.1	87.1	2,534.0	91.1
WORLD TOTAL	2,039.2	100.0	2,783.0	100.0

Source: *1946 World Oil Atlas*, Section 2 of *The Oil Weekly*, Vol. 121, May 20, 1946; *Minerals Yearbook, 1940*, U. S. Department of the Interior, Bureau of Mines, Washington, G.P.O., p. 1017.

d. THERMAL POWER COSTS IN THE UNITED STATES AND THE SOVIET UNION. A curious circumstance revealed in Map 1 is that, despite the relatively greater distances in the Soviet Union and its less well developed transportation system as compared with the United States, the range of electric power generating costs based on coal is narrower in the Soviet Union than in this country. Thus, while the cost of coal-generated power ranges from 4.5 to 7.0 mills per KWH in the United States, the range in the Soviet Union is from 4.5 to 6.0 mills. The Russian costs include a smaller charge per ton mile for transporting coal than the American costs. Since it was not uncommon during the prewar decade for the Soviet government to subsidize certain goods and services important to industrialization, including coal and railroad transportation, there are grounds for believing that the thermal power costs shown on Map 1 may be based, in certain parts of the Soviet Union, on coal priced below its real cost at the site of the power plant.[13]

We have, therefore, recalculated the Russian costs shown on Map 1, assuming that the same transportation rates prevailed as in the United States. On this basis we find that the costs shown for coal-generated electricity in the Soviet Union on Map 1 should, on the average, be adjusted upward, roughly as follows (in mills per KWH):

Costs on Map 1	Costs with revised transportation charges
4.5–5.0	no change
5.5	6.0–6.5
6.0	7.0–7.5

These adjustments would, therefore, result in a wider range of thermal power costs in the Soviet Union than in the United States.

The adjustment shown above may not take full account of the effect of government subsidies; it accounts for the subsidization of railroads but not for that of coal mining itself. At the time to which the available data refer, Soviet enterprises were undergoing a transition from heavy subsidies to a regime of being self-supporting; but key industries such as coal mining and railroad transportation are known to have been allowed a slower transition.[14] It is very likely, therefore, that the adjustment attempted above needs further correction upward, in order to answer the following question: In what regions of the U.S.S.R. would it be economically justifiable to construct atomic power plants,[15] thus withdrawing resources from coal-mining and railroad transportation?[16]

[13] See, for example, Alexander Baykov, *The Development of the Soviet Economic System*, Cambridge, England, Cambridge University Press, 1946, pp. 294 ff.

[14] *Management in Russian Industry and Agriculture,* by G. Bienstock, S. Schwarz, and A. Yugow, (A. Feiler and J. Marschak, eds.), New York, 1944, pp. 80–82. In 1936 the official journal on planning justified subsidies for coal as follows: "Low prices for coal, metal, tools and machines create additional incentives in order that machine techniques shall take root in all branches of national economy. But if at the same time cost of production is high, the maintenance of relatively low prices for goods produced by heavy industry is possible only through State subsidies." (M. Bogolepov, "On Sales Prices in Industry," *Planned Economy,* 1936, No. 5, pp. 76–77.)

[15] We are, of course, making no allowance for the possibility that the Russians as well as others will subsidize atomic energy plants because of their military significance. (See Chapter I, Appendix B.)

[16] It must be remembered, in addition (see Chapter II, A.1.d), that our figures of Russian costs in United States currency are based on an exchange rate of 19 cents per rouble, the rate established after the devaluation of the rouble in 1936. The electricity

e. Fuel Prices in 1946. As we noted earlier, the atomic cost estimates are based on prices in 1946, while Map 1 reflects fuel costs which prevailed before the war; if 1946 fuel prices had been used in their derivation,
the thermal generating costs shown on the map would have been considerably higher in many regions. But, as we stated before, it is questionable
how much permanent significance should be attached to 1946 fuel prices
in most parts of the world because of the acute coal shortage throughout
the world as an aftermath of the war. In other words, in most countries
coal prices in 1946 were probably unduly high compared with the general price level.

Although the United States did not fully escape the effects of the postwar fuel shortage, it is likely that 1946 coal prices in this country were in
better alignment with the 1946 general level of prices than in almost any
other country.[17] It is therefore quite significant to derive thermal power
generating costs for the United States, using 1946 fuel prices, for comparison with the estimated costs of atomic power. The increase in estimated
thermal costs over those based on prewar coal prices in the United States
might also provide some guidance on how much thermal costs elsewhere
in the world would increase over the costs shown on Map 1, if prices providing a more realistic approximation of postwar normal conditions were
used in their derivation. (See also Chapter II, B.)

As we have seen, all costs other than fuel for the hypothetical thermal
plant used in making the cost estimates shown on Map 1 are in terms of
1946 prices. When the prewar coal prices included in the generating cost
estimates are replaced by 1946 coal prices in the United States the estimated
total generating costs increase, on the average, by between 10% and 20%.
The shift in total generating costs (in mills per KWH) is about as follows:

Based on prewar coal price	Based on 1946 coal price[18]
4.5–5.0	5.0–6.0
5.5–6.0	6.0–7.0
6.5–7.0	7.0–7.5

costs in the U.S.S.R., expressed in United States currency, would appear much higher
if the pre-devaluation rate (50 cents per rouble) had been used. We have shown, however, that the post-devaluation rate was more appropriate for the purpose of judging
whether and where atomic power might compete with thermal power; the ratio between
power plant construction costs in the U.S.S.R. and in the United States corresponded to
the post-devaluation rate, rather than the pre-devaluation rate.

[17] It might even be argued that the average delivered price of coal in most places in
the United States was lower in 1946 compared with other prices than would be normal
over the longer-run postwar period, because the postwar series of increases in freight
rates, which increased coal freight rates on the average of about 43% as of October
1949, did not start until the middle of 1946. (See U. S. Interstate Commerce Commission, "Monthly Comment on Transportation Statistics," October 11, 1949.)

[18] Based on coal prices paid by power plants in various parts of the country (Federal
Power Commission, *Steam-Electric Plant Construction Cost and Annual Production Expenses, 1938–1947*).

In terms of the comparison with atomic power there is this major change: While with prewar coal prices most of the thermal power costs shown on the map were between 1 and 1½ mills less than the intermediate estimate of the cost of atomic power, with 1946 coal prices many of the thermal costs would fall at the intermediate cost estimate.

B. THE SIGNIFICANCE OF DIFFERENCES IN THE COMPOSITION OF TOTAL COSTS IN ATOMIC POWER AND CONVENTIONAL POWER

To this point, we have been concerned only with comparing the estimated *total* costs of generating atomic power and conventional power in each region of the world.   We want now to consider how differences in the composition of total costs may affect the future choice between atomic power and conventional power.

1. The Elements of Total Cost: Atomic Power and Conventional Thermal Power

In the last section of Chapter I we dealt briefly with certain differences in the composition of the total costs of atomic power and conventional thermal power which affect cost comparisons between them, as different assumptions are made about plant size and rate of capacity utilization.   A major difference was the relatively greater importance of fixed charges in the cost of atomic power compared with thermal power.   This difference implies, for example, that at the intermediate atomic cost of 6.5 to 7.0 mills per KWH (at a 50% rate of capacity use) fuel is an almost negligible item in cost, while in conventional thermal power at the same total cost fuel would constitute about 40% of all costs.   Again, at the same total cost, fixed charges (i.e. the current charges on plant and facilities) would constitute about 80% of atomic costs and about 50% of conventional thermal costs.

2. Trends in the Cost Components of Atomic Power and Conventional Power

If the component cost items whose relative importance is different in atomic power and conventional power should have different trends in the future, this would significantly affect our cost comparisons.   If, for example, there should be an upward trend in the cost of fuel compared to the cost of building and equipping power plants this would favor atomic power, and vice versa.

We are not able to predict what these trends will be.   The evidence of the past decade in the United States provides no clear indication of what the comparative cost trends have been in recent years, let alone a basis for

predicting future changes.   Between the prewar years and 1947 the average
cost per ton of fuel purchased by electric company power plants increased
by about 65% to 75%; during the same period the cost of building and
equipping steam electric stations (where cost changes are much more difficult to measure than for a relatively simple homogeneous product like
coal) is estimated to have increased by about 63%.[19]   Although coal costs
appeared to have risen somewhat more sharply than the cost of power
plants, we cannot be certain of this because of weaknesses in the indices of
measurement.

Two significant factors in the trend of coal costs are worth noting.
The one is the trend in miners' wages and welfare benefits.   Between
1939 and January 1949 average hourly earnings in bituminous coal mining more than doubled.[20]   Although this percentage increase in the
hourly earnings of coal miners does not differ significantly from the increase in the average hourly earnings in other industrial activities (as
measured, say, by the Bureau of Labor Statistics index of hourly earnings in all manufacturing), it is significant for our comparisons because
of the relatively large importance of the labor element in the cost of coal.
For example, wages constituted about 60% of the value of coal at the
mine.   We have not been able to derive a similar ratio for building and
equipping power plants.   However, wages in the engines and turbines industry, which produces a major part of power plant equipment, constitute about 25% of the value of the product; even including the wages embodied in the materials consumed in engine and turbine manufacture,
the ratio of wages to value of product does not exceed 40%.[21]   We might

[19] We consider the trend of costs in building and equipping conventional power plants
as the best approximation now available of the cost trends in the materials and equipment which will be incorporated in atomic power stations.

The sources of cost information are: for coal consumed by power plants, *Statistical Bulletin, 1947,* Edison Electric Institute, New York, June 1948; for the construction and equipping of power plants, an index prepared by Edwin H. Krieg, Chief Design Engineer, American Gas and Electric Service Corporation, "Progressive Engineering
Offsets Increased Costs of Steam-Electric Power Generation," *Civil Engineering,* April
1948, pp. 14–15.   Mr. Krieg's index, it should be noted, shows a considerably greater
cost increase than that indicated in an index prepared at the Stone and Webster Engineering Corporation, "Power Plant Construction Costs and Implications," by William
F. Ryan, *Power Plant Engineering,* August 1947, pp. 116 ff.

[20] U. S. Bureau of Labor Statistics, *Handbook of Labor Statistics,* 1947 edition,
Washington, G.P.O., 1948, and *Monthly Labor Review,* Vol. 19, November 1949.

[21] The ratio of wages to the value of coal was calculated from U. S. Bureau of the
Census, *Census of 1940, Mineral Industries, 1939,* Vol. I, Washington, G.P.O., 1944.
The ratio of direct wages to the value of engines and turbines was derived from the
1939 Census of Manufactures, Vol. II, Part 2.   The ratio of direct and indirect wages
(including wages embodied in materials consumed) to the value of engines and turbines
is based on figures supplied by Mr. Marvin Hoffenberg of the U. S. Bureau of Labor
Statistics; it is based on a table of inter-industry relationships in the United States prepared in the Bureau (in Appendix A of "Full Employment Patterns, 1950," Washington, 1946).   The ratio derived by Mr. Hoffenberg included both wages and salaries;
we estimated the wage component alone by applying data from the 1939 Census of
Manufactures.

expect then that an upward trend in wages relative to other prices, even if no steeper in coal mining than in other industrial activities, would favor the atomic power plant in the future.   On the other hand, we must not overemphasize the extent of the possible advantage.   The cost of coal constitutes only part of the costs of generating thermal electricity; in 1946 prices it accounts for about 30% to 40% of total generating costs. Furthermore, the wages embodied in delivered coal prices would consti- tute less than 60% of the total because of the transportation costs in- cluded.   Finally, increases in labor costs in coal mining might be offset by the greater mechanization of operations.   (See below, Chapter II, B.3.) The other trend worth noting is in the cost of transporting coal.   A series of postwar increases in railway freight charges has resulted, on the average, in a 43% increase in coal freight rates since the middle of 1946.   Only about half of this increase is reflected in the 1947 coal prices paid by power plants, on which the percentage increase cited above was calculated.[22]

The question of miners' wages is closely associated with the number of weekly hours worked and with the problem of providing an adequate number of coal miners.   Coal mining is arduous, dangerous work.   In the past it has been found difficult in some countries to recruit an ade- quate mine labor force at prevailing wage rates and working hours. For example, before the war France and Belgium had to recruit mine laborers from Poland and Italy.   The problem of recruiting miners has not eased since the war, and it will probably be increasingly difficult to recruit labor from Eastern Europe in the future.   A report from Bel- gium states: "The [colliery] situation is also complicated by the apparent reluctance of the postwar Belgian to work underground, with the result that little over a third of the hewers and less than half the other under- ground workers are of Belgian nationality. . . . "[23]   In the meantime, Great Britain, too, has experienced serious difficulties in recruiting enough miners to produce the required coal output.[24]   What is perhaps more serious over the long run is the increasing reluctance among British miners to have their sons enter coal mining.[25]

3. Technological Changes in Conventional Fuels

The trend in conventional fuel costs over the next few decades may be significantly affected by major changes in fuel technology.   Mines

[22] U. S. Interstate Commerce Commission, "Monthly Comment on Transportation Statistics," August 15, 1949; October 11, 1949, op. cit.

[23] The Economist, Vol. 157, September 10, 1949, p. 561.

[24] See, for example, Labor and Industry in Britain, Vol. 5, September–October 1947 (a monthly review published by the British Information Services).

[25] Mr. Herbert L. Matthews reports in the New York Times of June 15, 1948, after a visit to a British coal mine: "I asked the fathers among those leaning and crouching around me—there was no room to stand—whether they wanted their sons to take up mining.   The answer was an emphatic 'no' and it was a 'no' echoed from the Rhondda Valley through the Midlands to Fifeshire in Scotland."

may become more completely mechanized, even in the United States where mechanization has already proceeded further than in other countries. More drastic changes are also possible; these revolve around the demonstrated fact that coal can be converted into gas both above ground and underground, and that either natural gas or gas produced from coal can be converted into liquid fuels. Much has been written about the latter set of changes, which some observers believe to imply a "fuel revolution."[26]

A thorough analysis of the feasibility of the fuel revolution and its implications is beyond our scope. We want, however, to consider briefly how it might affect cost trends in conventional energy sources used in electric power generation. For this purpose we distinguish between two broad (but overlapping) lines of change: (1) those changes designed to convert "non-premium" fuels, such as coal, into fuels fetching a premium price, such as gasoline; (2) those changes designed to reduce the cost of mining or transporting coal.

The economic justification for converting solid and gaseous fuels to liquid form depends upon the expectation that the discovery of domestic petroleum resources will not keep pace with the growth of demand for liquid fuels. The objective is "to meet those liquid fuel requirements for which the user can afford to pay a premium over the price of coal on a heat-content basis."[27]

Although the conversion of coal to liquid fuel may eventually become economically feasible, other changes may come first.[28] According to Dr. Robert E. Wilson, Chairman of the Board, Standard Oil Company (Indiana):[29] "In the past much of our heavier petroleum residues have been burned by large power plants in competition with coal on essentially a BTU per dollar basis, and one of the first things to occur, if and when the petroleum supply situation tightens, will be the installation of equipment to convert these heavier residues into the lighter and more valuable liquid fuels (including gasoline). . . . In other words, the advent of catalytic cracking and the availability of hydrogenation are taking residual fuel oils out of the class of unavoidable by-products and giving them a definite and substantial value as a crude substitute. . . . when the crude price rises, they will gradually pass out of uses where they are in direct price

[26] See, for example, a series of articles entitled "Coal" in *Fortune,* Vol. 35, March and April 1947.

[27] Robert E. Wilson, "Oil from Coal and Shale" in *Our Oil Resources,* edited by Leonard M. Fanning, New York, McGraw-Hill Book Co., 1945, p. 214.

[28] The experience of the Pittsburgh Consolidation Coal Co., which closed its pilot plant for synthesizing gasoline from coal after 13 months of operation, provides some evidence (by no means conclusive) that gasoline from coal is not now economical. (*Business Week,* No. 1062, January 7, 1950, p. 26.)

[29] Wilson, *op. cit.,* pp. 214–215.

competition with coal. . . .   The burning of heavy fuel in large stationary power plants will be the first to go. . . . "

When the changes described by Dr. Wilson are no longer sufficient to meet our growing needs for the more valuable liquid fuels, the next cheapest method for deriving these fuels will be used, and so on.   From present cost indications natural gas would be the next cheapest source.   However, the relative scarcity of natural gas resources indicates that ultimately coal or oil-shale, of both of which we have very large reserves, will provide for at least a part of our liquid fuel requirements.[30]

We may now summarize the significance for future trends in conventional electric power costs of the technological changes just described. These changes are concerned with providing an increased supply of a fuel rarely used in power plants because it is much too expensive on a heat-content basis.   Their success will not reduce the price of this fuel to a point where it could be used in power plants; on the contrary, success depends on increases in its comparative price.   In this sense, these technical advances are irrelevant to our power cost comparisons.   However, to the extent that the emergence of new liquid fuels involves the "upgrading" of fuels now used in power plants—low grade fuel oils, natural gas, and perhaps coal to be absorbed by consumers other than power plants—it should result in an upward trend in their prices, possibly to the point of pricing oil and gas out of power-plant use.   This line of technical change, then, would tend to make conventional power plant fuels relatively more, rather than less, costly in the future.

The other general line of technological change has as its aim lower costs in winning coal from the earth, and possibly lower costs in transporting its energy content.   One way envisaged for reducing costs is through the more thoroughgoing mechanization of mines; the other is through the gasification of coal, which might be carried out underground or above ground, with the product transported to markets via pipeline.

As for the effects of mechanization, a highly optimistic estimate predicts that increased productivity might bring a return to prewar mine mouth coal prices, provided the cost of mine labor does not rise in the meantime.[31]   However, if miners' wages and welfare benefits keep pace with improvements in productivity—a much safer assumption than unchanged wage rates—this cost lowering obviously will not be achieved. Of course, regardless of its effects on costs, increases in productivity would be of the utmost importance in easing the problem of an inadequate mine labor supply.

[30] Eugene Ayres, "The Fuel Problem," *Scientific American*, Vol. 181, December 1949.
[31] "Coal: The 'Pitt Consol' Adventure," *Fortune*, Vol. 35, July 1947, p. 140.

As for underground gasification, the experts of the U. S. Bureau of Mines who have been conducting experiments with this technique (initially applied in Russia) reported in 1949 the following: "At this time . . . . it is not possible to do more than theorize on the economics of underground gasification. If, however, a producer gas having a heating value of 125 BTU per cubic foot can be developed by methods similar to those used in the experiment, the cost may well be as low as 14¢ per million BTU produced. This is equivalent to a coal cost of $3.50 per ton f.o.b. mine."[32] The equivalent coal cost is just about equal to the average mine value of coal in the United States in 1946. The BTU content per cubic foot is about ⅛ that of natural gas, indicating that the gasified coal is certainly not economically transportable. Major improvements would be required both in the cost of gasification and (especially) in the BTU content of the gas to make underground gasification economically feasible, except in those instances where certain coal deposits cannot be mined by ordinary means and there is a local market for the gas. The Bureau of Mines experts quoted above believe improvements are possible.

Mr. P. C. Keith, President of Hydrocarbon Research, Inc., is perhaps America's most enthusiastic advocate of this process. His best expectations (as stated in 1947) were to deliver coal gasified in West Virginia to New York City (and presumably to other market centers from this and other mining regions) at a price which equals the delivered price of coal in terms of energy contained; Mr. Keith believes that the cost advantage of transporting gas containing 1,000 BTU per cubic foot by pipeline rather than shipping coal by railroad might be sufficient to offset the extra processing cost (above ground) of obtaining a gas of such high BTU content from coal.[33] However, as with the proposals for converting coal to liquid fuel, the success of gasification may in fact depend on obtaining a premium price for gas as compared with coal because of its convenience in many uses.[34]

We conclude, therefore, that the technological advances envisaged for conventional fuels, although spectacular in some instances, provide no clear indication of how the comparative cost of power from conventional sources will be affected. Certain types of changes designed to increase

---

[32] "Laboratory and Field-Scale Experimentation in the Underground Gasification of Coal," by M. H. Fies and James L. Elder, Paper prepared for the United Nations Scientific Conference on the Conservation and Utilization of Resources, May 25, 1949, p. 15.

[33] "Coal: The Fuel Revolution," *Fortune,* Vol. 35, April 1947, p. 243.

[34] "Coal: The 'Pitt-Consol' Adventure," *op. cit.,* p. 143. See in addition the opinion of Commissioners Nelson Lee Smith and Harrington Wimberly of the Federal Power Commission: "Other than the prospects of being produced as a byproduct of other operations, economic conditions for gas manufactured from coal are not favorable until the price of gas fuel exceeds the price of coal." (*Natural Gas Investigation,* Docket No. G-580, Washington, G.P.O., 1948, p. 460.)

productivity in coal production and transportation (in the form of gas) may enable conventional energy to maintain or perhaps improve its 1946 relative price.   On the other hand, the possible development of synthetic liquid fuel production suggests that certain fuels now used in power generation, perhaps even including coal, may in time be rendered relatively more costly because of their new importance as sources of premium-priced fuels.

4. Significance for Different Countries of Differences in the Composition of Total Costs

Variations in the composition of total costs of atomic power and of conventional thermal power in respect to fixed charges and the cost of fuel may have a significant effect on economic comparisons in different countries.

a. INTEREST ON INVESTMENT.   Our cost comparisons have assumed the same rate of fixed charges in all countries.   It is possible, however, that the rate of fixed charges would be higher in economically underdeveloped countries than in the United States because the greater scarcity of capital in such countries would result in higher interest charges.   If the rate of fixed charges in certain countries should be higher than we have assumed, this would favor ordinary thermal power as compared with atomic power because of the smaller relative importance of fixed costs in total thermal power costs.[35]

The examples in Table 7, for two assumed investment levels in atomic power, indicate how our cost comparisons would be affected by changes in the interest rates.   In a country where the interest rate is, say, twice as high as our assumed rate, atomic power would cost between 1½ and 2 mills more per KWH of power generated than we have estimated. Therefore, while our estimates would show equivalent costs per KWH of power generated when the price of coal is from about $6.50 to $8.00 per ton, and the investment in atomic power is $215 per kilowatt, doubling the interest rate would yield equivalent costs when the price of coal is between $8.00 to $9.50 per ton, and so forth.

Although we must recognize that the interest rate might be higher in other countries than in the United States, it is important to note that the

[35] Cf., for example, Isard and Lansing, *op. cit.*   Two important exceptions to this generalization should be noted.   (1) In underdeveloped countries the provision of thermal electric power plants might also require a substantial investment in coal mines and railroads, some of which might be obviated by the use of nuclear fuel.   In this event the difference between the direct investment per kilowatt of capacity in the two kinds of plants would not measure the net capital cost advantage of the thermal plant.   (2) Compared with hydroelectric plants it is not at all clear that the investment per kilowatt in atomic power would be greater.   The capital costs of hydroelectric power vary greatly from project to project and are usually well above those in ordinary thermal power.   (On both these points see Chapter XIV, D.1.)

rate of fixed charges would not necessarily be larger by the exact amount of the increase in interest. The 11% rate of fixed charges we have assumed for the United States is made up of 3% for interest, 4% for depreciation and obsolescence, and 4% for miscellaneous charges. Taxes and franchise cost, which bulk large among the miscellaneous charges, chiefly reflect the legal and institutional setting in which American electric utilities operate. There is no fundamental reason why these charges should be the same in other countries. For example, given a public policy designed to encourage the development of an electric power industry, these charges might be considerably lower.[36]

TABLE 7. *Effects of different interest rates on comparisons between the costs of atomic power and ordinary thermal power.*

Assumed investment per kilowatt of capacity	Atomic power		Thermal power	
	Cost of electric power generated at a 50% rate of capacity use[a] (in mills per KWH)		Price of coal, which would yield the same KWH cost as in columns (2) and (3), at a 50% rate of capacity use[a] (in dollars per short ton)	
	with interest at 3%	with interest at 6%	with interest at 3%	with interest at 6%
(1)	(2)	(3)	(4)	(5)
$215	6.5–7.0	8.0–8.5	$6.60–$7.90	$8.15–$9.50
$315	10.2	12.3	$16.30	$19.50

[a] All costs other than interest are on the basis of the factors set forth in Chapter II, A.1.a.

b. FOREIGN EXCHANGE REQUIREMENTS. Certain countries do not possess adequate domestic energy resources to meet their future needs. These include not only underdeveloped regions, but also highly industrialized countries where there is little room for further expansion of energy production from domestic resources. Such countries will have to import fuel in order to expand their future power output. However, the importation of fuels in large amounts may be relatively costly because of the long-distance transportation often required, and, even more important, because such purchases, which must be paid for in foreign currency, may be difficult to finance, depending on the country's balance of payments. Charges for transportation are covered in the estimated power costs shown in Map 1. The additional problem of financing foreign purchases, and

[36] In this connection it is interesting to note that there is evidence that the Russians use a much lower rate of fixed charges in figuring the costs of hydroelectric power than is common in the United States. (See *Electric Power Development in the U.S.S.R.,* pp. 445 ff.) It could be argued that in using a lower rate the Russians do not measure their real costs—that they do not, for example, assign a proper interest rate considering their shortage of capital—but to argue so does not alter their established accounting procedures.

the possible advantages of atomic power in this respect, will now be considered.

Maps 2 and 3 indicate an extremely uneven distribution of conventional energy resources among the countries of the world. Both maps, of course, are based on known resources; with the future discoveries the pattern of resources may shift somewhat.[37] Moreover, the production of fuels shown in Tables 5 and 6 may increase substantially in certain areas that are now exploiting their known resources at only a fraction of their potential annual yield.

The possible discovery of new oil resources may be particularly important. New fields have been frequently discovered and centers of production have shifted markedly. Within the United States, for example, there has been a shift from the Pennsylvania fields to those in Texas, Oklahoma, and California. On a world scale the former dominance of the United States in world production is decreasing and the future will probably see the Middle East and Venezuela become increasingly important. Moreover, in 1949 large resources were discovered in the province of Alberta in western Canada, another example of a shifting situation.

The change in pattern applies to a lesser extent to coal mining. While more is known about the geologic existence of coal than oil fields, new coal fields are still being discovered, and old estimates of resources are being re-evaluated (although not always upward). Within the last two decades many new fields have been charted in the U.S.S.R., for example, and it is not unlikely that systematic surveys would produce similar results in other parts of the world. There is also a shifting in the relative importance of centers of production. Older areas retain their importance for long periods, but with the spread of industrialization, production in other areas increases. Production in the U.S.S.R. is rising rapidly, partly because of the development of hitherto unworked fields. In the Union of South Africa production has doubled within a recent decade, although it is still relatively small. It is quite possible for production in various Asiatic countries to rise far beyond present levels.

It would be unwise, because of these considerations, to conclude that all specific areas which do not at present have an adequate domestic fuel supply must in the future also lack one. In addition, electric power production might be based on low-grade fuels, such as peat, available in numerous regions. But although future development may alter the details of the picture, concentration of known resources shows that in the future, too, there will be many areas without adequate supplies of local fuel.

The selection of countries where domestic energy resources would be insufficient to meet future requirements depends upon a more detailed

[37] In this connection, see the final paragraph of footnote 11 of this chapter.

analysis of regions than we have made. However, some examples of countries which probably fall in this category are:[38] Argentina and Brazil with negligible fuel resources, and with many important population centers far away from potential water power; countries like Denmark or a North African group including Morocco, Algeria, Tunis, Libya, etc., with negligible fuel and water power resources; Hungary with fuel resources which are of low grade and costly to mine, and with negligible potential water power within its boundaries; countries like Italy, Austria, and Switzerland, with negligible fuel resources, whose power development in the past has been based mainly on water power, but where future water power development will be increasingly difficult because so much of the potential has already been developed; and perhaps countries like the United Kingdom where fuel supplies, formerly ample, may be difficult to maintain in the future, and where potential water power is negligible.

Atomic power may be extremely important to countries which would have to import conventional fuels. The international distribution of the resources of nuclear fuel is probably quite different than that of other energy resources. For example, Brazil is deficient in coal and other mineral fuels, but has extensive resources of thorium. How important these differences in resource distribution will be we cannot now say because of the paucity of information on resources of uranium and thorium.[39]

What may be even more important is that countries deficient in fuels will be able to buy uranium and thorium metal at a negligible cost compared to the cost of ordinary fuels. As we have seen, 1 lb. of uranium, perhaps costing about $50, would yield as much energy when fully consumed as about 1,250 tons of coal.

The size of the savings in foreign exchange resulting from the importation of uranium instead of coal depends on how many of the facilities for generating atomic power could be produced domestically. Great Britain, for example, might be expected to build and equip atomic energy plants domestically at an early date after the technology of such plants has been developed. Economically underdeveloped countries, on the other hand, probably would have to import much of the equipment required, perhaps for decades. But even for the latter class of countries substantial foreign

[38] We used two general criteria in selecting countries: (a) domestic resources of economically transportable fuels are not large enough, in our judgment, to satisfy potential requirements; and (b) there is either negligible potential water power, or so much of the potential has already been developed as to limit seriously future development, or there are many populated points too far away to be economically served by the available water power sites. No country is included if economically transportable fuels can be produced in sufficient quantities within the national boundaries to meet future needs, even if long-distance transportation from the point of production to important consuming regions is involved. In this case, of course, the costs of transporting the fuel would be covered in the estimates shown in Map 1.

[39] Cf. discussion in Chapter I, B.

exchange savings might result from the use of atomic power, as the following hypothetical calculation, using Argentina as an example, indicates:

(1) According to Map 1, Argentina's power costs in our hypothetical thermal power plant would be between 7 and 8.5 mills per KWH. If the investment in atomic power should be $215 per kilowatt[40] it would just equal the cost of the cheapest thermal power in Argentina, assuming 1937 coal prices of about $8 per ton at port cities.

(2) At $215 per kilowatt the investment in atomic power would be $85 per kilowatt more than the investment in ordinary thermal power. Let us assume that the entire difference consists of equipment which will be imported, and paid for through foreign borrowing. Assume, also, that interest is at 6% and that amortization is over a 25-year period. Total yearly charges per kilowatt of capacity then would be at a rate of 10% of $85, or $8.50.[41]

(3) On the assumption that power facilities will be operated, on the average, at 50% of capacity, a kilowatt of capacity would produce 4,380 KWH during the course of a year. At an extremely high thermal efficiency (10,000 BTU per KWH) about 1.7 tons of high grade coal would be consumed in a modern thermal plant in the production of 4,380 KWH. This amount of coal would have cost about $13.50 at 1937 prices.

(4) Therefore, the annual foreign exchange outlay entailed in the operation of the thermal plant would be $5.00 greater than in the atomic power plant per kilowatt of capacity. The advantages of the atomic plant in this respect would be reduced slightly by the need for importing uranium or thorium. Since, at $50 per lb. of uranium or thorium this would come to about 10 cents per 4,380 KWH, we place the advantage of the atomic plant at $4.90 per kilowatt of capacity.

(5) For thermal power to require as small an annual outlay of foreign exchange as atomic power under these circumstances, the price of coal would have to be $5.00 per ton at the port in Argentina.

The main elements of the foregoing example together with an example for another set of cost comparisons are summarized in Table 8. We may draw the following generalization: In countries which would have to choose between thermal plants requiring a continual importation of coal, and atomic power plants requiring a greater initial importation of facilities but negligible continual imports of fuel, the annual outlay of foreign exchange would be considerably less for the atomic plant, even though total costs in both types of plants, as we have estimated them, are equal. Or, put another way, while on the basis of a comparison of total

[40] Cf. Chapter I, C.2.
[41] We assume that the absolute decline in the interest payments as the loan is being amortized would over time be approximately balanced by the importation of certain replacement parts and operating materials.

costs, atomic power at \$215 per kilowatt of capacity could not compete with coal costing less than \$6.60 per ton, on the basis of a foreign exchange comparison it could compete with coal costing \$5.00 per ton. Again, while atomic power at \$315 per kilowatt of capacity could not, on the basis of total costs, compete with coal at less than \$16.30 per ton, it could on comparing only foreign exchange requirements, compete with coal at \$11.00 per ton. Of course, the advantage of atomic power in this respect would be increased to the extent that (a) the relative price of conventional fuel rises, and (b) not all the costs of constructing atomic facilities represent imports.

    c. CONCLUSIONS. We have found that differences in the composition of total costs may have quite important effects on economic comparisons

TABLE 8. *Atomic power and ordinary thermal power compared according to two criteria: total costs of production and foreign exchange requirements.[a]*

	Atomic power[f]		Thermal power[f]	
			Price per short ton of coal which would yield[d]	
Assumed investment per kilowatt of capacity	Total cost of electric power generated[b] (mills per KWH)	Annual outlay of foreign exchange per kilowatt[c]	Same KWH cost as in col. (2)	Same foreign exchange outlay as in col. (3)
(1)	(2)	(3)	(4)	(5)
\$215	6.5–7.0[e]	\$8.60	\$6.60–\$7.90[e]	\$5.00
\$315	10.2[e]	\$18.60	\$16.30[e]	\$11.00

    [a] On the assumption that a country would import fuel for thermal plants or facilities for atomic plants.
    [b] Assuming a 3% rate of interest for consistency with the estimates shown in Map 1; col. (3) assumes a 6% rate of interest.
    [c] Computed at 10% of the extra investment of atomic power compared with ordinary thermal power. Also includes 10 cents for imports of uranium to generate 4,380 KWH of electric power (see text for explanations).
    [d] Assumes that 1.7 tons of coal would be required to generate 4,380 KWH of electric power.
    [e] Estimated on the basis of the factors given in Chapter II, A.1.a.
    [f] Assuming power plant operations at 50% of capacity.

between atomic power and conventional power. On the one hand, a higher interest rate in economically underdeveloped countries would make our cost comparisons less favorable to atomic power. But in countries deficient in domestic energy resources, a calculation made strictly in terms of foreign exchange requirements, even with the higher interest rate, would make our comparisons more favorable to atomic power.

    In evaluating the relative weight of these two factors for future choices between atomic power and power from conventional fuels in different countries, we should keep these facts in mind. High interest rates favor the conventional plant; these rates are part of total fixed charges which include certain items determined by law and custom; as we pointed out earlier, higher interest rates in some countries may possibly be offset by lower rates on other fixed charge items than is customary in the United States.   On the other hand, the low cost of atomic fuel, which gives atomic power its foreign exchange advantage over ordinary thermal power, is primarily a result of the enormous energy content per unit of weight of uranium, a physical fact which is not likely to be offset by institutional differences among countries.

We would say, then, that for countries which are deficient in conventional energy resources, the possibility of foreign exchange advantages would weigh heavily in favor of atomic power, while, for underdeveloped countries with ample energy supplies, the scarcity of capital and consequent higher interest rates than in the United States might favor ordinary thermal power.   Even though a final answer to this crucial question cannot be supplied, perhaps something can be learned from the history of hydroelectric power, which, despite its generally heavier investment than thermal power, has been widely used even in underdeveloped regions.[42] It is significant that the savings of foreign exchange which would otherwise be spent on fuel imports is often offered as a major reason for the development of hydroelectric projects.[43]

In this chapter we have tried to find what determines the degree to which atomic power may compete, in a given geographical area, with power from other sources.   The relative level of costs, or the relative availability of the various sources for the generation of power, is one of the factors determining whether a given area may become a consumer of atomic power.   Other factors are those determining the demand for energy regardless of its source.   These factors are the purchasing power of the area's population, and the presence or absence of raw materials or markets for special goods and services that consume much electricity and can be sold within or outside the area.   As indicated in the Preface, Map 4 gives the geographical distribution of the world's population as a first, but very imperfect, approximation to the distribution of purchasing power. As to the specific energy-consuming industries, Part Two, to which we now proceed, will attempt to evaluate them as possible buyers of atomic power, primarily within the United States.

---

[42] Of course hydroelectric power has at least two important characteristics: (a) Much of the installation has an extremely long life which results in a relatively low rate of depreciation.   (b) It is often developed to serve a multiple economic purpose, i.e. irrigation, flood control, navigation, and power.   But remember that the atomic plant has also a multiple purpose, though not an economic but a military one!

[43] See, for example, press releases of the International Bank for Reconstruction and Development announcing loans for the development of hydroelectric facilities in Brazil and El Salvador.   (Releases No. 126 and 162, January 27 and December 14, 1949.)

Part Two
ATOMIC POWER IN SELECTED INDUSTRIES

CHAPTER III

The Industry Analyses: A Summary View

IN PART ONE of this book we confined our analysis to questions relating to the supply of energy. We inquired into the cost and other economic characteristics of atomic power and attempted, through comparisons with ordinary sources of energy, to determine whether atomic power might replace or supplement conventional energy resources in certain regions of the world. Now we will shift our attention from questions of energy supply to questions of energy utilization, in an attempt to determine what effects atomic power might have on a selected group of energy-consuming industries. The individual analyses which follow in Chapters IV to XII examine in some detail the importance of energy in particular industries, and the possible ways in which these industries might be affected by the development of commercial atomic power. This chapter provides a setting for these analyses and summarizes their major findings.

A. THE SETTING OF THE PROBLEM

1. The Purposes of an Analysis by Industries

A series of industry analyses such as we are about to undertake has two broad purposes. It should provide information for charting the course of individual industries under the impact of atomic power. That much is obvious. It should also help us to understand the general effects of atomic power on the entire economy. The answers to the questions we shall want to raise about the impact of atomic power on the economy of the United States or other countries—and, indeed, in some cases merely the posing of the proper questions on this subject—can, in good part, be found only in a detailed industry-by-industry analysis.

These chapters constitute no more than a beginning to the type of analysis which will have to be undertaken if we are to grasp the economic implications of atomic power. Their main value, perhaps, lies in the formulation of the appropriate questions for individual industries, and the construction of an analytic framework to handle these questions; in the tentative assembly and analysis of relevant economic and engineering data which permit the preliminary sorting out of the included industries into those which might and those which probably would not feel important effects as a result of atomic power; and in the insights gained about the general effects of atomic power on the economies of the United States and other countries.

2. *The Major Questions Raised in the Industry Analyses*

While we have tried to follow a uniform pattern in all of the industry analyses, we have had to modify the pattern in many different ways because each industry is in certain respects unique. The stress placed on different aspects of the analysis varies from one industry to the next, and the specific questions to which we seek answers are not always the same. Nevertheless, most of them have a common origin in three major questions which embrace our primary interests.

The first major question is obvious. Since atomic energy may be available at a lower cost than some of the energy sources commonly used today, we ask: Will industries in which energy is an important element in costs experience cost reductions as a result of its use?

The second question is related to the first one. In some industries where energy cost is important, atomic energy probably cannot be used without changes in production techniques. The only production processes in which atomic energy obviously can be used without any change are those which are now based on electricity, since it is generally assumed that atomic energy will appear commercially as electric power. Other processes now based on the burning of ordinary fuels (e.g. residential heating) or on particular chemical reactions (e.g. iron smelting) will require modification. For these industries we ask: Will atomic energy encourage the introduction of new production techniques?

A unique feature of atomic power is that it will not be characterized by the geographical variations in cost which are common for the energy sources used today. The reason for this is simple; atomic fuel has so great a concentration of energy per unit of weight that we may say that for practical purposes it is a weightless fuel with cost-free transportation. Geographical variations in power costs may, of course, still arise from such factors as differences in the size of central electric stations, or in the costs of constructing central stations of a given size, and so forth; but these variations are expected to be small, and possibly short-lived, compared with the varying costs for coal transportation and the natural differences governing the cost and availability of water power.

The likelihood that the cost of atomic power will tend to be geographically uniform suggests the third major question to be considered in subsequent chapters. In some industries the availability of low-cost fuel and/or power, or the availability of an unusually suitable fuel, has been highly important in the selection of factory sites. The question arises, therefore, whether under the influence of virtually equal-cost atomic power the locational pattern in these industries would change in favor of new sites possessing advantages not held by areas near low-cost fuel and power.

The analysis by industries centers, then, on three primary questions: (1) the possibility of cost reductions in industries which today consume much energy per unit of product; (2) the possibility that low-cost atomic power might encourage process changes in which methods based on atomic energy would substitute for the direct fuel-burning or chemical methods used today; and (3) the possibility of changes in the location of some manufacturing activity as a result of the greater geographical equalization of power costs. The effects which atomic power might have on individual industries would often fall within two of these general categories, since a process change might occur in conjunction with either one of the others.

Two additional points should be noted. In comparing a new production process based on atomic power with a method in current use, and in comparing a new production site with an existing production site, we use total costs as the relevant measure unless otherwise specified. Part of the total cost with existing methods and in existing locations represents current charges on a fixed investment. Since the investment cannot be recovered, these costs must be disregarded in any comparison of existing plants with new plants. In actual practice, therefore, when we speak of changes in location or in process, we usually have in mind a situation where new facilities are to be constructed to take care of increased demand, or to replace worn-out existing facilities.

The reader should be aware, too, that our relocation calculations (unless otherwise specified) assume that the major cost items other than power are the same in the hypothetical new site as in the old. This is done because figures on these other costs at new locations generally are not available and cannot reliably be estimated except through an intensive study of a particular site, a task considerably beyond our resources. The sites we select in our examples should not, therefore, be taken as the precise places in which new production locations might develop, since other costs may be unfavorable there. Instead, they are meant to illustrate the kinds of regions to which production might move, other costs being neutral.

3. *The Significance of the Industry Analyses for More General Economic Questions*

The economy-wide effects of the development of atomic power are analyzed in Part Three, where relevant examples drawn from the industry analyses are discussed. Here we want merely to indicate a few of the questions whose importance transcends the individual industry analyses to help orient the reader whose primary interest is in the industry chapters.

The general economic questions which the individual industry analyses may help us understand can be grouped most conveniently into two cate-gories: (1) What are the resource-saving effects of the commercial use of atomic power? and (2) What new impulses to economic development might it generate? The two categories are not mutually exclusive since the economic-development impulse may appear first as a resource-saving effect. We distinguish between the two in order to indicate at what points we think the developmental impulse may be particularly important.

a. RESOURCE SAVING. Let us try to see, first, what the individual in-dustry analyses can tell us about the resource-saving effects of atomic power in the entire economy. One less-detailed way to estimate the resource-saving potentialities of atomic power—and this is a method which has frequently been used—is to take the total amount of electricity pro-duced in the United States and to estimate what might be a reasonable unit cost reduction as a result of atomic power. The multiplication of total electricity production by unit cost reduction is used to make an esti-mate of cost saving (i.e. resource saving) in the entire economy.

But this approach to the measurement of resource saving has a serious shortcoming. That is, the level of demand for atomic energy may greatly exceed what might otherwise be the level of demand for more expensive electricity, particularly as a result of induced process changes permitting nuclear power to become the major energy source in industrial processes where higher-cost electricity cannot now successfully compete. The resource-saving calculation which would have to be made on this account would depend on the extent to which the change in process re-duced production costs in the affected industries.[1] To be sure, our pre-liminary investigations do not provide a definitive measure of resource-savings, but they do give valuable insights into the possible cost-reducing effects of process change.

b. ECONOMIC DEVELOPMENT. Just where important impulses to eco-nomic development at home and in other countries may be felt as a result of atomic power is difficult to foresee. There may, for example, be major possibilities in the effects which the innovation will have on individual industries. Unusual cost reductions in one industry may set the stage for rapid growth in that industry and other related industries. There is the possibility, too, that process changes brought on by atomic power may give rise to new industrial activities designed to meet the needs for novel types of industrial equipment, new and untried engineering skills, etc. Undoubtedly

[1] For a fuller discussion, see Chapter XIII, A.1.

the full impact of atomic power will not be experienced until these new industrial activities emerge and provide the processes, factories, and machines
which can fulfill the inherent possibilities of low-cost atomic power.

Our industry analyses will provide some indication of the possible importance of these effects.   Process change, in particular, emerges as an
important possibility in many of the industries we consider; it seems reasonable to believe that atomic power will provide a fertile field for the development of new types of industrial activities which might contribute significantly
to continued economic progress in the United States and elsewhere.

Atomic power might also bring into play at least two distinctly new factors which are potentially of great importance to world economic development.   Both of these are regional in character and our evaluation of both
depends on an analysis by industries.   The first is the development of new
production centers in some industries, which appears possible on the basis
of our analysis.   If such new locations developed, the economic effect would
often extend beyond the particular industry concerned, especially if the
region involved were one in which human resources had been under-utilized
in the past.   The initial impulse provided by the establishment of an industry in such regions might set in motion a chain reaction leading to higher
per capita income in the area, primarily as a result of the more effective
utilization of local labor.   In this case the increased mobility of the fuel
resource might serve, in effect, as a partial substitute for the lack of mobility
of the workers in the region who have remained despite the scarcity of
employment opportunities.   A study of the possible importance of such
instances would be a fruitful field for further investigations on the economic
impact of atomic power.

There are many countries throughout the world, such as Brazil, Argentina, Italy, and perhaps Great Britain, in which indigenous energy resources are so severely limited as to constitute a serious impediment to future
economic development.   Atomic power may be important to such countries
for two reasons: (1) they may have abundant resources of atomic fuel which
could take care of their future energy needs, e.g. Brazil has sizable deposits
of thorium; (2) even if they do not have resources of atomic fuel, they will
probably be able to buy it in its natural form, i.e. thorium or uranium ores
and concentrates, at only a slight outlay of foreign exchange compared to
the cost of its fuel equivalent in coal or petroleum.[2]   As we have already
pointed out, the fuel component of the cost of atomic power is negligible.
Capital charges are the important element in costs, and it is possible that
either in the near future, as in Great Britain, or eventually, as in economically underdeveloped countries, much of the plant and equipment would be
supplied from internal resources.   In effect, this would remove fuel from

[2] For a fuller discussion, see Chapter II, B.4.b.

that class of products which cannot under any circumstance be produced in adequate amounts from domestic resources. It is difficult to know how important this change would be without knowing the future prospects for world trade. But there are some grounds, at least, for believing that the physical and economic factors (i.e. foreign exchange difficulties) which otherwise result in an extremely uneven distribution of energy supplies among the nations of the world could be greatly modified by the advent of commercial atomic power.

If atomic power will be available where other fuels are scarce, either as direct heat from the reactor or as electricity, the question arises: To what uses can this type of energy be economically applied? In particular, it is important to know whether it can be used in certain industries which are of key importance in economic development, e.g. iron and steel, railroads, etc. Our industry analyses provide the beginnings of an answer.

4. The Selection of Industries

The industries covered in our analysis are aluminum, chlorine-caustic soda, phosphate fertilizer, cement, brick, flat glass, iron and steel, railroad transportation, and residential heating. In selecting these industries we were guided by the three major questions discussed above which are central to the industry analyses.

Because of our interest in whether cost reductions are possible as a result of atomic power, these are industries in which the cost of energy is an important element in total production cost. Thus, while for all manufacturing in the United States energy cost is about 2.3% of the factory value of products, the relative importance of energy cost in the industries covered in our analysis is, in all cases, well above this average figure.[3] This is evident in Table 9, which summarizes the relevant figures.

Since we have wanted to examine the possible relocation effects of atomic power, we have included some industries in which energy resources have had a locational influence. The industries falling in this category are aluminum, flat glass, and iron and steel. The study of locational effects is also important for phosphate fertilizers in which the use of atomic electricity might encourage the growth of a production process with somewhat different locational determinants than the process commonly used today.

Finally, because of our interest in process changes, we have included certain industries in which atomic power can be used only with the introduction of such changes. The term "process change" is used in a specialized sense to mean the use of atomic energy, either as direct heat or as electricity, in an industry where electricity is not now the primary energy source. We

[3] Average for all manufacturing was derived from the 1939 Census of Manufactures by Lincoln Gordon for "Power and Fuels," *op. cit.*

TABLE 9.   *The cost of energy as a percentage of the cost of production in selected industries.*[a]

Industry	Energy cost as per cent of cost of production	Remarks[b]
Aluminum	20	Includes only electricity used in reducing alumina to aluminum, and refers to reduction plants located near low-cost hydro-electric power. Electricity cost is expressed as a percentage of the estimated mill cost of producing pig aluminum.
Chlorine-caustic soda	8	Refers to chlorine and caustic soda produced jointly in the electrolytic process, and expresses electricity cost (at 3.5 mills per KWH, a typical prewar rate for this industry) as a percentage of the prewar manufacturers' price of the two products.
Phosphate fertilizer	33	Refers to double superphosphate fertilizer produced from elemental phosphorus smelted in an electrothermal furnace using electricity costing about 3.25 mills per KWH (assumed cost of atomic power). Electricity cost is expressed as a percentage of the estimated cost of producing the double super-phosphate fertilizer. Phosphate fertilizer today is generally produced by a chemical process in which energy cost is insignificant.
Cement	15–26	Refers to both fuel and power consumed in cement manufacture and assumes coal costing $2.00–$6.00 per ton (assumed representative range of prewar coal costs for cement mills in various parts of the United States).
Brick	20	Refers to both fuel and power consumed in brick manufacture, and shows energy cost as percentage of factory value of products for the entire industry in 1939.
Flat glass	7–10	Refers to both fuel and power consumed in glass manufacture and shows energy cost as percentage of factory value of products in representative production centers.
Iron and steel	12	Refers to both fuel and power consumed in blast furnaces, steel works, and rolling mills, and shows energy cost as a percentage of factory value of products (excluding duplication) for the entire industry in 1939.
Railroads	8	Refers to all energy consumed and shows energy cost as a percentage of total operating expenses on railways in 1939 and 1945.

[a] Excludes residential heating because in this instance energy is not used in the production of another good or service, but is, in effect, the good being produced.

[b] For a complete explanation of the meaning and derivation of these percentages, see the individual industry chapters.

define process change in this way because if atomic power appears as electricity, it can be used without any change in method wherever electricity is now used.   For example, it would make no difference in aluminum reduction whether the source of electricity is a nuclear reactor or a waterfall. However, for those industries which obtain energy from coal or some other fuel (or, like phosphate fertilizer, are based on a chemical process), it would usually make a great difference if their energy source were changed to electricity.   A production technique which is novel in at least some respects would be required.   The difference might be less drastic where the nuclear reactor is used as a source of heat rather than of electricity, but even here quite important changes in production technique might be required. For example, the substitution of nuclear heat for coal in residential heating would require the establishment of district heating systems to replace the home furnace.

The industries in which process change, thus defined, is an important subject for study are phosphate fertilizer, cement, brick, flat glass, iron and steel, railroad transportation, and residential heating. With the exception of phosphate fertilizer, which is produced chemically, coal, oil, or gas are of great importance in these industries. The amount of fuel consumed by them in 1939 in terms of its energy equivalent in tons of bituminous coal, and the amount of coal consumed as such, are shown in Table 10. Coal consumed as such in these industries amounted (in 1939) to 66% of all bituminous coal produced in the United States and to about 60% of the bituminous and anthracite coal combined. All the fuel consumed in these industries took more than 40% of the nation's total energy supply from mineral fuels and was the equivalent of between 80% and 90% of the nation's coal supply.

Three users stand out in the total picture: iron and steel, railroads, and residential heating. It would be important to study these industries if for no

TABLE 10. *Fuel consumed in selected activities in which fuel, not electricity, is the primary energy source.*

Industry	All fuel consumed in 1939, total in bituminous coal equivalents	Coal consumed as such or as coke
	(thousands of tons)	
Cement manufacture[a]	7,300	5,200
Brick manufacture[b]	3,100	2,200
Flat glass manufacture[c]	1,000	300
Iron and steel manufacture[d]	65,700	55,000
Railroad transportation[e]	90,100	74,700
Residential heating[f]	189,000	124,000
TOTAL	356,200	261,400
TOTAL AS:		
Per cent of all bituminous coal produced	90	66
Per cent of all bituminous coal and anthracite coal produced	80	59
Per cent of all mineral fuels produced	42	

[a] Derived from *Minerals Yearbook, 1940, op. cit.*, chapter on "Cement."

[b] Derived from *Census of Manufactures, 1939,* U. S. Bureau of the Census, Washington, G.P.O., 1942, report on "Brick and Hollow Structural Tile."

[c] *Ibid.*, report on "Flat Glass."

[d] *Ibid.*, report on "Blast Furnaces, Steel Works and Rolling Mills." Adjusted to exclude intra-industry duplication.

[e] Derived from *Statistics of Railways in the United States, 1939,* U. S. Interstate Commerce Commission, Washington, G.P.O., 1941, p. 65.

[f] Derived from *Minerals Yearbook, 1940, op. cit.* p. 846. Includes, in addition to residences, the fuel used in offices, hotels, schools, hospitals, and probably stores. It may be more accurate to refer to this category as non-industrial space heating.

other reason than their towering importance as energy consumers. Obviously, if atomic power could compete with fuel in these industries, there might be enormous effects on fuel production and associated industries. Again, if the use of atomic power becomes economically feasible in one or more of these industries, it could be very important in the industrialization of those countries where conventional fuels are scarce, and where atomic power might be more freely available. Although this question will receive no further mention in this chapter, we should not lose sight of it.

The decision to include a particular industry in the analysis has been based on the factors mentioned above. Several industries for which we made preliminary analyses (ferro-alloys, copper, lead, zinc, pulp and paper, ocean transportation, chemical nitrogen, and some phases of agriculture) have not been included in the report because of the impossibility of assembling complete enough data, as well as shortages of time and personnel, and so forth. These and other industries which are not discussed in this report would undoubtedly repay detailed analysis. In addition, the industries covered here have not been investigated in nearly as much detail as would be desirable. We hope that others will carry forward this necessary work.

5. *The Cost of Atomic Power*

Our analysis of the possible economic effects of atomic power starts from a set of ideas concerning the cost and other economic characteristics of atomic power. Some of these have been mentioned earlier in this chapter in connection with specific questions. Others which are particularly relevant to the industrial analysis will be briefly stated in this section. They have been abstracted from the analysis and data shown in Chapter I, and relate mainly to the cost of producing atomic power. Although costs will be expressed in terms of electricity, some of the industrial applications studied assume the direct use of nuclear heat. Equivalent heat costs are used in these cases.

According to the estimates derived in Chapter I, the cost of generating commercial atomic electricity might fall into one of the following classes (assuming 1946 prices, and electric central station operations at 50% of capacity):

> 4.0–4.5 mills per KWH
> 6.5–7.0 "   "   "
> About 10 "   "   "

We have evaluated these estimates in terms of an approximate time scale of commercial use. The highest figure we take to represent the level of cost which would be typical of the first commercial plants producing atomic power. The lowest figure, on the other hand, we consider to represent the minimum cost at which atomic electricity would ever be produced by techniques considered likely at the present time. We believe costs would fall this low, if ever, only after the commercial atomic power industry has had many years of experience. The intermediate cost figures we consider to represent an approximation of the level of costs which might prevail in those atomic power plants which will be built, say, 5 to 10 years after the first installations. These plants would incorporate the improvements based on the lessons learned in constructing and operating the earliest plants but would still fall short of the highly efficient designs which would incorporate the advances of a good many years of commercial operation. For the purposes of the industry analyses, only the low and intermediate estimates are significant; as long as the highest cost prevails, there are not likely to be any important industrial effects.

It must be understood that the cost figures are only rough estimates. Therefore, even though we believe that they can best be understood as applying to different stages in the commercial use of atomic power, it is also possible that any one of them could materialize as the real long-run cost of producing atomic power, depending on how the many unknown factors shape up in the process of development now under way.

The firmest figure is probably the estimate below which atomic costs are not expected to fall, for this is based on the assumption that the investment per kilowatt in atomic electricity would ultimately be as low as the investment in ordinary thermal electricity. The only remaining difference, in this event, would be in the cost of the fuel "burned" in the two types of plants.

This minimum cost figure (with the modifications noted below) is of great importance in our analysis. The essential question we ask in our industry studies is: How cheap must atomic power be in order to bring about certain effects such as locational changes, process changes, and cost reductions? We then ask whether the cost required is above, below, or at the borderline of the lowest conceivable cost of atomic power. It is in this connection that the minimum cost estimate is used. It enables us to find out what effects could *not* follow under any circumstances; and gives us a sense of the nature and scope of those changes which might take place. Thus it helps to define the general area within which the changes resulting from atomic power would fall.

Additional important characteristics of atomic power can be seen when the estimate of generating costs is shown in terms of its component items. For this purpose we take the two lower cost classes, since only they are significant in the industry analyses. Their breakdown into fixed charges, fuel, and labor, maintenance, and supplies follows (in mills per KWH) :[4]

[4] These figures do not agree exactly with those in Chapter I because they are adjusted to totals which have been rounded to the closest 0.5 mill.

	Minimum estimate	Intermediate estimate
Total cost:	4.0–4.5	6.5–7.0
Fixed charges	3.5	5.4
Fuel	0.002–0.02	0.002–0.02
Labor, maintenance, and supplies	0.5–1.0	1.1–1.6

Two major factors are revealed in this breakdown of estimated costs: (1) fixed charges dominate in production costs; and (2) the cost of fuel is of negligible importance.   The fuel item considered here is the day-to-day consumption of natural uranium or thorium.   These consist almost wholly of materials which are not fissionable in their natural state but which have the rare property of being converted into fissionable substance when exposed to fission in the nuclear reactor.   The reactor probably will require, in addition to the natural uranium, the investment of an initial stock of some pure fissionable substance (plutonium, uranium 235, or uranium 233) in order to begin operations.   This will be relatively costly, but it figures only as an item in the initial capital cost per kilowatt of capacity and is covered in the fixed charges on investment.   We assume that the reactor will replenish its supply of such materials through its own operations by converting feed materials (natural uranium or thorium) into fissionable substance.

These factors have at least two implications for the industry analyses. The first pertains to the question of the percentage of capacity at which the atomic power station operates.   The costs shown are for operations at 50% of capacity, a reasonable average figure for total electricity production. Since fixed charges are so important, however, substantial cost reductions per KWH are possible if capacity is more fully utilized.   Many of the industries we consider in our discussion will be characterized by an almost continuous and unvarying energy demand, so that they will use a kilowatt of capacity almost 100% of the time.   For these industries, lower estimated KWH costs reflecting the fuller utilization of capacity are used in our analyses.   Total costs per KWH at 80% and 90% of capacity (100% of capacity is rarely attained) as compared with costs at 50% are listed below (in mills per KWH):

Percentage of capacity	Minimum estimate	Intermediate estimate
50	4.0–4.5	6.5–7.0
80	2.7–3.2	4.6–5.0
90	2.5–3.0	4.2–4.6

A second important implication is that the cost of producing atomic electricity is likely to remain very much the same over the lifetime of an existing atomic power installation.   This would sharply distinguish atomic electricity from ordinary thermal electricity if the upward trend in the cost of conventional fuel continues, since rising fuel costs would result in higher electricity costs.   In this respect the atomic installations will resemble hydroelectric plants since for them also fixed charges are the dominant element in costs.

All of the figures we have shown so far are estimated costs for the generation of atomic power.   Because many of the industries considered in the analysis will normally buy electricity from central electric plants, sometimes located at a considerable distance from the consuming factory, there will be additional costs for the transmission of electricity.   Such transmission costs are extremely difficult to estimate, particularly because the distances involved cannot be predicted.   We assume that, in general, intensive power-using industries will try to locate close to the central electric plant in order to avoid excessive transmission charges, but this factor must always be balanced against other cost factors.

We have chosen not to make explicit allowance for transmission costs in most of our industry analyses.   Instead we allow for transmission implicitly by assuming that the border line of realizable atomic costs will fall somewhere between the minimum and the intermediate estimates, and in no case at the estimated absolute minimum.   We have, in effect, a cost zone of about one-half to one mill per KWH (at 90% of capacity) into which we relegate transmission along with other uncertainties surrounding the cost estimates.   The exception to this procedure among the electricity-consuming industries is railroad transportation.   Here we make explicit allowance for transmission because of the obvious need in this industry for moving electricity over extended distances.

B. MAJOR FINDINGS OF THE INDUSTRY ANALYSES

The following summaries report those findings of our industry studies which we consider most significant, as they apply to the United States.   In our detailed industry discussions and elsewhere in this report, we attempt, wherever the results warrant it, to weigh the implications of the American analysis for other countries.   Unfortunately, we could not carry over into brief summary statements these extensions of the primary argument.

The reader should be aware that we have often found it difficult within the limits of a brief summary fully to convey the tentative nature of our conclusions, or to provide adequate information on the shortcomings of the primary data.   These questions are dealt with in greater detail in the individual industry chapters.

1. Aluminum

The aluminum industry consumes an extremely large amount of electricity per ton of product.   The electricity input—about 18,000 KWH per ton—is the largest among the industries covered here.   The size of the per unit power requirements has forced the industry to seek sources of extremely low-cost electricity for aluminum reduction works. These have usually been near hydroelectric power. The orientation to cheap power has meant that additional costs have been incurred in the production and marketing of aluminum both for the transportation of raw materials (2 tons of bauxite per ton of alumina, 2 tons of alumina per ton of aluminum) to production centers, and for the transportation of pig aluminum to consuming centers.

We have considered three primary questions. (1) Could atomic power reduce costs in existing production locations? (2) Could it bring production sites closer either to aluminum markets or to bauxite deposits as a result of a greater geographical equalization of electricity rates? (3) Could it be important to the continued growth of the aluminum industry by providing a cheaper future source of electric energy than would otherwise be available?

We find that lower costs at existing sites are not likely to result from atomic power, since the industry is already using energy sources which produce electricity more cheaply than appears possible for atomic power plants.

The possibility of relocating aluminum production to two new sites was considered: (1) Dutch Guiana, which provides the bulk of bauxite consumed in the United States; and (2) Mobile, Alabama, which is closer to bauxite resources and also closer to important markets than some existing aluminum plants, and which is now an important center of alumina production (an intermediate stage in making aluminum from bauxite). Assuming that existing aluminum production sites will continue to provide a growing supply of equally low-cost power in the future as in the past, relocation to the new centers would require atomic power rates of about 3.5 mills per KWH or less in Dutch Guiana, and 3.25 mills or less in Mobile. Since such rates are on the borderline of estimated realizable atomic costs, we have concluded that while locational shifts might occur as a result of atomic power, they would not bring about major reductions in aluminum prices.

Whether atomic power could provide a cheaper future source of electric energy than ordinary power depends in the main on four variables: (1) the estimated future demand for aluminum if relative metal prices remain as they are today; (2) the future availability of water power sites with favorable natural conditions; (3) the future costs of constructing hydroelectric developments; and (4) the future costs of alternative fuels that might be used for generating electricity. These variables can be combined into the following composite picture. Estimates of future aluminum demand indicate that something like 3.5 million additional kilowatts of capacity might be required to satisfy the growth in aluminum output between 1945 and 1970. This amount, which is about 80% of the total generating capacity of all federally-owned hydroelectric plants in 1946, would have to be made available even though unused water power sites may, in general, have less favorable natural conditions than those already exploited. Furthermore, the cost of constructing hydroelectric developments will be higher than in the prewar decade. Because of higher construction costs alone, alternative sources of electricity of 3.9 mills per KWH or less in the Gulf Coast area would provide cheaper aluminum in important markets than additions to hydroelectric capacity at existing sites.

While rates of 3.9 mills might make ordinary thermal electricity plants fueled by natural gas suitable for aluminum production, there are several reasons why these plants would be less effective in attracting aluminum production than atomic power:

(1) The future cost and availability of natural gas is uncertain, while atomic power costs should be pretty much the same over the lifetime of an installation. This is important because rising fuel costs, and the resultant higher electricity rates, could be disastrous for aluminum producers with their very high power consumption per unit of production.

(2) Potentially, atomic costs can fall below costs possible with natural gas except when the latter is an almost cost-free by-product of oil field operations for which there is no other market.

(3) Thermal power stations based on natural gas would be locationally inflexible as compared with atomic power. Natural gas happens to be cheap on the Gulf Coast, but this is only one of many sites at which aluminum might be produced. Sites even closer to important markets might have lower aluminum production and transportation costs than Mobile, particularly if aluminum production from fairly ubiquitous low-grade clays becomes technically feasible. On this basis, the New York area, or other important market centers, might attract aluminum production with atomic power, but not without it.

We conclude, therefore, that atomic power may be of great importance in supporting the future expansion of aluminum output in the United States, and if so, it would undoubtedly be in locations other than those favored by aluminum producers today.

2. Chlorine-Caustic Soda

Chlorine-caustic soda is, like aluminum, an electrolytic industry, but it differs sharply from aluminum in how it might be affected by atomic power. Its power requirements per unit of product (about 2,850 KWH per 2 tons, 1 each of chlorine and caustic soda), although large by comparison with the ordinary run of industries, are relatively moderate when compared with aluminum reduction. The weight loss involved in passing from the raw material (salt) to the final products is negligible compared to that involved in passing from bauxite to aluminum. As a result, producers of chlorine-caustic soda are generally located in the vicinity of their markets rather than of cheap water power or of salt deposits, if these are distant from markets, and there is no reason to expect atomic power to change the essential locational pattern. The main question, therefore, is whether atomic power will provide cheaper electricity in those places where producers are located to serve an already existing market.

The general conclusion we reach is clearly implied in the essential nature of the industry. In order for production to be market-oriented, it must be widely dispersed, and if it is widely dispersed, producers will encounter varying power costs. In some, though not all, of the desirable production sites, atomic power might provide electricity more cheaply than existing sources. The precise locations in which costs might be reduced through the use of atomic power are difficult to select. The data available to us indicate that the general level of power costs in major production centers in Michigan, Rhode Island, and New Jersey is such that atomic power may be cheaper, but this listing is at best inadequately supported by the available data. A rough calculation indicates that a cost reduction of no more than about 5% may reasonably be expected in those places where atomic electricity is cheaper than existing sources.

3. Phosphate Fertilizer

The cost of energy is unimportant in the manufacture of phosphate fertilizer, but it could become important with a change in the production process. The production methods currently used are based on phosphate rock and sulfuric acid; the production method which could be used requires electricity in large amounts in place of sulfuric acid. However, the sulfuric acid method could be replaced by the electric process only if the cost of electricity were very low. Could atomic electricity be cheap enough to encourage the adoption of the new process based on electricity?[5]

In the new method, the power requirements per unit of product are great: about 13,000 KWH per ton of elemental phosphorus, which may be compared with 18,000 KWH per ton in the reduction of aluminum. The raw material requirements per ton of product are also great: about 7 tons of phosphate rock per ton of elemental phosphorus as compared with approximately 4 tons of bauxite per ton of aluminum. Aluminum production demonstrates that if electricity requirements per unit of output are sufficiently large, it often pays to ship weight-losing raw materials over very long distances for processing at low-cost water power sites. A second major question is, therefore, whether atomic electricity near the phosphate rock would yield a cheaper product than low-cost water power located at some distance from the phosphate rock mines.

[5] The analysis assumes the production of elemental phosphorus to be used in making double superphosphate fertilizer. Quite different conclusions might apply to the production of elemental phosphorus for other uses.

To answer these two questions we studied the position of the greatest phosphate rock-producing region in the United States, the deposits in Florida. Our analysis disclosed that economies of transportation would probably now favor sites near phosphate rock for the production of phosphate fertilizer by the sulfuric acid method, although formerly this was not so. The first comparison was, therefore, between estimated costs of producing phosphate fertilizer by the sulfuric acid method and by the electric method in Florida. This disclosed that electricity rates of about 3.5 mills per KWH or less would justify the electric furnace method in Florida. Since atomic electricity might be produced at this cost, we concluded that atomic power might encourage the use of the electric method in place of the sulfuric acid method, while electricity costs in Florida based on conventional fuels might be too high to justify the use of the electric method.

The second comparison was between electric smelting in Florida and electric smelting of Florida rock in the Tennessee Valley region. The latter region was considered as favorable a location as would exist for the low-cost electric smelting of Florida rock outside the ore-producing region. This disclosed that, as a result of the extra transportation costs of shipping the rock to low-cost hydroelectric energy, atomic electricity in Florida at 3.5 mills per KWH probably would be cheaper than electric smelting elsewhere even with lower-cost hydroelectric power.

The analysis disclosed that there would be no dramatic reductions in the cost of phosphate fertilizer as a result of atomic power. The choice between methods would probably be made on the basis of quite small cost differentials. The importance of atomic power lies in its ability to tip the balance in favor of the electric method under circumstances in which electricity from conventional sources might prove too costly to justify its use.

4. Cement

In cement production much energy is consumed per unit of output, mainly as fuel to be burned in the kiln rather than as electricity. Our analysis considers two possibilities for replacing ordinary fuels by atomic energy: (1) the use of atomic electricity in a process involving the substitution of electric kilns for fuel-fired kilns; and (2) the use of extremely hot gas from the nuclear reactor to be fed directly into the cement kiln. Neither method is feasible with present techniques, nor is it known whether the technical difficulties involved can be overcome. Our analysis is, in a sense, an attempt to determine whether the possible savings in the cost of producing cement by the new methods make it worthwhile to attempt to solve these technical problems.

Two characteristics of cement manufacture are important in evaluating the possible role of atomic energy. (1) Fuel and power constitute an important element in costs; for example, with coal at $4 per short ton, they come to about 20% of the total cost of production. (2) Production sites are usually determined by the market. As a result, fuel and power costs vary widely at different production locations. Nuclear reactors might therefore offer cost-reducing possibilities in cement manufacture in some places. The important question is: How cheap must atomic energy be in order for these two untried and unperfected methods to compete with ordinary fuel in cement manufacture?

Our analysis of comparative energy costs suggests that the electro-thermal method of cement manufacture based on atomic electricity probably could not compete except where coal costs $12 per ton or more. Since this is an unusually high coal price for large-scale users in normal times, we conclude that atomic energy in the form of electricity would not be important in cement manufacture. On the other hand, the nuclear reactor as a source of hot gas might compete with coal costing from $4 to $6 per ton; since many regions in the United States are likely to have coal prices within this range, we conclude that atomic energy applied in this form might be quite important to the American cement industry if the engineering problems involved in its use could be solved.

The analysis is complicated by another factor. Gas at the extremely high temperature required in cement manufacture can be piped only over very short distances. This may mean that cement mills would be able to use hot gas from nuclear reactors only if the reactor were located close to the kiln. Unless additional uses are found for the reactor in the vicinity of the cement mill, this might limit the use of nuclear reactors to mills large enough to require units within an (as yet unknown) economic size range. On this basis a large portion of the cement industry might be excluded from the potential benefits of the use of atomic energy.

5. Brick

Brick resembles cement in several respects. Energy constitutes a substantial element of production costs; in 1939 it accounted for about 20% of the factory value of products of the brick industry. Market orientation of production is even more pronounced than in cement manufacture because economies of scale in brick production are not as important as in the cement industry. As in that industry, therefore, energy costs probably will vary widely depending on the level of fuel costs in the region where the particular brick factory is located.

The resemblance does not end here. The energy requirements in brick manufacture, like those in cement production, are chiefly for fuel to burn in the kiln and only secondarily for electricity. Without making new calculations for brick, we assume, therefore, that atomic energy will be economically feasible only if the nuclear reactor can be used as a source of hot gas for direct use in the kiln. However, the usual brick plant is extremely small in comparison with cement mills, almost certainly too small to require a nuclear reactor of its own within an economic size range. Because of this, we conclude that even if the nuclear reactor could provide gas hot enough for brick manufacture, its importance would be severely limited by the small size of brick plants, unless there were additional large-scale uses for the reactor in the immediate vicinity.

6. Flat Glass

Although fuel and power are less important in the total cost of producing flat glass than in most of the other manufacturing industries covered in this report, major production centers have been located near fuel supplies. The fuel used in the United States is almost invariably natural gas, with glass furnaces in both old and new gas-producing regions. Our analysis of the possible use of atomic power in this industry is based on the application of electric glass melting furnaces which are new and as yet not wholly perfected, but for which engineering estimates of cost and performance are available.

We have considered two questions about the possible effect of atomic power on glass manufacture: (1) Could it reduce production costs in present sites? (2) Could it cause the industry to move out of the gas fields into new production centers located closer to markets? The second question arises because the raw materials other than fuel used in glass production are widely distributed, so that a greater geographical equalization of energy costs might make production closer to market economically desirable.

As for present production sites, it appears likely that atomic power used in the electric glass melting furnace could reduce costs in many of the older gas-producing regions in Central and Eastern United States, but not in the newer flush gas-producing regions of the Southwest. However, for the older regions it appears probable that electric furnaces using coal-generated electricity could also effect some reductions in costs, although not quite so much as might be possible with atomic electricity. But as further analysis shows, the possibility of cost reductions in present locations is of questionable significance.

Turning to the question of possible changes in production sites as a result of atomic power, we encounter the following paradox. Even though furnaces using natural gas today are located in gas-bearing regions, figures on the comparative costs of transporting both natural gas and glass suggest that it would be cheaper to pipe gas to locations near consuming markets and manufacture glass there. It appears likely, therefore, that with the expansion of the pipeline network for gas distribution in the future, market areas will grow in importance as centers of glass production, at the expense of gas-bearing regions.

Furthermore, if the electric glass furnace should be introduced, it too would encourage production close to market. Indeed, our data indicate that these furnaces might very well be cheaper than furnaces using piped gas in such important market regions as New York and Chicago, even if the electric furnace used electricity generated from coal. On the basis of this evidence, we conclude that glass production probably will move to market regions even in the absence of atomic power.

There is no evidence that the trend to market orientation in glass production would be measurably strengthened by the advent of atomic power. The cost advantages seem clearly on the side of the market for furnaces using other sources of energy (either natural gas or coal-based electricity), and the most that atomic power could do would be to reduce production costs somewhat in the new locations. Our data suggest that these cost reductions would be quite moderate, possibly in the neighborhood of 2% of glass production costs.

7. Iron and Steel

Iron ore smelting today is completely dependent on coal in the form of coke. For this purpose coal acts both as a source of heat and as a chemical agent. Energy (not necessarily from coal) is also necessary in the later stages of steel manufacture including steel refining and rolling. The over-all fuel economy involved in combining coke ovens, blast furnaces, steel furnaces, and rolling mills has resulted usually in creating works where all of these operations are carried on at a single site. When these operations are taken together, it becomes clear that coal is important not merely as a source of blast furnace fuel but that it also has a powerful locational influence on the entire integrated operation. Iron and steel manufacture takes 15% of all the bituminous coal produced in this country (chiefly for blast furnace use), and also consumes large amounts of other fuels in steel furnaces and rolling mills.

Could atomic power alter this situation? It would appear that it could bring about changes in one or both of the following ways: it might either replace coke as a blast furnace fuel and thereby release the entire operation from its fundamental dependence on coal; or it might separate the steel furnace and rolling mill from the blast furnace and produce important effect only on the former operations, which can be based on electricity with present-day technology. We consider both possibilities in our analysis.

As to the separation of steel furnaces and rolling mills from the blast furnace, the analysis discloses the following possibility. Atomic electricity might make the electric melting of steel scrap worthwhile in important steel-consuming centers which do not today have steel plants or which produce much less steel than they consume. In other words, atomic electricity rates might be low enough so that, together with the transportation savings possible from locating production close to market, they would make such furnaces competitive with integrated steel works elsewhere.   This development is rendered much more likely by the recent announcement of the Republic Steel Corporation that continuous casting has been successful on an experimental basis.   In this method only one step is needed for converting molten steel to semi-finished shapes, so that the massive and expensive equipment needed for making ingots and billets is eliminated.   Taken together with low-cost atomic electricity, this might open the way to decentralized steel production in such important steel-consuming centers as New York, Boston, and St. Louis.

Scrap supplies would set a definite upper limit to steel production by this method.   Although the annual supply of scrap is not likely to exceed 30 to 40 million tons at most for a long time to come, the economic limit to electric scrap melting probably would be considerably lower, since integrated steel works, which will continue to use open-hearth furnaces operating on a mixed charge of scrap and pig iron, will compete for the limited scrap supplies. Thus, while this development probably could bring about some decentralization of steel production, the bulk of steel production in the United States would continue to be organized on the present basis.

Could electricity replace coke as a blast furnace fuel?   We find that there is on the horizon one technique which might be economically feasible in the United States.   In this method iron ore would be reduced to sponge iron by using hydrogen as the reducing agent rather than carbon.   This new process, just emerging from the laboratory, is considered to hold promise by the research metallurgists who have experimented with it.

Atomic electricity could be used to produce hydrogen by the electrolysis of water.   Could iron be produced at a lower cost by this method than by the coke blast furnace in the present centers of steel production?   Could the new method result in new iron production sites which would provide lower-cost iron in important markets than present locations?   To answer these questions we constructed estimates of the comparative cost of iron reduction by the two methods.   Obviously these comparisons are extremely rough since the hydrogen method has not yet been tried commercially.

Our calculations indicate that if the price of coking coal is about $8 per ton or more, atomic electricity used in the hydrogen process would, on the most favorable cost assumptions, be on the competitive threshold of the coke blast furnace.   On this basis we find that existing steel production centers will continue to find the coke blast furnace a less costly source of iron than hydrogen furnaces based on atomic electricity.

Present steel centers are favorably situated with respect to coal. In a process based on atomic electricity the major raw material influencing location would be iron ore, and we therefore have attempted to determine whether the hydrogen process based on atomic power near iron ore would be cheaper than the coke blast furnace elsewhere. Here we find that the iron ore region of Northern Minnesota, which produces the bulk of the iron ore consumed in the American steel industry, might be able to deliver iron at a lower cost in the Chicago-Gary market than steel plants located in this area, and could almost match costs in the Pittsburgh region.

This particular result rests on assumptions which may prove invalid, particularly with respect to the relative costs of transporting iron ore and sponge iron. However, it points to the possibility that atomic electricity might bring the hydrogen reduction process at new production sites within economic reach.

If the hydrogen process based on atomic electricity should be adopted, it would have major implications for the location and size of steel-producing facilities. By-product fuel economy would no longer be of great importance since the hydrogen reduction process would not yield anything like the volume of hot gases exhausted by the coke oven and blast furnace. This change would weaken one of the most important factors which has historically made for the locational integration of all operations from the coke oven through the rolling mill. If the new hydrogen process is used, it might prove possible to reduce the iron ore at the ore site, since iron ore would then be by far the bulkiest raw material, and to place the later stages of production near the market for steel products. Furthermore, it appears possible that sponge iron operations could be successfully set up in a region to exploit ore resources of approximately $\frac{1}{10}$ the size needed for economic investment in a blast furnace. What might happen, therefore, is that in time iron reduction would undergo much decentralization with hydrogen sponge iron plants developing in various regions, based on local reserves of iron ore.

Under these circumstances some integrated operations on a considerably reduced scale might return through the back door. For as the economic size of plants were reduced, smaller markets could be served by a single plant and the market and ore sites might draw closer together. Of course, the chances for such a development would be enormously improved by the perfection of continuous casting and other small-scale finishing techniques, since they would scale down considerably the size of the units required for converting the iron into certain basic shapes.

8. Railroad Transportation

While energy costs account for about 8% of railway operating costs, other operating expenses are also affected by the type of railroad motive power used. Aggregate fuel consumption by American railroads constitutes somewhat more than 10% of the country's total consumption of mineral fuels.   The energy consumed today is derived primarily from coal, although the trend in recent years has been in favor of diesel oil.   This trend may reverse itself if oil costs tend to become very high and if the recently developed coal-burning gas turbine locomotive compares favorably with the diesel locomotive.   Electricity as a source of railway motive power accounts for only a very small percentage of the industry's total energy consumption.

It is conceivable that nuclear reactors could be used directly to propel locomotives, but we do not consider this likely because of a combination of technical and economic factors.   Our main question is: Could atomic electricity encourage a significant increase in railroad electrification in place of fuel-fired locomotives?   Our basic comparison is between electric locomotives using electricity generated in atomic central stations, and diesel locomotives, which are taken to represent the best in modern fuel-fired locomotives.   The analysis has considered only main-line locomotives traveling about 200,000 miles per year.   In order to bring the subject within manageable proportions, we undertook highly simplified comparisons from which only very general conclusions can be drawn.

The most important variables affecting the choice between diesel and electric locomotives are the comparative costs of diesel oil and electricity, and the density of railroad traffic.   Considering diesel oil costs in various parts of the country and the prevailing density of railroad traffic in the same regions, the electricity rates required to render electrification economically feasible were in most cases found to be below the minimum estimates of future atomic costs.   For three regions (the Great Lakes, Central Eastern, and Pocahontas regions, as classified by the Interstate Commerce Commission), railroad traffic appeared sufficiently dense to justify possibly a fair amount of electrification, but only if atomic electricity could be made available at the lowest estimated future cost, while diesel oil costs remained at current high levels.

Since significant expansion in railroad electrification based on atomic power probably could not be justified anywhere in the United States unless atomic electricity were produced at the minimum estimated future cost, the industry probably will not undergo extensive electrification as a result of atomic power.   First of all, it is rather doubtful that atomic costs will fall to the very minimum.   Secondly, if its costs were that low, atomic power might, according to our studies, replace coal in many of its current uses. Since coal tonnage constitutes about ⅓ of the total freight originated on American railways, there is at least a serious question whether as a result of the decline in coal shipments the density of railroad traffic would not in many cases fall below the minimum necessary to justify electrification. Therefore, the general conclusion seems justified that railway motive power in the future, as in the past, will continue to be based on diesel oil and coal, the latter fuel possibly to be used in radically new locomotives.

9. Residential Heating

Space heating consumes annually a greater amount of energy than any of the industries covered by our analysis. In 1945 the fuel consumed for this purpose accounted for almost 20% of the total energy supply from mineral fuels and water power.[6] By and large, the heating of residences is not performed by an existing industry but by the individual householder or his landlord. How would nuclear heating affect this important energy-consuming category? Could it bring about the growth of a district heating industry using central nuclear reactors as its energy source? Or could it substitute for conventional heating methods in some other way?

In the other activities covered in our analysis we have considered atomic power primarily in the form of electricity. We do not consider electricity for space heating purposes because (1) electric heating with ordinary resistance heaters is at a decided disadvantage in comparison with direct heat from the same thermal source, chiefly because of the energy loss involved in converting heat into electricity, and (2) the only electric heating device which could redress the energy loss involved in producing electricity, the heat pump, is today much too expensive for common use. Nor do we deal with the possibility that nuclear reactors whose primary product is electricity will supply direct heat as a by-product, because at best this could usually satisfy only a small part of a locality's heating needs. We confine our analysis to the possibility of district heating based on centrally-located straight heat-producing nuclear reactors. Heat produced by the experimental nuclear reactor which began operating at Harwell, England, in July 1948 is reported already being used for space heating, but only within the confines of the Harwell establishment.

The most important variables affecting the cost of residential district heating are population density and weather conditions. In general, a combination of high population density and cold winters would be required to bring district heating into competition with conventional heating methods. This is mainly because there are decided economies of scale in the laying of underground pipe, so that a length of pipe to carry a load, say, four times that of a smaller capacity pipe of the same length can be laid at about a 50% increase in cost. Furthermore, since fixed charges dominate in costs, unit costs are reduced substantially in any size system as capacity is more fully utilized.

Our analysis shows that the concentrated heating demand which would probably be required to bring nuclear district heating costs down to a competitive level exists in very few American cities. Some cities in which conditions seem satisfactory for nuclear district heating are Boston, Buffalo, Chicago, Milwaukee, New York, and Newark, Paterson, and Jersey

---

[6] This total includes the heating of schools, hotels, offices, hospitals, and probably stores, but the analysis is confined to residential uses.

City, New Jersey. In all these cities the average density of population is over 13,000 persons per square mile. The lowest average density exists in Milwaukee, but the winters there are the coldest among the cities named. The others represent varying combinations of exceedingly high population density (New York) and quite cold winters (Buffalo). But all of them are marked by both characteristics to a considerable extent. Very cold winters alone, as in the case of Duluth, for example, would not be a sufficient condition for the success of nuclear district heating, nor would high population density, as in some cities with mild winters.

For several reasons this listing of cities is highly tentative, and at best, merely illustrative of suitable locations. Actually, the suitability of any locality for district heating could be determined only through an engineering survey of that locality. In the absence of this, we have assumed certain typical values for the cost of laying pipes underground, the nature of the buildings to be heated, and so forth, which may be inappropriate for some or all of the cities shown. This is a basic technical problem with which we were unable to cope.

There is, in addition, a basic statistical difficulty which might have been overcome, but only with the expenditure of an amount of work completely beyond our resources. This difficulty arises from the use of city-wide averages on population density as a guide to the selection of localities suitable for nuclear district heating. City-wide averages are inappropriate because of sharp differences in population density among districts within a city. The cities we list undoubtedly contain some districts where the density is too low to justify district heating, while many other cities not listed include certain districts where population density would permit nuclear district heating.

Because of the shortcomings of our analysis, we draw only two very general conclusions. (1) Nuclear district heating might be available at a cost low enough to compete with conventional heating methods in some localities, the heating market thus served accounting for a sizable part of the nation's space heating requirements. This may be seen from the fact that the cities listed above include about 10% of the population of the United States. In terms of fuel consumed for space heating purposes their percentage importance would be much greater, since a large part of the population of the United States lives in mild regions where relatively small amounts of fuel are used for space heating. (2) The substitution of nuclear district heating for conventional methods would not involve large cost savings. Instead, in most cases where nuclear heating would be competitive, its costs would be in the same general range as conventional methods, but nuclear heating might prevail because of its convenience.

CHAPTER IV

Aluminum

BAUXITE, the ore from which aluminum is derived, is a highly localized raw material. The United States, for example, must depend on distant foreign deposits for most of its requirements. Since approximately 4 tons of this ore are needed to produce 1 ton of aluminum metal, sizable economies in transportation probably could be effected if aluminum were produced near the bauxite deposits; but the production of 1 ton of pig aluminum from alumina requires about 18,000 KWH of electricity for electrolysis. Hence, the reduction of aluminum is almost invariably carried out at the site of cheap electricity, even though this involves transporting the bauxite to the cheap power and then transporting the pig aluminum from there to consuming centers.

Would atomic power reduce the costs of producing aluminum simply by providing cheaper electricity than would otherwise be available? Would atomic power bring about savings from reduced transportation costs by making it possible to locate aluminum reduction works, powered by atomic electricity, closer to bauxite deposits, aluminum markets, or sites combining both advantages?

These two questions will guide our analysis. We will begin with a brief description of how aluminum is produced. Then we will consider the possible effects of atomic power on costs and production locations in the United States. In a third section we will discuss whether atomic power might play a major role in increasing the aluminum output of this country. We will attempt to apply our conclusions, finally, to other countries. The materials presented do not pretend to predict all the possible directions which aluminum production might follow in the future. We believe, however, that the analysis will reveal a number of quite new factors which might operate with the coming of commercial atomic power.

A. PROCESSES OF PRODUCTION

There are two steps in the conversion of bauxite to aluminum metal. In the first stage, alumina is precipitated from bauxite ore which has been ground and dissolved in a hot caustic soda solution. In the second stage, aluminum metal is derived from this alumina by electrolysis, i.e. electric current is passed through a bath of alumina and cryolite, which separates the oxygen from the alumina to yield aluminum metal.

The production of alumina and aluminum requires the following inputs of materials and electricity per ton of output (the amounts are approximate):

To make one ton of alumina (from bauxite)	To make one ton of aluminum (from alumina)
2 tons of high grade bauxite	2 tons of alumina
¾ ton of coal (or its equivalent in another fuel)	1,200–1,600 lbs. of carbon electrodes
132 lbs. of soda ash	50 lbs. of cryolite
125 lbs. of lime	70 lbs. of aluminum fluoride
	18,000 KWH of electricity

Because of these requirements, aluminum is usually produced in the United States close to exceptionally cheap sources of electricity, while alumina plants usually are located at some point between the bauxite mine and the site of the aluminum reduction plant. The energy required to obtain alumina is so small compared to the amount needed for aluminum reduction (it would take approximately 9 tons of coal to produce the electricity required in reducing the alumina for 1 ton of aluminum, compared to only 1½ tons to produce the alumina) that the reduction of bauxite to alumina is economic in regions where energy costs are too high to attract aluminum plants.

With its enormous energy requirements per unit of output, the reduction of alumina to aluminum is the operation which might be seriously affected by the introduction of atomic power. According to the data in the following section, typical costs of producing aluminum metal are around $200 per short ton, when power costs about $36 per ton. The electricity required in the electrolytic reduction process (9 KWH per lb.) would account for the following percentages of total cost at varying power rates per KWH:

Electricity cost (mills per KWH)	Cost of aluminum per ton[1]	Power as a per cent of total cost of aluminum
2	$200	18
4	236	31
6	272	40
8	308	47
10	344	52

It is obvious from these figures that low electricity costs are the lifeblood of the aluminum industry.

[1] The figures in this column were derived by varying the cost of power while holding all other costs constant.

B. EFFECTS OF ATOMIC POWER ON COSTS AND PRODUCTION SITES

1. The Importance of Electricity and Transportation in Production Costs

A breakdown of the costs of producing aluminum in typical plants in the United States is shown in Table 11. The two sets of data correspond to the lowest and highest estimated costs in plants owned by the Aluminum Company of America. Some of the previously indicated characteristics of the industry are clearly shown here. The costs of electricity account for almost 20% of the total costs, while the costs of transporting the

TABLE 11. *Costs of producing pig aluminum. Estimated postwar mill costs at typical U. S. plants, with special reference to the costs of power and transportation of bauxite and alumina.*[a]

	Dollars per short ton of pig aluminum	Per cent of total cost
Power for electrolytic reduction of alumina to aluminum	$28.80–37.80	*19.1–19.7*
Transportation of bauxite and alumina	16.52–24.40	*10.9–12.7*
a. Bauxite from mine to alumina plant	6.22–7.92	*4.1–4.1*
b. Alumina from alumina plant to electrolytic aluminum plant	10.30–16.48	*6.8–8.6*
Other costs	105.78–129.76	*70.0–67.6*
a. In alumina production	36.78–46.28	*24.3–24.1*
b. In aluminum production (excluding cost of alumina)	69.00–83.48	*45.7–43.5*
TOTAL	$151.10–191.96	*100.0–100.0*

Source: Adapted from data in "Aluminum Plants and Facilities," *Report of the Surplus Property Board to Congress, September 21, 1945, op. cit.*

[a] Data refer to Alcoa plants only; the lower figures are for that Alcoa plant (Badin, North Carolina) with the lowest estimated postwar costs, the higher figures are for that Alcoa plant (Vancouver, Washington) with the highest estimated postwar costs. Actually, the report shows a range of costs for each plant; the figures are, therefore, the lower Badin costs and the higher Vancouver costs. The estimated total postwar costs of all other Alcoa plants fall within the range shown, although for single items other plants may have costs above or below the range in the table.

Alcoa costs, as estimated by the Surplus Property Board, exclude all intra-company profits. Hence, the high degree of vertical integration characteristic of Alcoa results in lower costs than would otherwise be the case. Thus, for example, shipping costs for bauxite exclude profits since the steamship company is an Alcoa subsidiary, while the freight on alumina presumably includes profits to the railroad carrier. It is well to note, too, that these estimates of postwar costs are considerably below those in *Aluminum*, by N. H. Engle, H. E. Gregory, and R. Mossé, Chicago, Richard D. Irwin, Inc., 1945; we have preferred to use them because they appear to be more recent.

bauxite and alumina to suitable reduction sites account for between 11% and 13% of total costs.   It is noteworthy, too, that of the total transportation costs, the larger share is for the movement of alumina to electrolytic reduction plants; in the case of the higher cost plant, this cost was more than twice the cost for transporting the bauxite.   Finally, we should realize that power plus transportation account for only ⅓ of the total costs.   Obviously, then, major variations in "other costs" could break a production site which was favorable from the standpoint of power costs or transportation.   In this analysis we will assume that these costs are constant in order to highlight the variables which are our main concern.

2. The Possibility of Cost Reductions in Present Sites

Since power accounts for roughly 20% of the total cost of producing pig aluminum, substantial cost reductions could be effected if power costs were reduced.   But we have seen that aluminum reduction is already carried out in proximity to cheap power; this is reflected in the average power rates implied in Table 11, which range from 1.6 to 2.1 mills per KWH.[2]   The estimated atomic costs shown in Chapter I were under no circumstances lower than or even as low as these rates, and the estimates considered most reasonable were substantially higher.   We can, therefore, conclude that atomic power probably will not result in reduced power rates for the aluminum industry in present locations, or in any decrease in aluminum costs.

3. The Possibility of Locating Aluminum Reduction Plants at New Sites

a. CLOSER TO RAW MATERIALS.   Could the need for transporting bauxite and alumina long distances be reduced through the use of atomic power?   The Alcoa plants represented in Table 11 derive virtually all of their aluminum from bauxite mined in Dutch Guiana.[3]   The bauxite is shipped to the United States for conversion to alumina; the costs in Table 11 are based on the assumption that this operation is performed in the new Alcoa plant at Mobile, Alabama.   The alumina produced in Mobile is then shipped to plants close to cheap power for the electrolytic reduction

[2] The Surplus Property Board estimates of postwar power costs in plants which can compete in today's markets exceed the 2.1 mills per KWH rate in a few instances.   The outstanding case of such operations is at Jones Mill, Arkansas, which may be considered a rare case of existing raw material-oriented production.   The operation is based on Arkansas bauxite, and the electric power is generated from natural gas.   Reserves of Arkansas bauxite are quite limited and, for reasons set forth in a subsequent section, electricity based on natural gas may have only limited significance for the aluminum industry.   The Jones Mill operation is, therefore, not considered to be an important exception from the standpoint of our analysis.

[3] Our attention has been called to the fact that some of Alcoa's production is based on Arkansas bauxite.   However, this does not seriously affect Table 11, which for simplicity assumes Dutch Guiana as the sole source, because it is by far the major source.

to pig aluminum. According to the data in Table 11, if all other costs remained equal, as much as 11% to 13% of the cost of production could be saved by locating aluminum reduction works at the bauxite deposits, and as much as 7% to 9% of the cost by locating aluminum reduction works at Mobile. However, this assumes that atomic power would be available at about 2 mills or less per KWH; according to our estimates, such low costs of atomic power are unlikely.

On the other hand, the costs of production in Dutch Guiana or in Mobile would be less than in present locations if atomic electricity could be made available at less than about 3.5 or 3.0 mills per KWH, respectively.[4] Electricity rates of 3.0 to 3.5 mills per KWH do come within the range of estimated atomic costs, although rates of less than 3.0 mills involve extremely optimistic assumptions. We find, then, that rates which are on the borderline of realizable atomic costs might make new production locations possible in the aluminum industry, but that they probably would, at best, permit only slight decreases in the present costs of aluminum production.

b. CLOSER TO MARKET. The advantages or disadvantages of relocating aluminum plants cannot be determined simply by an analysis of production costs; we must also consider the comparative costs of transporting pig aluminum to market from the different locations. These costs are particularly important in aluminum because production sites based on low-cost power are often in regions far removed from aluminum markets.

TABLE 12. *Aluminum markets and aluminum production centers: estimates by regions.*

Geographic region	Probable percentage of U. S. postwar market for aluminum[a]	Probable percentage of U. S. aluminum reduction capacity after the war
Northeast	30–35	15–20
Southeast	1–3	35–40
Pacific Northwest	1–3	30–35
Southwest	5–7	10–15
Midwest	53–62	0
TOTAL (midpoint of range)	100.0	100.0

Source: *Aluminum,* by N. H. Engle *et al., op. cit.,* Table 102, p. 348.

[a] Market is in terms of consumption of metallic aluminum, not of aluminum made up into finished products. The estimates should be used only as a rough approximation; events in recent years may have changed the details.

The lack of geographic correspondence between market and production centers is evident from the data in Table 12. For example, the Southeast and the Pacific Northwest, each with roughly 35% of the estimated postwar aluminum reduction capacity of the United States, are estimated to

[4] These rates would equalize production costs at the old and new locations. The savings in transportation would just balance the increase in electricity costs.

account together for 2% to 6% of the American postwar market for aluminum. The Northeast, with 30% to 35% of the estimated postwar market, has 15% to 20% of aluminum reduction capacity; and the Midwest, with 53% to 62% of the postwar market, has no aluminum reduction plants. This means that aluminum from the Northwest and the Southeast will be shipped to the Midwest and the Northeast for manufacture.

Of the two new location possibilities we have considered, Mobile appears to have a transportation advantage over present sites in so far as particular markets are concerned. To serve the New York-Philadelphia market, for example, costs of ocean transportation from Mobile probably would be less than the lowest transportation costs from the Pacific Northwest or from Alcoa, Tennessee. Data in a recent study of the aluminum industry indicate that the cost of transporting a short ton of aluminum from both these places to the Middle Atlantic seaboard is roughly $20 per ton.[5] We can roughly estimate that the cost of ocean shipment from Mobile to New York would be approximately 25% cheaper.[6]

The possible savings in transportation to the New York market from Mobile compared with the Pacific Northwest or Tennessee is, therefore, $5.00 per ton, or 0.25 cents per lb. of aluminum. Such a saving, the equivalent of 0.28 mills per KWH of electricity consumed in aluminum production, increases to about 3.25 mills per KWH the maximum atomic power rate in Mobile (previously estimated at 3.0 mills) which would render aluminum produced in that region for the New York-Philadelphia market cheaper than that produced in certain existing sites. This makes the location of aluminum reduction plants in Mobile more likely, and broadens the possible extent of cost reductions.[7]

C. ATOMIC POWER AND THE EXPANSION OF ALUMINUM PRODUCTION

In the preceding analysis we have stressed the part played by power and transportation costs in determining the total cost of producing alumi-

[5] Engle *et al., op. cit.,* Table 103, p. 350. The lowest costs from the Pacific Northwest were calculated on the assumption of ocean transportation via the Panama Canal. The costs from Tennessee were apparently based on the lowest quoted of all rail rates.

[6] The distance from Mobile to New York is 1,700 nautical miles as compared with 5,900 nautical miles from Portland, Oregon, to New York. Although the former distance is roughly ⅓ the latter, the cost of transportation is estimated to be approximately ¾ because of the markedly lower ton mile rates for greater distances. If the relative rate of shipping aluminum from Mobile to New York is lower than this, the advantage of Mobile over the Pacific Northwest obviously would be greater.

[7] The importance of the costs of transporting pig aluminum to market suggests that if atomic power makes aluminum reduction economically feasible in Mobile, for instance, other locations, even closer to market, might benefit at least as much. We have not analyzed this possibility, but believe it deserves study. Preliminary calculations suggest that just about as good a case could be made for New York, for example, as we have made for Mobile.

num metal.   Since we found that the rates at which aluminum plants now purchase power are below the probable future costs of atomic power, we judged it unlikely that reductions would occur simply through a cheapening of power.   The possibility of cost reductions through transportation savings as a result of relocation was only slightly more promising, even though some new sites might begin to produce aluminum as a result of atomic power.   However, the analysis was based on the assumption that atomic power would have to compete with the extremely low-cost hydroelectricity that the aluminum industry uses today.   This assumption is examined in the present section, which considers the future expansion of aluminum production.

1. Growth of Aluminum Demand

We have not attempted in this analysis to construct an estimate of the future demand for aluminum in the United States.   But it is important to note that with the steadily increasing demand for aluminum, the role of atomic power in the industry might be considerably greater than under the conditions we assumed in the preceding analysis.   The average annual production of primary aluminum in this country grew roughly fourfold in the 15 years between 1910–1914 and 1925–1929, and more than sixfold in the 15 years from 1925–1929 to 1940–1944.   The average level of production of new metal in 1940–1944 (547,000 short tons) reflects, in part, the very high output of the war years.   But this does not necessarily vitiate the peacetime importance of the production increase, since the output in 1947 was about 520,000 tons—not far below the average for the years 1940–1944.

If the use of light metals is still in its infancy, as is frequently asserted, the demand for them may be expected to continue to increase rapidly. One authority has expressed the view that aluminum "should gradually work its way up to second place in weight production" among the metals, and that "it would seem probable that it will be a strong contender for second place by 1970."[8]   On this basis the total production of aluminum in 1970 might be in the neighborhood of 2 million tons, which was the level of production of the second place metal, copper, in 1944.   A potential level of aluminum demand in 1960 of between 1.90 and 3.15 million tons was estimated by the Bonneville Power Administration, which assumed that 1.7 and 2.7 million tons respectively would be new metal.[9]   These estimates, it should be observed, imply rates of increase in output over the next

[8] Zay Jeffries, Chemical Department, General Electric Company, in "Metals and Alloys of the Future," *American Metal Market*, Vol. 54, March 28, 1947, p. 7.
[9] Unpublished manuscript on the aluminum industry by Samuel Moment, prepared for the Columbia Basin study of estimated industrial development and power requirements in the Pacific Northwest, Bonneville Power Administration, August 15, 1947.

15 to 25 years which do not differ very much from the rates of growth during earlier periods.

2. Power Requirements of Increased Aluminum Production

If the above estimates of the possible growth of aluminum demand are correct, the demand for electricity in aluminum reduction will grow from roughly 10 billion KWH in 1947 to between 31 billion KWH and 49 billion KWH in 1960 or 1970. The average increase would be approximately 30 billion KWH, which at 100% load for aluminum plants would require an additional 3.5 million kilowatts of electric-power capacity.

Can an additional 3.5 million kilowatts of power be made available at costs as low as those enjoyed by the aluminum industry today? The answer depends essentially on our ability to provide large-scale additions to our hydroelectric capacity which will generate electricity as cheaply as our lowest-cost producers today. To determine this point precisely would require detailed studies of potential water power sites which are beyond the scope of this study. The magnitude of the problem is suggested by the following comparisons: 3.5 million kilowatts constitute about 25% of the total installed hydroelectric capacity for public use in the United States in 1946, of which only a small part produces exceedingly cheap water power; or, again, in 1946 the generating capacity of all federally-owned hydroelectric plants, which includes most of the low-cost electricity generated in multiple-purpose projects, was about 4½ million kilowatts.[10] Therefore, although we should not conclude that cheap electricity based on water power will not be available to satisfy the future needs of the aluminum industry, we must realize that the future supply of very cheap water power may fall short of the estimated needs of the aluminum industry. Aluminum producers may well be forced to pay higher power rates for an important portion of their new electricity requirements.

The low electricity rates charged the aluminum industry may not apply to future installations for still another reason. They are based on construction costs which prevailed when the particular water power projects were built—the prewar decade in the case of the Pacific Northwest and the Tennessee Valley. Since that time, construction costs have shown a steep rise. Between 1939 and 1946, for example, the cost of equipping hydroelectric installations, as measured for a selected group of major items of equipment, increased by about 50%.[11] On this basis a project where electricity rates of 2 mills per KWH were possible with prewar construction costs might, if built in 1946, be forced to charge rates of about 3 mills per KWH. This is of course purely illustrative, but we should remember that

[10] *Statistical Bulletin, 1946,* Edison Electric Institute, New York, July 1947, p. 14.
[11] M. G. Salzman, "Design, Construction and Operation Control Rising Costs of Hydro Power," *Civil Engineering,* Vol. 18, April 1948, p. 23.

our atomic costs have been estimated as of the later period. So far as future competition between atomic power stations and new hydroelectric plants is concerned, the costs which prevailed for both in 1946 are probably the most appropriate measure of comparison.

How to insure that a supply of cheap electricity will continue to be available to the aluminum industry has, indeed, already become a matter of some concern. The *Engineering and Mining Journal* of June 1948[12] states in an editorial that much higher costs are expected in future hydroelectric developments, and that the competition for even this higher-cost power will be severe as regions like the Pacific Northwest become more populous and more highly industrialized. The editorial continues: "Meanwhile, it is reported that the aluminum producers are considering Texas, with its abundant gas supply and accessibility to South American and Caribbean bauxite, as an appropriate location for new capacity. However, other industries are also looking toward the gas of the Southwest, including manufacturers of synthetic gasoline and added pipeline outlets. Thus, there are at least some grounds for wonder if requirements for aluminum can be met in the future without a price increase to assure the producers of their power needs."

As the cost of power increases at those sites which had hitherto possessed only the advantage of cheap power, other production sites closer to raw materials and/or markets obviously will become more attractive. Just how cheap power would have to be at the latter sites in order for aluminum producers to favor them over water power sites for locating new plants cannot be determined until the cost of hydroelectric power from new dams is known. By way of example, we may note that if hydroelectric power were to cost aluminum reduction works 3 mills per KWH at the higher-cost production site in Table 11, all other costs remaining equal, production still would be cheaper at Mobile as long as electricity there cost less than 3.9 mills per KWH. Atomic power rates below 3.9 mills per KWH at 100% load are well within the range of minimum realizable atomic costs estimated in Chapter I.

Of course, if the cost of hydroelectric power rises as the supply of very cheap sites is exhausted, the attraction of other production sites will tend to increase whether or not atomic power is developed. The editorial we have just quoted indicates that this is already beginning to take place. For example, electricity can be produced for less than 3.9 mills at 100% load in modern steam plants based on very cheap fuel. Production locations on the rim of the Mexican Gulf might therefore increasingly attract aluminum plants because cheap electricity based on natural gas would be available from nearby wells. There are, however, the following reasons for believing that certain special characteristics of atomic power may make

[12] Vol. 149, p. 68.

it more effective than cheap conventional fuels in drawing aluminum production away from hydroelectric sites.

a. The cost of atomic electricity will not vary greatly from place to place, while the cost of electricity based on conventional fuels varies with the distance from the source of fuel supply. As a result, cheap atomic power would permit a higher degree of raw material or market orientation than would generally be possible with other fuels. For example, although electricity based on natural gas would permit production in the Gulf Coast region with the advantage of greater proximity to bauxite and certain markets, atomic power might as easily permit production even closer to market, in the New York area, for example.

b. Cheap electricity based on a fuel like natural gas might fail to attract aluminum plants because of uncertainty as to its future cost. The costs of atomic power, like the costs of hydroelectricity, would probably consist almost wholly of fixed charges on plant and equipment, which would remain fairly constant over the power station's lifetime. The cost of electricity based on natural gas will fluctuate with variations in the cost of that fuel, and could rise steeply in a particular region if production declined as local resources were exhausted. The uncertainty as to future fuel costs applies also to coal; for with coal, too, costs may rise as natural conditions get worse and as miners' wages increase. Since variations in the costs of electricity are of crucial importance in the economics of the aluminum industry, there will be less willingness to relocate production on the basis of a source of electricity for which costs may rise than on the basis of an electricity source whose costs probably will remain fairly constant. The force of this consideration has already been felt in the early attempts to establish aluminum reduction plants in Texas. A note in the *Engineering and Mining Journal* states: "The only obstacle to the selection of Texas is the cost of natural gas and the reluctance of gas producers to commit themselves to any long-term agreement for large quantities of their product."[13]

c. Atomic electricity costs could conceivably fall well below the costs of ordinary thermal electricity. Ultimately, as we indicated in Chapter I, atomic costs may reach the level of costs in an ordinary thermal plant operating with an almost costless fuel; the costs of thermal electricity based on conventional fuels cannot drop that low except in the rare case of plants located in oil and gas fields which are burning gas which otherwise would be wasted.

3. *Aluminum Production from Ores other than Bauxite*

The expanded aluminum production of the future may be based in part on aluminum-bearing raw materials whose occurrence is more common

[13] Vol. 149, May 1948, p. 106.

than bauxite. Fairly extensive experimental work on the recovery of alumina from kaolin and other clays has been carried on in recent years; some of the experiments show that aluminum could be produced from these more abundant materials at an increase in aluminum costs of about $40 per ton.[14]

If the recovery of alumina from such clays should prove economically feasible, part of the industry might change from using highly localized raw materials to using fairly ubiquitous raw materials. These clays might be so evenly distributed, for example, that there would always be a good chance of finding them close to cheap hydroelectric power. Then the costs of transporting bauxite and alumina, so important in aluminum production today, could be avoided, and the transportation savings on raw materials possible with atomic power might no longer be important.

No close cost comparisons can be made at this time between alumina production from clay and from bauxite. We can, however, make two generalizations about how the possible effects of atomic power on the aluminum industry would be altered if ubiquitous clays should be used. First, the locational effects of atomic power, if aluminum production is based on ubiquitous clays, would be limited to securing a higher degree of market orientation. Electricity costs based on atomic power in the vicinity of markets might here exceed cheap hydroelectricity by no more than the cost of transporting the finished product to market. Since these transportation costs are often sizable (as in transporting aluminum from the Pacific Northwest to New York) the locational advantages made possible through atomic power might be quite important. Our second generalization is that the pressure on cheap hydroelectric power implied in the growth of aluminum production will not be eased by the use of clays; if hydroelectric costs rise, this will tend to strengthen the attraction of market-oriented atomic power.

D. APPLICATION OF THE ANALYSIS TO OTHER COUNTRIES

We can draw at least three broad conclusions about the implications of atomic power for the American aluminum industry. The first relates to the possibility of reducing the cost of aluminum through the use of atomic power. We consider it unlikely that the cost of power at present production sites would be reduced or that significant savings would result from relocating production closer to bauxite and/or aluminum markets. Our

---

[14] Based on a comparison of the estimated postwar costs of alumina production at Mobile ("Aluminum Plants and Facilities," *op. cit.*, p. 115) with the estimated costs for a new process using kaolin. (U. S. Department of the Interior, Bureau of Mines, R. I. 4069, "Recovery of Alumina from Kaolin by the Lime-Soda Sinter Process," by Frank J. Cservenyak, May 1947, p. 41.)

second conclusion is that atomic power might encourage new production sites closer either to bauxite deposits or to aluminum markets, even though this would result in only slight reductions in aluminum costs. Finally, we believe that atomic energy might help provide electric power for a greatly expanded aluminum industry, since the water power available for some of its future electricity needs will be more costly than that used today.

Our negative conclusion regarding possible cost reductions and the tentative character of our second and third conclusions attest to the relative abundance of cheap water power in this country. Electricity is made available to aluminum plants at their present locations so cheaply that further reductions of power costs through the use of atomic energy are most unlikely. Furthermore, savings in transportation costs that might be realized through the orientation of production closer to bauxite and/or markets would apparently in large part be outweighed by the higher costs of electricity, based on atomic energy, at these new locations compared with existing sites. Also, we know that the opportunities are not exhausted for water developments similar to those providing cheap electricity today, even though construction costs have risen considerably.

Clearly, therefore, the applicability of the preceding analysis to other countries which now produce aluminum, or may in the future do so, depends primarily on their supply of cheap water power. Another important factor is the relative cost of transportation in these countries. If their transportation charges are higher than in the United States, cheap water power as opposed to production sites close to raw materials or markets will be relatively less attractive than here, and vice versa. This in turn will be determined in great part by whether they use more or less of a relatively cheap transport medium such as river and ocean shipping than in the United States.

In the absence of a detailed country-by-country analysis, we have to fall back on quite general indicators of the position of various aluminum-producing countries in respect to the possible importance of atomic power. The general indicator which we have used here is the availability of cheap electricity based on water power according to the electricity cost map (Map 1) in Chapter II. This map is subject to especially serious limitations regarding hydroelectricity, for which cost data generally are not available.

On this basis, we find that among the countries other than the United States which were important prewar aluminum producers cheap hydro-electricity was available as follows:[15]

---

[15] For this purpose a cost of 4 mills per KWH has been used as the dividing line between cheap and dear hydroelectric power. Since the load factor for most of these sites is not known to us, it cannot be determined whether this constitutes electricity as cheap as that consumed by aluminum producers in the United States.

Countries with cheap hydroelectricity		Countries without cheap hydroelectricity
Canada	Norway	France
Italy	Switzerland	Germany
Japan	U.S.S.R.	United Kingdom

Two important qualifications should be noted. The first is that the countries in the right-hand column may have some low-cost water power sites for which data are not shown in the energy cost map; France, for example, has a good deal of water power, some of which may be quite cheap. Also, we must remember that countries may not need to rely on their own electricity for aluminum production; France or Germany, for example, might import electric current from Switzerland for aluminum production.

Our tentative conclusion is, therefore, that atomic power would be more likely to reduce aluminum costs in France, Germany, and the United Kingdom than in the United States; while Canada, Italy, Japan, Norway, Switzerland, and the Soviet Union, like our own country, would not experience such cost reductions. For the Soviet Union, with its vast distances and relatively poor transportation system, lower costs might be possible if aluminum plants were located closer to bauxite deposits, but even this is debatable because of the exceedingly cheap hydroelectricity available at the Dnieper Dam, the site of much of the Soviet prewar aluminum production.[16]

As in the United States, atomic power may play a more important role in these countries as they expand their aluminum production. Among the countries shown, Canada and Norway probably possess sufficient cheap water power to support a great increase in their aluminum output.[17] Italy, on the other hand, appears to be pressing against the upper limit of its potential water power; to expand aluminum output substantially it probably would have to rely on more expensive power.[18] Switzerland and Japan represent a less clear-cut case since they have unexploited power resources, but there is reason to believe that these are relatively costly to develop.

[16] Krzizhanovsky Power Institute, *Electric Power Development in the U.S.S.R.*, *op. cit.* The Dnieper rates of about ½ mill per KWH (as shown on the energy cost map) would render transport of raw materials and finished products over extremely long distances feasible. The electric power costs we show for the Soviet Union depend on our rouble-dollar conversion rate. (See Chapter II, A.1.d.)

[17] Since this chapter was written, there have been reports of plans of the Aluminum Company of Canada for constructing aluminum reduction works costing between $200 and $300 million in British Columbia, site of low-cost hydroelectric power. (*New York Times*, Sept. 11, 1949, Section III, p. 1, col. 6.) Aluminum produced here might also compete in markets in the United States, although presumably no more favorably than existing aluminum plants in our Pacific Northwest.

[18] It is possible that in Italy some of the prewar production of aluminum was based on relatively dear hydroelectricity, since much of the country's hydroelectricity is more expensive than 4 mills per KWH.

In the Soviet Union, prewar literature on long-range planning stressed the close relationship between the development of electric-process industries such as aluminum and the development of the country's water power. An official source stated, for example: " . . . in the future stages of construction the location of electricity-consuming industries will be largely tied up with hydroelectric development."[19]   The same source confidently predicted that the water power to support this growth was available, singling out especially the extremely cheap water power in the southeastern reaches of the Soviet Union, particularly along the Angara, Irtysh, and Yenisei Rivers and their tributaries.   Since large power developments are projected for these rivers, with estimated future costs as low as between ½ and 1 mill per KWH,[20] we can assume that atomic power could be of little importance in providing for future growth, even if it made possible a better orientation of production to markets or raw materials.[21]

Despite these long-run proposals to utilize cheap water power, recent additions to aluminum capacity appear to have been made near bauxite deposits in the Urals in the vicinity of Sverdlovsk and Krasnovishersk.[22] These deviations from the long-range plans, if such they are, may have been in response to wartime requirements for a quickly expanded output, which temporarily ruled out the use of water power resources whose development would consume much time.   In addition, the long-distance transportation from regions possessing cheap water power to fabrication plants may have been difficult to provide because of wartime pressures.   Electricity in this region is now based on fuel rather than water power.   If the location of aluminum production in this region is based on enduring strategic or economic considerations, atomic power might possibly provide cheaper electricity than would otherwise be available.[23]

[19] *Electric Power Development in the U.S.S.R., op. cit.*, p. 364.
[20] *Ibid.*, pp. 466, 470.   Major cities in the regions of planned development are Semipalatinsk and Krasnoyarsk.
[21] These production centers would be located roughly 800 to 1,500 miles from possible fabrication centers in the Urals.   According to our analysis, producing regions in the United States with less favorable power rates than are projected for these regions probably could deliver to markets at even greater distances more cheaply than could atomic-based production in the consuming region.
[22] *Business Week*, No. 958, January 10, 1948, pp. 97–98, based on *Map of the Fatherland, 1917–1947,* by N. N. Mikhailov.
[23] Even if new aluminum plants are not built in this region, the plants already constructed will probably continue to produce, and they might benefit from atomic power.

CHAPTER V

Chlorine and Caustic Soda

CHLORINE and caustic soda are obtained together in a process based on the electrolysis of salt (sodium chloride) in solution. The cost of electricity for these products, as for aluminum, is a fairly important element of production. However, the electricity required per unit of output is considerably less than for aluminum; and there is almost no weight loss involved in extracting the two products from salt, as compared with the sizable weight loss when bauxite is made into aluminum.

These differences limit the potential role of atomic power in the chlorine and caustic soda industry as compared with the aluminum industry. Because of the smaller electricity requirements per unit of output, the industry today is not always drawn to very low-cost electricity. As a result of the absence of substantial weight loss in processing, possible raw material transportation economies are of little concern. As the subsequent analysis will show, possible cost reductions in the industry from the use of atomic power will generally be limited to the cheapening of power in present production sites.

Our analysis opens with a brief description of how chlorine and caustic soda are obtained. A second section discusses the possibilities of reducing costs in the American industry. The final section sets forth some general conclusions.

A. PROCESSES OF PRODUCTION AND FACTORS IN THE LOCATION OF PLANTS

The passage of electric current through a solution of common salt in water separates the salt into chlorine and sodium. The chlorine is released as a gas, and the sodium reacts with the water to produce caustic soda (sodium hydroxide). The reaction also releases hydrogen, which may either be discarded or used in the production of other compounds.

This process, the electrolysis of salt in solution, is the only way chlorine is produced commercially. In extracting 1 ton of chlorine, a little over 1 ton of caustic soda is made. Caustic soda is also produced commercially by an alternative process in which soda ash (sodium carbonate) is causticized with lime. The relative importance of the latter process has been declining in recent years because of the growing demand for chlorine; it appears likely that the quantity of caustic soda produced chemically will experience an absolute decline.[1]   This discussion is therefore confined to the electrolytic process in which chlorine and caustic soda are co-products.

In this electrolytic process, the input of approximately 2 tons of salt and 2,850 KWH of electric power (2,450 KWH for electrolysis, 400 KWH for illumination and mechanical and other operations) yields about 1 ton each of chlorine and caustic soda.[2]   Since almost the entire weight of the salt is recovered in the two finished products, there is no obvious advantage in terms of transportation costs in locating production near the raw material rather than near the market.   Because higher rates generally prevail for transporting the finished products than for transporting the salt, the transportation cost advantage appears to be with the market.[3]   Furthermore—and this is a very important factor—salt, although not a ubiquitous resource, occurs quite widely; it can therefore often be found relatively near possible markets for its derivatives.

The attraction of very cheap power sites far from markets does not appear to be great.   The reason for this is apparent from the following relationships: (1) Since about 2,850 KWH are required per 2 tons of products (1 ton of chlorine and 1 ton of caustic soda), a difference of 1 mill per KWH in electricity costs results in a saving of only $2.85 for the combined products.   (2) At rail transportation costs of 1.0 to 1.5 cents per ton mile for the products, extra distances to cheap power of no more than 100 to 140 miles would overcome an electricity cost advantage of 1 mill per KWH; 200 to 280 miles of additional transportation would cancel a power cost advantage of 2 mills per KWH, etc.   Since the difference in generating costs between low-cost hydroelectricity and thermal electricity at a high load factor (as we estimate cost) is usually not greater than about 2 or 3 mills per KWH in the United States, transportation of raw materials over great distances to cheap power, such as is found in the aluminum industry, is obviously uneconomic in the production of chlorine and caustic soda. Therefore the location of production close to very cheap power will probably be feasible only where such power sites are found within a relatively short distance of the market.

[1] *Chemical Engineering*, Vol. 55, February 1948, p. 107.
[2] Unpublished manuscript prepared by the Bonneville Power Administration.
[3] For rail distances of 600 miles and above, a study by the Federal Coordinator of Transportation shows rates per ton mile for chemicals in tanks about 25% higher than for salt. (*Freight Traffic Report*, Vol. III, Washington, 1935, pp. 157, 159.)

It has been pointed out to us that the transportation of solid salt, in some instances, may not be economical even when relative freight rates are lower on salt than on chlorine and caustic soda.   This is because it is more economical to feed the salt to the electrolytic cells in the form of brine; the cost of preparing salt for shipment and of producing a brine from the shipped salt may be enough above the cost of producing a brine at the salt deposit (by introducing water underground through drilled wells) to cancel the cost advantage of transporting salt rather than chlorine and caustic soda.

B. POSSIBLE ECONOMIC EFFECTS OF ATOMIC POWER

1. The Importance of Power Costs

Data on the costs of producing electrolytic chlorine and caustic soda are not available to us. This absence of published cost estimates perhaps is explained by the fact that most plants produce a large number of additional chemical products based either on the primary products or on other salts in the brine. This makes it difficult to isolate the cost of the primary products.[4] Even without such data we can indicate the importance of power costs by showing them as a percentage of the selling price of the product. The following tabulation, for instance, shows the cost of electric power as a percentage of the prewar price of the two products, at varying costs of power. The calculations assume the consumption of 2,500 KWH per short ton of caustic soda and 1,770 pounds of chlorine (equivalent to approximately 2,850 KWH per short ton of chlorine), and are based on a price of 2.7 cents per pound for caustic soda (76% solid) and 2.15 cents per pound for chlorine.[5]

Cost of electricity (mills per KWH)	Percentage of total price
2.0	4.6
2.5	5.8
3.0	6.9
3.5	8.0
4.0	9.1
5.0	11.5
6.0	13.8
7.0	16.1

It is obvious from these figures, which probably encompass the range of power costs for electrolytic chlorine and caustic soda plants throughout the United States, that the cost of power is a fairly important factor in total costs. Even at rates below 3 mills per KWH the cost of power constitutes up to 6% of the price of the products, compared with an average cost of power for all manufacturing of only about 2%.

2. Possible Cost Reductions From the Use of Atomic Power

Whether atomic power will reduce power costs in this industry, and if so by how much, depends on the price at which power is now made available to electrolytic chlorine and caustic soda plants. The Federal Power Commission report just cited indicates that representative producers in the industry have energy costs of about 3.5 mills per KWH.[6] Atomic power,

[4] Cf. *Columbia River and Minor Tributaries*, Vol. I, prepared by the U. S. Army Corps of Engineers, pub. by the 73rd Congress, 1st Session, as H. D. No. 103, Washington, G.P.O., 1933–1934, p. 293.

[5] Tabulation adapted from Federal Power Commission, *Power Requirements in Electrochemical, Electrometallurgical, and Allied Industries*, Washington, G.P.O., 1938, p. 14, using more recent power requirements than were assumed in that study.

[6] *Ibid.*, p. 14.

on favorable assumptions, could be made available to industries with a high load factor at this rate, according to the estimates in Chapter I. (American electrolytic chlorine and caustic soda plants in the aggregate are estimated to have operated at about a 75% load factor in 1938.)[7] On this basis, although atomic power might be used, it would not result in substantial cost reductions.

We can assume that in numerous electrolytic chlorine and caustic soda plants in the United States electricity costs more than 3.5 mills per KWH. The industry was shown earlier to possess characteristics making for a high degree of market orientation. This has resulted in a fairly wide dispersion of plants to satisfy markets in various parts of the country, with new plants operating as markets develop in new regions.[8] Some of the plants are known to have access to sources of cheap hydroelectric power, e.g. the producers at Niagara Falls. Others depend on thermal electricity, and presumably have energy costs commensurate with the regional price of fuel.

The data at our disposal do not enable us to provide specific illustrations of places where atomic power might be a cheaper source of electricity to producers in this industry. When we examine the electricity cost map (Map 1) together with a map of chlorine-caustic soda plants,[9] we find that some important centers for chlorine and caustic soda production are in regions where atomic electricity might provide cheaper power at a high load factor than conventional fuels. These include sites in Michigan, New Jersey, and Rhode Island (on the basis of prewar coal prices), and possibly even West Virginia (on the basis of current coal prices). But these findings are of doubtful accuracy because the electricity cost map (and similar calculations) indicates only the general level of electricity costs in particular regions. This might differ considerably from the energy cost to specific firms, who might avail themselves of limited supplies of low-cost power existing in the region (e.g. from scattered water power developments) or who might obtain low-cost energy through favorable arrangements with local public utilities, etc. This does not invalidate our general conclusion that atomic power could provide cost reductions in many specific regions.

One final point must be stressed. Although atomic electricity might reduce power costs somewhat in this industry, the possible savings probably would not be greater than about 5%, as may be seen by referring to the figures shown above. If, for example, 5 mill power (a reasonable figure at 75% load for thermal electricity in various parts of the United States) should be replaced by atomic electricity at 3.5 mills per KWH, the reduction in production costs would be about 5%.[10]

[7] *Ibid.*, p. 68.
[8] *Ibid.*, p. 69.
[9] *Ibid.*, p. 65.
[10] Derived on the assumption that costs of production are 75% of the prices used in the earlier tabulation.

C. SOME GENERAL CONCLUSIONS

Our analysis indicates that atomic power may be important to chlorine and caustic soda plants because, since they tend to be located close to markets, their power costs vary considerably among regions. Some of these costs are probably higher than our lower estimates of the future costs of atomic power. But we must remember that the industry has not sought very cheap power sites largely because power consumption per unit of output is moderate in comparison, for example, with aluminum. This means that the effect of small reductions in power costs on total production costs is relatively insignificant.

Also, since production will continue to be oriented to markets, atomic power could not by itself encourage new production sites in this industry in the way that we considered possible in the aluminum industry. Rather, atomic power would probably be used only in places which already provide a market because of the concentration there of other industries consuming chlorine and caustic soda or their derivatives.

These conclusions seem to apply also to countries other than the United States, where production of electrolytic chlorine and caustic soda probably shows the same tendency to locate close to markets. Since power costs in industrial centers vary more throughout the world than within the United States, there should be numerous places where atomic power might provide cheaper electricity.

CHAPTER VI

Phosphate Fertilizers

THERE are two varieties of phosphate fertilizers in common use: ordinary superphosphate and double (or triple) superphosphate.[1]   In the processes now employed in the United States to produce this fertilizer, the otherwise insoluble plant food content of phosphate rock is made available as a soil nutrient by treating the rock with sulfuric acid.   Phosphate rock and sulfuric acid are the primary materials; only insignificant amounts of energy are used.   But the importance of energy will increase greatly in the future if new techniques of fertilizer manufacture involving the use of elemental phosphorus become important.   Elemental phosphorus is commercially produced in this country by smelting in electric furnaces, an operation which requires large quantities of electricity per unit of product.   A major purpose of this chapter is to determine whether in producing phosphate fertilizers atomic power would be cheap enough to encourage the use of the electric furnace in place of the chemical process based on sulfuric acid.

We will also be concerned with questions of location when the electric and chemical processes are used.   The chemical process is used to produce different concentrations of phosphate fertilizer, with plant location depending mainly upon the type of fertilizer produced.   As we will show later, with the less concentrated variety (ordinary superphosphate) sizable economies in transportation are possible if producers are located close to fertilizer markets; with the more concentrated variety (double superphosphate) economies in transportation usually are possible if producers are located near the phosphate rock deposits.   This is even more true in the production of elemental phosphorus, so that the increasing use of this method would be almost certain to favor new production sites near phosphate rock regions.

Our comparison of the electric furnace method for producing elemental phosphorus with the sulfuric acid method for producing double superphosphate will involve both production costs and transportation costs.   The transportation advantage implied in the use of the electric furnace will be important to us both for its effect on the cost of phosphate fertilizer and because it may encourage new production centers.

The discussion opens with a brief description of the production proc-

[1] Certain other varieties, including ammonium phosphate and liquid phosphoric acid, are used in quite small amounts.

124

esses important to our analysis, with special emphasis on their locational significance.   This is followed by an analysis of the possible effects of atomic power both in widening the use of the electric furnace and in encouraging the growth of new production locations.   A final section considers some of the implications of the analysis for countries other than the United States.

A. PROCESSES OF PRODUCTION AND FACTORS IN THE LOCATION OF PLANTS

1. The Sulfuric Acid Process

Almost all the phosphate fertilizer used in the United States is produced by the action of sulfuric acid on phosphate rock, which makes the phosphorus pentoxide (P_2O_5), the plant food constituent of the rock, available in a form which the plants can absorb.   The P_2O_5 is made available in either of two concentrations: (a) as ordinary superphosphate, which usually contains between 16% and 20% P_2O_5; or (b) as double superphosphate, which usually contains between 40% and 50% P_2O_5.[2]   While the ordinary superphosphate is produced merely by treating phosphate rock with sulfuric acid, double superphosphate is obtained by first making phosphoric acid by a process similar to that used to make ordinary superphosphate, and then using the acid thus produced to treat additional phosphate rock.

Ordinary superphosphate is by far the more important of the two varieties of phosphate fertilizer, accounting for about 90% of the total American production of superphosphate fertilizers (in terms of P_2O_5 content).   Although the absolute output of superphosphate has greatly increased in the past fifteen years, its relative importance has not noticeably changed.[3]   The proponents of double superphosphate, a fairly new product compared to ordinary superphosphate, argue that its relative importance will grow owing to the greater economies in transportation and packaging per unit of plant food content which result from the use of material in a more highly concentrated form.   They point out that such cost reductions will mean real savings to the farmer, especially since experiments show that equivalent quantities of available P_2O_5 in double superphosphate form give approximately the same increases in yield as are obtained from ordinary superphosphate.[4]

[2] The latter variety is also referred to as triple superphosphate or concentrated superphosphate.

[3] Based on data in an unpublished manuscript on the phosphate fertilizer industry by Lewis G. Prichard, Henry O. Parsons, Robert R. Stewart, and Roscoe E. Bell, prepared for the Columbia Basin Study of estimated industrial development and power requirements in the Pacific Northwest, Bonneville Power Administration, June 1947.

[4] A. L. Mehring, "Double Superphosphate," United States Department of Agriculture, Circular No. 718, Washington, G.P.O., December 1944.   Some of the factors which have retarded the growth of double superphosphate appear to be that: (1) tech-

The determinants of plant location differ for the two types of superphosphate. In the production of the ordinary superphosphate, transportation costs are economized by shipping phosphate rock to producing plants located close to markets. This is true mainly because the phosphate rock has about twice the P_2O_5 content of ordinary superphosphate per ton. Therefore, to produce 1 ton of ordinary superphosphate containing 16% P_2O_5 might require about ½ ton of rock and about ⅛ ton of sulfur (to produce ½ ton of 70% sulfuric acid). Since (a) the total weight of both is less than the weight of the final product and (b) the transportation rates per ton mile on each of the two raw materials are below the rates on the fertilizer, production locations close to market are desirable.

In obtaining double superphosphate, on the other hand, transportation costs are usually economized by locating the plants near the raw materials. For, as we have seen, double superphosphate has about 1½ times the P_2O_5 content of phosphate rock per ton. To produce 1 ton of double superphosphate containing between 40% and 50% P_2O_5 might require about 1½ tons of rock and about ¼ ton of sulfur (to produce about 1 ton of 70% sulfuric acid) so that the total weight of both is about 75% more than the weight of the final product.[5] The smaller weight of the final product might be offset if, for instance, the transportation rates were twice those on the raw materials, but available data on rates indicate that this is generally not the case.[6]

2. The Electric Furnace Process[7]

The electric furnace smelts phosphate rock to produce elemental phosphorus. In addition to its other uses, chiefly in the manufacture of chemicals, elemental phosphorus can be used in the preparation of double superphosphate fertilizer. This involves first that the elemental phosphorus be made into phosphoric acid, and second that phosphate rock be treated with this phosphoric acid. Essentially the second step is the same as in the

---

nical developments leading to production of more highly concentrated fertilizers are of relatively recent origin; (2) change to a new product is slow even if its superiority is demonstrated; and (3) the machinery for applying fertilizer in the field has been designed for distributing ordinary superphosphate. (*Columbia River and Minor Tributaries, op. cit.*, p. 189.)

[5] There is an implicit assumption here (not necessarily crucial to the conclusion reached) that the sulfur is found close to the phosphate rock. This is based on conditions in the United States where sulfur is produced almost wholly in Texas and Louisiana, across the Gulf of Mexico from Florida, where most of our phosphate rock is mined.

[6] See Table 13.

[7] The analysis in this chapter is confined to elemental phosphorus used in the production of double superphosphate. Quite different conclusions might follow with respect to its use in other (non-fertilizer) products.

sulfuric acid process we have just described; the sole difference is in the manner of producing the phosphoric acid. While in the former method it is made by applying sulfuric acid to phosphate rock, in the electric furnace method it is produced by burning elemental phosphorus and absorbing the resulting phosphorus pentoxide in water.

If phosphate rock is used to produce elemental phosphorus rather than double superphosphate, the way is open to new transportation savings. Instead of converting the elemental phosphorus to double superphosphate at the phosphorus production site, both elemental phosphorus and phosphate rock can be shipped to locations closer to the fertilizer market for the final step in production. To produce 1 ton of double superphosphate about 283 lbs. of elemental phosphorus and 900 lbs. of phosphate rock would be needed, and since the combined weight of the two raw materials is little more than half the weight of the final product it would probably pay to transport them rather than the double superphosphate fertilizer. The amount saved would depend on the comparative transportation rates on the several items, which we shall examine later.[8]

The production of 1 short ton of elemental phosphorus in the electric furnace requires the following inputs of materials and electricity (the amounts are approximate):

phosphate rock	7 tons
coke	1.5 tons
silica pebble	2 tons
electric power	13,000 KWH

Coke is required as a reducing agent and silica as a flux. Since so many tons of rock are required per ton of final product, it is clear that economies of transport will usually dictate that production be near phosphate rock deposits, provided electricity rates there are not very much above those at an alternative site. If a low-cost water power development is nearby, the transportation of even large amounts of weight-losing raw materials might be justified.[9] This question will be considered later.

3. Transportation Costs of Superphosphate Fertilizer: A Summary

Table 13 will be a useful reference in connection with our subsequent analysis. It summarizes many of the transportation relationships we have just discussed, and puts them in readily comparable cost terms. A comparison of columns (3) and (4) for the first two categories illustrates the

[8] See Table 13.

[9] As, for example, in aluminum production (Chapter IV). Note, however, that aluminum reduction requires more electricity per ton of product, and that the weight loss from bauxite to aluminum is not nearly so great. Phosphate rock could be shipped to cheap water power over only a fraction of the distances common in aluminum production.

advantage of production near markets in the first case, and near raw materials in the second case.   Examination of column (5) shows the great transportation economies per ton of phosphate plant food if double superphosphate is transported rather than ordinary superphosphate.   Finally, a comparison of columns (3) and (4) for the last two categories illustrates the additional transportation advantage in producing elemental phosphorus near the rock and transporting it (plus the necessary phosphate rock) closer to market for conversion to double superphosphate.

In summary, therefore, we can conclude that if ordinary superphosphate is used, producing locations close to market are economically justified. Since ordinary superphosphate is more expensive to transport (and, by inference, also to package, market, etc.) than double superphosphate, presumably the popularity of double superphosphate, a relative newcomer, will grow at the expense of ordinary superphosphate.   In obtaining double superphosphate with sulfuric acid, production sites near phosphate rock are generally justified.   Even greater transportation economies are possible if double superphosphate production is based on elemental phosphorus, for then it is less costly to transport the elemental phosphorus (plus the necessary phosphate rock) to production sites close to market, than to transport double superphosphate fertilizer.

B. POSSIBLE ECONOMIC EFFECTS OF ATOMIC POWER

In order to simplify the discussion as much as possible, we have confined the following analysis to Florida phosphate rock, which in 1947 accounted for about 75% of all phosphate rock mined in the United States.   With respect to Florida rock, we will consider two questions: (1) Would atomic power be cheap enough to make the electric smelting of phosphate rock in Florida competitive with the sulfuric acid process for double superphosphate production?   (2) Would it be cheaper to use atomic power in Florida for the electric smelting of Florida rock than to send the rock to low-cost hydroelectric centers elsewhere?

We are limiting our analysis in this way because of the following major assumptions: (1) The competition of ordinary superphosphate need not be considered because transportation and other economies will insure the ever-increasing importance of double superphosphate as compared to ordinary superphosphate.   (2) The competition of electrically-smelted Tennessee rock using TVA power need not be considered because Tennessee resources are severely limited and cannot support the future growth of the industry.   The TVA itself recommended that Florida rock be used to relieve the strain on Tennessee's resources.[10]   (3) The competition of electrically-

---

[10] *Annual Report of the Tennessee Valley Authority, 1944,* Washington, G.P.O., pp. 10–11.

TABLE 13. *Transportation requirements and costs of superphosphate fertilizer by different concentrations and different production processes.*[a]

(1) Final product (P_2O_5 content in parentheses)	(2) Major raw materials consumed per ton of final products (in approximate amounts)	(3) Representative costs of transporting raw materials required per ton of final product 500 miles by rail	(4) Representative costs of transporting final product 500 miles by rail	(5) Representative cost of transporting 1 ton of P_2O_5 in form of final product 500 miles by rail[b]
Ordinary superphosphate (16% to 20%)	Phosphate rock ½ ton Sulfur ⅛ ton	$2.50	$5.85	$32.50
Double superphosphate by sulfuric acid process (40% to 50%)	Phosphate rock 1½ tons Sulfur ¼ ton	7.00	5.85	12.19
Double superphosphate from elemental phosphorus (40% to 50%)	Elemental phosphorus 283 lbs. Phosphate rock 900 lbs.	3.75	5.85	12.19

[a] All of the figures are approximate. They are based mainly on data in "Transportation Costs as They Affect New Phosphorus Industries in the West," by Roscoe E. Bell and Donald T. Griffith, Western Phosphate Fertilizer Program, June 1947; and "Transportation Costs of Sulphur," by Roscoe E. Bell, Pacific Northwest Coordination Committee, April 1948. Both reports have been issued as mineographed releases of the U. S. Department of the Interior. The freight rates used throughout were in effect prior to July 1, 1946.

[b] Column (4) divided by the average P_2O_5 content of the final product.

smelted phosphorus based on the enormous phosphate rock deposits of far Western States (Idaho, Wyoming, Utah, and Montana), so far largely unexploited, need not be considered because, even on the basis of assumptions generally favorable to this region, phosphorus produced there probably would not be able to compete with eastern production in a very large part of Southern, Eastern, and Central United States (including the largest phosphate fertilizer-consuming regions).[11]

1. Comparative Costs of the Sulfuric Acid Process and the Atomically Powered Electric Furnace in Producing Fertilizer Materials in Florida

Atomic power will not be important in the future production of phosphate fertilizer in Florida unless it can provide fertilizer at least as cheaply as that produced with sulfuric acid. In order to determine how cheap atomic power would need to be, we have brought together some estimates of comparative production costs in the electric and sulfuric acid processes. The data for the two processes are from different sources, and although both sets of figures belong (roughly) to the prewar decade, they are for different years. Because the data were not specifically designed for such a comparison, and because they are estimates, they should be considered only as a general guide to comparative costs.

To place the comparison on a uniform basis, all figures are expressed in terms of 1 ton of P_2O_5 in the form of double superphosphate. Using data from two main sources we find that during the prewar decade all costs other than sulfuric acid (in the chemical process) and other than electric power (in the electrothermal process) per ton of P_2O_5 in double superphosphate form produced in Florida would have been approximately the same—roughly $30 per ton.[12]

The decisive factor in the choice between the electrothermal process and the chemical process, therefore, would be the comparative costs of sulfuric acid and electric power. The two processes require 2.3 tons of 50° Baumé (about 62.2%) acid and 4,670 KWH of electric power respectively per ton of P_2O_5 in double superphosphate form, and approximately equal costs of production for the two methods would result, given the following costs of sulfuric acid and electric power:

[11] Bell and Griffith, *op. cit.*, pp. 19 ff.

[12] The main sources of cost information are "TVA Estimates Favorable Costs for Concentrated Superphosphate," H. A. Curtis, A. M. Miller, and J. N. Junkins, *Chemical and Metallurgical Engineering*, Vol. 43, November and December 1936; and *Columbia River and Minor Tributaries, op. cit.*, pp. 170–173. The figures actually work out to $28 per ton in the chemical process and $30 per ton in the electrothermal process.

If the price of sulfuric acid (50° Be.) is (dollars per ton)	The electricity rate which will equalize production costs is (mills per KWH)
$6.50	3.2
7.00	3.4
7.50	3.7
8.00	4.0
8.50	4.2
9.50	4.7
10.00	4.9

The most reasonable estimate available to us of the 1946 price of 50° sulfuric acid in Florida is between $7.00 and $8.00 per ton. On this basis atomic power rates would need to be between 3.5 and 4.0 mills per KWH for the costs of production by the electrothermal process to equal the costs of the chemical process.[13] For an industry operating at a high load factor (such as 90%), these atomic rates might be realized according to the estimates of Chapter I.

The calculations of the preceding paragraph are, of course, only approximate and do not permit us to predict whether atomic power will be important to the development of phosphate fertilizer production (or elemental phosphorus production) in Florida. However, they do seem to support the belief that, given sulfuric acid prices within the range shown, atomic electricity might be sufficiently cheap to make the comparative costs of production of the two methods so close that the choice between them would become a matter of indifference to be decided by other factors.

One factor which might prove decisive under this set of circumstances is that the cost of transporting the product to market might be cheaper per unit of P_2O_5 for the electrothermal process. As we noted earlier, the electric furnace method produces elemental phosphorus while the sulfuric acid process does not. The opportunity exists in the former process to ship the elemental phosphorus to a region close to market for conversion to double superphosphate, while with the chemical process the double superphosphate fertilizer would be the most economical form in which to transport the material. Table 13 indicates that the savings per ton of double superphosphate if elemental phosphorus were used might be about $2.10 for a distance of 500 rail miles. This is the equivalent of about $4.20 per ton of P_2O_5 in the form of double superphosphate.[14]

[13] A 1946 acid price is used for comparability with the atomic costs, which were estimated as of 1946. The price is based on data in *Census of Manufactures, 1947,* U. S. Bureau of the Census, Washington, G.P.O., 1949, section on "Industrial Inorganic Chemicals."

[14] Note that this assumes that the final production of double superphosphate from elemental phosphorus takes place elsewhere. If costs of conversion there are substantially higher than in Florida it might not prove desirable to take advantage of the savings in transport.

These calculations show that atomic power would not have a decisive role in producing cheaper superphosphate fertilizer.   If we assume 3.5 mill atomic power and $7.50 sulfuric acid, the savings per ton of P_2O_5 in a market 500 rail miles distant would be about $4.20 on a total cost of about $50.00, and proportionately larger savings would accrue over longer distances.   But production costs are only a fraction of the eventual price of phosphate plant food to the farm consumer, which might run to about $125 per ton of P_2O_5 content.[15]   In terms of retail prices, therefore, atomic power in Florida, on the basis of our figures, would only bring about reductions of 3% or a little more, depending on the transportation distances involved.

We should not dwell on the importance of these cost reductions, considering the large element of uncertainty in the figures on which they are based.   But they reinforce the view that the electric method would not become important in Florida unless atomic power were made available at a cost of around 3.5 mills per KWH or less.   Florida, it should be remembered, has relatively high-cost electricity on the basis of conventional fuels. Map 1 in Chapter II indicates electricity costs of approximately 6 mills per KWH at 50% plant factor, which would make possible rates of about 4.25 to 4.50 mills per KWH to an industry with a high load factor.   At this level of electricity costs, the electric furnace could compete in the production of superphosphate fertilizers only if sulfuric acid were about $8.50 to $9.00 per ton of 50° acid.   This could mean either that production in Florida would be based on the sulfuric acid method rather than the electric furnace, or that Florida would send out its phosphate rock for electric smelting in regions with lower-cost electricity.

We conclude, therefore, that while atomic power probably will not result in important reductions in the price of double superphosphate fertilizer, it may prove important in stimulating the use of the electric method as opposed to the method using sulfuric acid.   Whether it will help Florida keep electric smelting plants which might otherwise be moved to cheap power sites, we will consider in the next section.

2. *The Comparative Costs of Smelting Florida Rock in Florida and Elsewhere*

Can cheap hydroelectricity elsewhere attract electric phosphorus smelting away from Florida even at the extra cost of shipping 7 tons of phosphate rock per ton of elemental phosphorus?   Cheap hydroelectricity may be available in the future in the Tennessee Valley and in the Pacific Northwest, among other regions.   We shall not consider the Pacific Northwest in this analysis in view of the opinion cited above that elemental phosphorus pro-

---

[15] Based on OPA retail ceiling prices in selected cities as summarized in an unpublished manuscript of the Bonneville Power Administration.

duced from local rock in that region probably could not compete in large sections of the United States with Florida production.[16]   Obviously, production burdened with the added costs of long-distance rock transportation would be at an even greater competitive disadvantage.   We direct our attention, therefore, to the relatively nearby Tennessee Valley region.

Perhaps the question is best approached on the basis of some purely hypothetical calculations.   The production of 1 ton of elemental phosphorus requires about 13,000 KWH of electricity.   Hence a difference of 1 mill per KWH in electricity cost makes a difference of $13.00 in the cost of producing 1 ton of elemental phosphorus.   The production of 1 ton of elemental phosphorus also requires 7 tons of phosphate rock, and the cost of transporting such rock over distances of 500 miles or more is about 8 mills per ton mile. Hence a power cost differential of 1 mill per KWH would be offset by the additional transportation of phosphate rock for about 225 miles.

Low-cost Tennessee Valley electricity might be found at a distance of about 750 miles from the Florida phosphate rock fields.   The cost of transporting 7 tons of phosphate rock to this region is, therefore, the equivalent of about 3.2 mills per KWH of electricity consumed per ton of elemental phosphorus.   This means that a production site in the Tennessee Valley must have electricity rates of less than 0.3 mills per KWH to attract electric smelting away from Florida if, for example, atomic rates of 3.25 to 3.5 mills per KWH prevail in Florida.   To attract electric smelting away from Florida when electricity rates there are about 4.25 to 4.5 mills per KWH (a minimum estimate on the basis of conventional fuels), rates in the Tennessee Valley region would need to be less than 1.3 mills per KWH.   The rates necessary in the Tennessee Valley to draw Florida rock for electric smelting, whether Florida used atomic electricity at 3.25 to 3.50 mills per KWH or conventional electricity at 4.25 to 4.50 mills per KWH, are well below the power rates possible in that region.   A reasonable rate in the Tennessee Valley is, perhaps, 3 mills per KWH,[17] and on this basis production in Florida would be cheaper at electricity rates of about 6 mills per KWH or less.

Our tentative conclusion may be stated in the following way: If electricity generated from conventional fuels in Florida is economically feasible in the electric smelting of phosphorus for phosphate fertilizer as compared with the sulfuric acid process there, it will probably also be sufficiently cheap to prevent lower-cost electricity regions elsewhere from attracting Florida rock for electric smelting.   Atomic electricity would not be important, in this case, as a possible means of bringing processing plants to Florida which otherwise would migrate elsewhere.

[16] See above, Chapter VI, B.

[17] Estimated on the basis of data in "Aluminum Plants and Facilities," *op. cit.,* Appendix 8, pp. 82–83.

C. GENERAL CONCLUSIONS

In applying the tentative results of the American analysis to other countries, it is important to know the conditions under which the phosphate rock is found.   Does a country, for example, have both low-cost hydroelectricity and phosphate rock within a single region, as the United States has in Tennessee and some of its far western states?   Where this combination exists atomic electricity probably will be unimportant to the phosphate fertilizer industry.   Does the country, on the other hand, have important reserves of phosphate rock in regions devoid of low-cost electricity, as is true of Florida?   It is here that the use of atomic power might prove important.

The enormous North African deposits in Tunisia and Algeria strikingly resemble those in Florida.   In this region, as in Florida, the question arises whether the cost of phosphate fertilizer would be lower with atomically-powered electrothermal furnaces than with the chemical process.   The closeness of estimated costs for the two processes in the United States suggests that important advantages through the use of atomic power are unlikely here, especially in view of the proximity of large deposits of sulfur in Italy and pyrites in Spain.   Possibly, however, the electric method might be preferred because of transportation economies.

But this is only one instance.   Important amounts of phosphate rock are produced and consumed in many countries of the world.   Probably in most cases the sulfuric acid process, because of its comparative simplicity, will limit the importance of electrothermal operations based on atomic power, especially in view of the general availability of sulfur and pyrites and the relative ease with which they are transported.   One major exception to this generalization must be made.   An advantage of the electrothermal process in special cases is its ability to use grades of phosphate rock which are unsuitable for the chemical process.[18]   The possible importance of this advantage may be judged from the Russian experience that the costs of producing concentrated fertilizers from low-grade phosphorites by the electric method are considerably lower than by alternative processes.[19]   With raw materials of this kind atomic power might offer important advantages through possible reduction of power costs or, if production is already oriented to cheap power far from the phosphate rock, through a possible reduction in the cost of transporting raw materials to processing sites.

[18] *Columbia River and Minor Tributaries, op. cit.,* p. 169.
[19] *Electric Power Development in the U.S.S.R., op. cit.,* p. 363.

CHAPTER VII

Cement

WHETHER atomic energy will prove of use to the cement industry is considered here chiefly for two reasons. First, fuel and power costs are important in the total cost of cement production, so that possible reductions in these costs through the use of atomic energy may be quite significant to the industry. Second, the total amount of fuel consumed in making cement is very great. Therefore the availability of a new source of energy may be of concern to countries with poor fuel resources planning programs of modernization and industrialization which require much cement for construction.

In this chapter we will consider whether there is a reasonable possibility that atomic energy may be used in the production of cement. We will indicate only briefly the technical problems involved, emphasizing chiefly whether the use of atomic energy would be justified on the basis of its cost as compared with the cost of conventional fuels.

The chapter begins with a section on how cement is produced, including questions of raw material requirements and plant location. This is followed by an analysis of the comparative costs of providing process heat and power from coal and from atomic energy. A final section sets forth some general conclusions.

A. THE PROCESS OF PRODUCTION AND FACTORS IN THE LOCATION OF PLANTS

Cement is ordinarily produced by heating a properly proportioned mixture of limestone and clay to extremely high temperatures. Energy is required in the kiln to sustain temperatures of between 2,400° and 2,800° Fahrenheit. This is by far the major energy use in cement manufacture, and is at present satisfied through the burning of fuel. Electricity is used both to pulverize the raw materials before they are fed into the kiln and to grind the resulting pellets, known as "clinkers," into cement.

To produce one barrel of cement weighing 376 lbs., about 600 lbs. of limestone and clay are needed. The carbon dioxide lost by the raw materials during the heating process accounts for the weight difference between what is fed into the kiln and the final product. The fuel required for process purposes varies widely, depending chiefly on the efficiency of the kiln. A representative prewar figure was about 120 lbs. of coal per barrel of cement, although kilns with the highest thermal efficiency required considerably less.[1]   In fact, reasonable estimates predict that coal consumption in modern efficient plants might soon fall to about 60 lbs. per barrel.   Some existing plants have already been reported to burn not much more than 60 lbs. of coal per barrel of cement in their kilns.[2]   The electric power required in cement manufacture, chiefly for grinding, averages about 20 KWH per barrel.   The coal equivalent of the combined fuel and power requirements in a modern mill on the basis of 1 lb. of coal per KWH might be as low as 80 lbs. per barrel of cement.

Since the raw materials other than fuel required in cement manufacture are to be found almost everywhere, production sites are located in all regions using large amounts of cement.   As the economies of production favor plant locations close to major markets, it is obvious that cement mills will have widely differing energy costs, depending on the level of fuel costs in each region.

B. COMPARATIVE FUEL AND POWER COSTS OF COAL AND ATOMIC ENERGY

1. The Importance of Fuel and Power Costs

Although fuel and power are not in general sufficiently important to determine the location of cement plants, their cost is a major item in cement manufacture.   Table 14 indicates the percentage of total production costs which might be accounted for by fuel and power with varying prices of coal.   It is noteworthy that even if coal costs as little as $2.00 per ton, fuel and power might amount to 15% of the total production costs, while with coal costing $6.00 to $8.00 per ton, this figure would rise to 25% and 30%.

2. Comparative Costs: Coal and Atomic Energy

There are two general ways in which atomic energy could be applied to cement manufacture.   One would be through the use of atomic electricity not only for those operations which today require electric power, but also for heating the kiln.   The second method would be to use atomic electricity only for those operations which employ electricity today, and to use hot gases from the nuclear reactor as a substitute for conventional fuels in the kiln.

a. USING ATOMIC ELECTRICITY FOR ALL OPERATIONS.   The technology of cement manufacture would have to be modified before atomic electricity could replace conventional fuels in the kiln.   The only cements to be made

---

[1] Fuel Efficiency in Cement Manufacture, by N. Yaworski, V. Spencer, and others, W.P.A. National Research Project, Report No. E-5, Philadelphia, 1938, p. 32.

[2] Ibid., p. 62.   See also the small coal requirements (about 65 lbs.) for a new plant in Sweden, a country with poor coal resources, in "Sweden's Modern Cement Plant," by Bror Nordberg, Rock Products, Vol. 49, January 1946, p. 78.

electrothermally on a commercial basis so far are certain special aluminum cements produced in France and Switzerland by stationary electric furnaces. Ordinary cement is not produced in stationary furnaces but in rotary kilns. There are serious technical difficulties in adapting electric heating to the quite different operating conditions of a rotating furnace. It is impossible to know whether this can be done and, if so, at what cost, at least until a pilot plant has been placed in operation. We shall assume nevertheless that the technical problems can be solved, and compare the energy cost of the new method with that of the conventional method.

TABLE 14. *Fuel and power as a percentage of the total cost of cement production with varying coal prices.*[a]

Price of coal ($ per short ton)	Fuel and power costs as a per cent of total costs
$2.00	15.1
4.00	20.4
6.00	25.7
8.00	29.8
10.00	33.9

[a] Costs other than fuel and power are based on 1929 data in U. S. Tariff Commission, *Report on Cement*, No. 38, 2nd Series, Washington, G.P.O., 1932. Any change in these costs since that time will obviously affect the percentages shown, and these data should, therefore, be taken to indicate the general magnitude of fuel and power costs rather than their precise relative importance. The costs represent averages for plants which appear to have produced about 40% of the total cement output in the United States in 1929. There are indications that the plants not covered in this sample had similar production costs.

[b] Based on an assumed input of 60 lbs. of coal for burning in the kiln and 20 KWH of electric power. The cost of electric power is estimated from the cost of coal on the basis of the factors specified in Chapter II, A.1.a, under the assumption that the electric facilities are used at 50% of capacity. The calculation assumes that no electricity is produced from waste heat from the kilns.

This comparison is made in Table 15, which shows at what costs atomic electricity would have to be made available in cement manufacture to equal fuel and power costs at varying coal prices. The calculations assume that 60 lbs. of coal and 20 KWH of electricity are required in the conventional method as against the equivalent of a total requirement of 165 KWH of electricity in the electrothermal method. The coal requirements assume a very high level of thermal efficiency in the kiln. This has been done for the purposes of our comparison because the electrothermal method would be feasible (if at all) only where coal is very expensive and, therefore, where efficient burning methods would probably be employed. The electricity requirements in the electrothermal method have been estimated from the amounts consumed under similar conditions in electrothermal ceramic firing. It should be understood that the comparison is only between energy costs.

In the absence of information on other costs (including fixed charges) we are, in effect, assuming them the same in both methods.

The comparisons in Table 15, restricted as they are to energy costs, strongly suggest that atomic electricity will not encourage the development of electrothermal methods of cement manufacture unless it turns out to be unexpectedly cheap while coal becomes unusually expensive. Before World War II there were very few populated regions where the price of coal exceeded $10 per ton; yet electrothermal operations could not compete even with $10 coal unless power were available at 2.8 mills per KWH or less. According to our estimates, atomic electricity at this cost, while not impossible, is highly unlikely.

TABLE 15. *Fuel and power in cement manufacture: costs based on coal compared with costs based on electricity.*

If coal cost per ton is	Fuel and power per barrel will cost[a] (in cents)	Electrothermal kiln will equal cost in col. (2) at following electricity cost[b] (in mills per KWH)
(1)	(2)	(3)
$2.00	16	0.97
4.00	23	1.4
6.00	31	1.9
8.00	38	2.3
10.00	46	2.8
12.00	53	3.2
14.00	61	3.7
16.00	68	4.1
18.00	76	4.6
20.00	83	5.0

[a] Based on an assumed input of 60 lbs. of coal for burning in the kiln, and 20 KWH of electric power. The cost of electric power is estimated from the cost of coal on the basis of the factors specified in Chapter II, A.1.a, under the assumption that the electric facilities are used at 50% of capacity. The calculation assumes that no electricity is produced from waste heat from the kilns.

[b] Derived from data shown in "Electric Heating Systems in Furnaces for Ceramic Firing," by P. Gatzke, *Ceramic Industry*, Vol. 27, October 1936, p. 269. According to these figures approximately 145 KWH of electricity would be required for heating the kiln per barrel of cement. To this has been added 20 KWH for the operations which are today electrified. The electricity consumed in the kiln has a smaller BTU value than 60 lbs. of coal. This is explained by the closer heating controls made possible by electric operation.

Only if postwar normal coal costs are between $12 and $20 per ton in some regions is there a reasonable possibility that atomic electricity could be of use in the industry. There probably will be few (if any) such regions, and their cement needs may not be sufficiently great to warrant the research necessary to develop electrothermal cement kilns. Even in this case the use of atomic electricity probably would be economical only if the cement mills operated at from 75% to 100% of capacity; for reasons discussed later, it is far from certain that such an even level of cement mill operations could be sustained.

b. NUCLEAR REACTORS AS A SOURCE OF DIRECT HEAT AND ELECTRICITY. It is apparent that, for reasons discussed in Chapter I, it would be a difficult engineering problem to operate nuclear reactors at high enough temperatures to supply heat to cement mills. There would also be problems of finding suitable materials for pipes to carry the intense heat to the cement kiln.

As in the case of the electrothermal cement kiln, we shall assume that the engineering problems involved in the use of nuclear heat in cement kilns can be solved. We must emphasize that they appear much more difficult than those associated with the development of electric kilns. In the remainder of this section we will compare the cost of energy by this method with the costs based on conventional fuels, to determine the possible economic importance of atomic energy applied in this form to the cement industry.

TABLE 16.   *Fuel and power in cement manufacture: costs based on nuclear reactor as source of both direct heat and electricity compared with costs based on coal (all atomic costs expressed in terms of equivalent electricity cost at 50% plant factor).*

		Cost of direct heat and power from nuclear reactor will equal cost in col. (2) if atomic electricity can be produced at following costs at 50% plant factor (in mills per KWH)[b]		
If coal cost per ton is	Fuel and power per bbl. will cost (in cents)[a]	Assumption 1: Both heat facilities and electric facilities operate at 50% of capacity	Assumption 2: Heat facilities operate at 100% of capacity; electric facilities at 50% of capacity	Assumption 3: Both heat facilities and electric facilities operate at 100% of capacity
(1)	(2)	(3)	(4)	(5)
$2.00	16	3.6	3.7	4.2
4.00	23	4.3	4.8	5.6
6.00	31	5.1	6.2	7.2
8.00	38	5.8	7.3	8.6
10.00	46	6.6	8.7	10.2
12.00	53	7.3	9.8	11.6
14.00	61	8.1	11.2	13.2

[a] Based on an assumed input of 60 lbs. of coal for burning in the kiln, and 20 KWH of electric power. The cost of electric power is estimated from the cost of coal on the basis of the factors specified in Chapter II, A.1.a, under the assumption that the electric facilities are used at 50% of capacity. The calculation assumes that no electricity is produced from waste heat from the kilns.

[b] Based on an assumed input of 800,000 BTU of direct heat and 20 KWH of electricity. We have assumed the BTU requirements to be roughly the same as in kilns burning coal. This implies the same thermal efficiency as in fuel-fired kilns, which might prove incorrect; we consider this relatively unimportant in light of the roughness of the comparisons in all respects. The calculation assumes that no electricity is produced from hot exit gases from the kiln. The calculation is based on a fixed relationship between the cost per BTU of energy per hour generated in the reactor and the cost per KWH of electricity, for each of the three cases assumed.

Table 16 summarizes the comparative costs of energy in cement manufacture from coal and from nuclear reactors supplying both direct heat and
electricity.   The figures on atomic costs are all expressed in terms of the
equivalent cost of atomic electricity at a 50% plant factor, but they are based
on separate calculations of the cost of process heat and of electricity from the
nuclear reactor under the following three assumptions:

1. That both the heat facilities and the electric facilities operate at an
   average of 50% of capacity;
2. That the heat facilities operate at 100% of capacity, and the electric
   facilities at an average of 50% of capacity;
3. That both the heat facilities and the electric facilities operate at
   100% of capacity.

It is difficult to say which assumption is most realistic with respect to the
percentage of capacity at which production will normally take place.   The
assumption that all facilities might operate at only 50% of capacity is based
on the customary existence of pronounced seasonal variations in cement
output.   This is, of course, because construction activity out-of-doors is
ordinarily highly seasonal.   Materials are sometimes affected by adverse
weather conditions, and certain types of activity, e.g. the laying of roads or
highways, are almst impossible in bad weather.   On the other hand, technical progress has overcome many of the difficulties associated with adverse
weather, and trade practices producing a seasonal pattern are also slowly
being changed.[3]   A diminishing seasonal pattern in construction, combined
with attempts to regularize production of cement through stockpiling, may
succeed in some instances in bringing about operations approaching 100%
of capacity.   We may note that in 1942 the entire United States Portland
cement industry operated at close to 75% of capacity.[4]   This possibility
provides the basis for Assumptions 2 and 3.   These two assumptions differ
only in that Assumption 2 allows for the additional possibility that the
electricity load may be irregular even at capacity operation.

The comparisons shown in Table 16 indicate that from an energy cost
standpoint, the nuclear reactor used to supply both heat and electricity
might compete even with relatively low-cost coal.   Under Assumptions 2
and 3, the nuclear reactor might be competitive with coal costing as little as
$4.00 per ton, provided atomic costs fall somewhere between the minimum
and intermediate estimates of Chapter I.   There is, of course, a stronger
possibility of cost competition with $4.00 coal under Assumption 3 than
under Assumption 2; in the latter case atomic costs falling almost within the

[3] See Simon Kuznets, *Seasonal Variations in Industry and Trade,* New York, National
Bureau of Economic Research, 1933, pp. 149–151.
[4] *Minerals Yearbook, 1945,* U. S. Department of the Interior, Bureau of Mines, Washington, G.P.O., 1947, p. 1232.

minimum range would be necessary to equal coal costing $4.00 per ton. Under Assumption 1 atomic power might be competitive with coal costing about $6.00 per ton. During the prewar period when the average value of coal f.o.b. mines in the United States was about $2.00 per ton, there were many regions where the costs were in the neighborhood of $4.00 per ton, and a few where the cost to large coal users was as high as $7.00 per ton. It appears, therefore, that the nuclear reactors might be a cheaper source of energy than coal in many regions, even on the basis of prewar coal prices. With coal costs in 1946 as high as $4.00 to $5.00 per ton in many important coal-producing regions, cement production in most parts of the United States might well be cheaper with the nuclear reactor as an energy source.

Of course, all of this is highly conjectural. As we stressed earlier, there are formidable technical difficulties to overcome before the nuclear reactor could be used in the way we have assumed. The calculations seem to show, however, that a solution of these problems holds greater promise for the cement industry than an attempted changeover to electrothermal methods based on atomic electricity. We will return to these considerations in the concluding section of this chapter.

We should note one final problem. Because of heat losses in transit, it would be necessary—even if all other technical problems were settled satisfactorily—to locate cement kilns close to the nuclear reactor. This might mean that the cement mill would need its own nuclear reactor. The question would then arise whether the mill was large enough to require nuclear facilities above an as yet undetermined minimum economic size.

In this connection, the following relationship is useful: For a plant with an annual capacity of 500,000 barrels of cement, a nuclear reactor of from 6,000 to 7,000 kilowatts would be required to provide both direct heat and electricity. Thus, plants of varying capacities would require nuclear reactors of the following sizes:

Annual capacity (barrels of cement)	Nuclear reactor[5] (in kilowatts)
500,000	6,000– 7,000
1,000,000	12,000–14,000
1,500,000	18,000–21,000
2,000,000	24,000–35,000
5,000,000	60,000–70,000

These figures may be compared with figures on the capacity of cement plants in the United States shown in the following listing for the year 1945:[6]

[5] These estimates assume that when producing at full capacity, the cement kiln would operate continuously, while the electrical load might be at either 50% or 100% capacity. The latter variation accounts for the range shown.

[6] Minerals Yearbook, 1945, op. cit., p. 1233.

Annual capacity (in barrels)	No. of plants
less than 1,000,000	38
1,000,000– 1,999,000	84
2,000,000– 2,999,000	21
3,000,000–10,000,000	11
TOTAL	154

Information is not available on the minimum economic size of nuclear reactors, and it is even difficult to make a reasonable estimate. The atomic cost estimates mentioned above in our discussion of Table 16 presumably relate to an optimum-sized reactor which we assume to be approximately 75,000 to 100,000 kilowatts. Only the largest cement plants would require reactors of this size. As size declines, cost will probably increase, although we cannot say how much. The possibility must be recognized, therefore, that the cost comparisons discussed above may be inapplicable to cement mills of less than 5 million barrels, and particularly so with respect to the large group of plants, by far the majority, with annual capacities of less than 2 million barrels. This could mean either that nuclear reactors might not prove economical for large sections of the cement industry in the United States and other parts of the world, or else that other uses for reactor heat and power in the vicinity of the cement mill would have to be found. A theoretical third possibility, that cement mills locate themselves near atomic power stations supplying electricity to a city or region, is not considered generally feasible because this would often require additional transportation of limestone and clay, which would increase costs substantially.

C. SOME GENERAL CONCLUSIONS

We have seen that cement manufacture has two characteristics which render atomic energy potentially important: (1) fuel and power constitute an important element in costs, and (2) production sites are almost invariably determined by market considerations. As a result, fuel and power costs vary widely according to the price of fuel in particular producing regions. Nuclear reactors, which will probably be characterized by relatively uniform energy costs in different regions, might offer cost-reducing possibilities in cement manufacture.

Two possible ways of applying atomic energy in this industry were suggested: either in electrothermal kilns or through the use of hot gases from the nuclear reactor. Both require the investment of time and money in applied research, since neither is feasible on the basis of known methods. Of the two ways, the heating of gases to high temperatures in the nuclear reactor for use in the kiln probably offers the most difficult engineering problems, but if technically feasible it also holds a much

greater promise of commercial success.   Our cost comparisons show that
the nuclear reactor as a source of hot gas might compete with coal costing
between $4.00 to $6.00 per short ton.   This means that large sections of
the world with coal prices above these figures might benefit from the
application of atomic energy.   Since the electrothermal method probably
could not compete with coal costing less than $12.00 per ton, its use would
be highly restricted.

Aside from the possibility of cost reductions through the use of atomic
energy, in some regions atomic energy might facilitate the production of
cement in much larger quantities than would otherwise be possible, for
cement production is a voracious consumer of fuel.   It has been estimated
that in the United States in 1927, a prewar peak year of cement production, approximately 13½ million tons of coal equivalents were used to
provide fuel and power for the industry.[7]   It is therefore possible that
countries with poor fuel resources might experience difficulty in paying
for the foreign coal needed to support a high level of cement production.
In 1947 Brazil, for example, had total imports of 1.7 million tons of coal.
With a population that year of approximately 45 million persons, it is
obvious that a level of per capita cement consumption approaching that
in the United States would have required coal imports, for this purpose
alone, exceeding the country's total coal imports for the year.

Since atomic fuel may be more readily available throughout the world
than conventional fuels, it may encourage the growth in cement output
essential to large-scale modernization programs in countries handicapped
by a scarcity of coal.   Furthermore, as we have seen, domestic production
based on atomic energy might not be at a comparative cost disadvantage
if direct heat from the nuclear reactor could be used in the cement kiln.

[7] Yaworski, Spencer, et al., op. cit., p. 16.

CHAPTER VIII

Brick

THE characteristics of brick manufacture which will determine the role of atomic power in the industry are in many ways like those of cement. Production is based on a widely distributed raw material, common clay in the case of ordinary brick. Fuel is the other important ingredient, and since the amount consumed weighs considerably less than the finished brick, plants are generally located close to markets rather than to fuel.

The cost of fuel and power is an important part of the total cost of brick production. For the entire American industry the cost of fuel and power in 1939 came to about 20% of the factory value of products.[1] Because plants are widely scattered near brick markets, energy costs vary widely to individual manufacturers according to the level of fuel costs in the particular region. Under these circumstances, a source of energy characterized by small geographic variations in cost is of potential importance in certain regions.

The major portion of the industry's energy requirements is for fuel used in the kilns for baking brick. The cement analysis shows that for this kind of energy requirement atomic electricity could be of importance only where the cost of conventional fuel is unusually high.[2] Atomic power probably could be used on a broad scale in this industry (as in cement) only if it proved possible to use hot gas from the nuclear reactor in the brick kiln. First of all, of course, the technical problems already mentioned in our discussion of the cement industry would have to be overcome.

However, one basic characteristic of brick production, which differentiates it from cement, militates against the direct use of hot gas from the nuclear reactor. While cement is for the most part made in large establishments, brick is usually produced in relatively small plants. As a result, the energy requirements of an ordinary brick plant could be satisfied by a nuclear reactor considerably below the probable minimum eco-

---

[1] *Census of Manufactures, 1939, op. cit.* The data are for the Census industry "Brick and Hollow Structural Tile," which produces mainly brick.

[2] This is corroborated by P. Gatzke in "Electric Heating Systems in Furnaces for Ceramic Firing," *op. cit.,* p. 267. The author states: "For the present at least, the electric furnace will probably remain a laboratory device as far as the crude ceramic industry is concerned. I cannot imagine that the crude ceramic industry can economically fire its products electrically, because the value, per unit weight, of crude ceramic ware, . . . is much too low." This generalization clearly applies to brick production.

nomic size of operation.[3] For example, the average annual capacity of brick plants in 1925 (a relatively good year for the industry) was about 10 million bricks.[4] The energy needs of this size plant could be satisfied by a nuclear reactor of about 1,000 kilowatts, which is almost certainly too small for economical operation. The largest brick plant in that year had an annual capacity of about 140 million bricks, for which a nuclear reactor of about 14,000 kilowatts would be sufficient. The energy requirements of even the largest plant might be well below the minimum size for the economic operation of nuclear reactors.[5]

Therefore it appears likely that atomic power would be important to the industry only if there were other uses for the nuclear reactor in the vicinity of the brick plant. Cement production offers one such possibility. A moderate-size cement mill, according to our estimates, might require a nuclear reactor of about 20,000 kilowatts. If a brick plant were established close by, it could use heat and power from the same reactor.

The combination of cement and brick production is an attractive one because, in good part, both serve the same general market. Raw materials required for the two products can also usually be found within a small area. But the possibilities are limited by the fact that the making of brick is much more widely scattered than the making of cement. In 1939, for example, there were in the United States 800 establishments producing brick and hollow structural tile, as compared with 160 making cement.[6] Unless the economics of brick production should, through the influence of atomic power and/or other technical developments, shift sharply in favor of larger-scale producing units, only a small proportion of the total industry might benefit from sharing the use of atomic power with cement mills.

Conversely, cement mills which do not themselves require economically-sized nuclear reactors are unlikely to be shifted into that category by being combined with brick-producing plants. Only very large brick plants could have a significant effect on the size of the nuclear reactor required. These would be found in proximity to extremely large centers of population such as New York and Chicago, which are probably large enough markets for cement to require nuclear reactors above the minimum economic size.

[3] It is significant in this connection that the mechanization of brick production apparently has been retarded by the predominantly small scale of brick manufacture, because many machines are efficient only in large plants. (M. E. West, *Brick and Tile,* W.P.A. National Research Project, Report No. N-2, Philadelphia, 1939, p. 11.)

[4] *Ibid.,* p. 14.

[5] As the cement discussion indicated, the question of size is important for nuclear reactors providing high temperature process heat because transmission beyond short distances is not feasible. Hence, factories using high temperature reactor heat will probably need to have their own reactors.

[6] *Census of Manufactures, 1939, op. cit.*

CHAPTER IX

Flat Glass

FLAT GLASS manufacture is included in our analysis chiefly for two reasons.[1] First, energy cost is a fairly important element in the total cost of production. Energy is required primarily as fuel to heat the glass furnace; it is worth inquiring whether atomic electricity would be a cheaper source of furnace heat than the fuels ordinarily used. Second, the fuel most widely used in glass manufacture in the United States is natural gas; the cost and availability of this fuel seem to have had an important bearing on where glass plants are located. Would atomic power tend to bring about changes in plant location which would reduce the delivered costs of glass in important markets?

The chapter opens with a brief discussion of the production process, the major raw materials, and the selection of plant sites. This is followed by an analysis of the possible economic effects of atomic power on glass manufacturing through its influence on fuel and power costs and on the location of manufacturing centers. A final section sets forth some general conclusions.

A. PROCESSES OF PRODUCTION AND FACTORS IN THE LOCATION OF PLANTS

For our purposes we need indicate only a few essentials of glass manufacture. The major raw materials, outside of fuel, are silica sand, soda ash, lime, and cullet (broken glass). These are melted in a furnace at temperatures of about 2,500° to 3,000°F. The molten glass passes from the furnace to machines which roll the glass to the desired thickness. The sheets are then conveyed to an annealing oven (lehr) in which the glass passes through descending temperatures until it emerges at the temperature of the outside air. Finally, the sheets are ground and polished (in the case of plate glass) and cut to appropriate size.

Gas, which was first used in glass manufacture in the latter part of the nineteenth century, has become of major importance because of the over-

---

[1] In our highly developed industrial economy, the glass industry produces a large variety of products. The *Census of Manufactures, 1939, op. cit.,* classifies all glass production into three groups: "flat glass," "glassware, pressed or blown," and "mirrors and other glass products made of purchased glass." This chapter is confined to the first of these groups. The main products of flat glass plants are window glass and plate glass, both widely used in building construction. Plate glass, in addition, finds a major market in automobile manufacture.

all economies in its use resulting from its cleanliness, the ease with which the temperature can be controlled, and the development of highly efficient furnaces based on gaseous fuel.[2]

Where flat glass factories are situated today has been heavily influenced by where natural gas was to be found. The important production centers of Western Pennsylvania, West Virginia, and Ohio owe their predominance in good part to the early discovery there of natural gas. The subsequent discovery of natural gas in southwestern United States made this area another important center in the glass industry.[3] Mainly because of the large investment required in flat glass manufacture, there are very few plants in the industry, but all of these are in or near natural gas-producing areas. During the mid-1930's more than 75% of the total output of sheet glass was from 8 factories and more than 95% of the total output of plate glass from 6 factories. All of these factories were in the vicinity of natural gas fields.[4]

Although the importance of natural gas in determining where glass factories are to be located is borne out by the facts about plant location, there is an element of paradox in the attraction exercised by this fuel. Location of factories in gas-producing regions is made possible by the fact that the other raw materials required in glass production are rather widely distributed. As a result, they can usually be found close to gas-bearing regions. By the same token, such materials should be available near important glass markets. It is to be expected, therefore, that glass manufacture would be located close to natural gas only if it is more costly to transport the natural gas required in glass production than it is to transport the glass. This does not appear to be the case, as the following figures indicate: A ton of finished glass produced in a modern plant requires approximately 23,000 cu. ft. of gas for fuel and power. This amount of gas can be transported for 100 miles, at a representative load factor, for about $0.55 to $0.83, while the cost of transporting 1 ton of glass (plus the weight of the container) 100 miles is reported at $1.60 to $2.20.[5]

Since it seems to cost less to transport the gas required in glass manufacture than to transport the finished flat glass, why hasn't the market exercised a stronger influence on plant location? The explanation may

[2] U. S. Tariff Commission, *Flat Glass and Related Glass Products,* Report No. 123, Second Series, Washington, G.P.O., 1937, p. 20.

[3] *Ibid.,* p. 20.

[4] *Ibid.,* pp. 41, 93.

[5] Gas transportation costs from Federal Power Commission Natural Gas Investigation, "Problems of Long-Distance Transportation of Natural Gas," November 1947, p. 55. Glass transportation costs estimated from Tariff Commission, *Flat Glass . . . , op. cit.,* pp. 71, 109; and from published and unpublished data of the Interstate Commerce Commission.

be that the long distance transportation of natural gas is a relatively recent development; even today, the network of gas pipe-lines does not cover all the important centers of population.[6] This has had a twofold effect. First, potential gas consumers have been limited to those regions in which gas could be obtained. Second, because of their inability to move their total output away from the producing areas, gas producers in the past have been forced to sell locally at extremely low prices. Thus, gas prices in producing areas have tended to be considerably below the price (minus the cost of transportation) in distant consuming regions, where the demand for natural gas often outweighs the supply, and where, therefore, gas prices reflect its value for those uses in which gas is a preferred fuel, e.g. for household uses. As a result the indicated transportation differentials may not truly measure the difference between the gas price to glass producers in gas-bearing regions and more distant regions.

If this hypothesis is correct, the recent growth of the pipe-line network for low-cost gas transportation should have weakened the attraction of gas-producing regions for new glass plants. A recent authoritative statement on glass plant location tends to support this belief without isolating the factors at work: "In the past, cheap fuel was the deciding factor in the location of glass factories, but as far as can be judged from the new factories built in recent years, nearness to desirable markets is now the most important consideration in choosing a site."[7] We must be cautious about projecting this trend into the future. Gas is a wasting resource, and the demands placed upon it appear to be increasing; for example, it is considered a possible raw material for synthetic gasoline.[8] The time may come when the potential needs will far outweigh the supply, and under these circumstances rational use (or local patriotism) may necessitate that first priority be given to demands within the gas-producing regions. This development (or even its anticipation) may again strengthen the attraction exercised by the gas-producing regions if glass production continues to be based chiefly on natural gas.

B. POSSIBLE EFFECTS OF ATOMIC POWER ON PRODUCTION COSTS AND PLANT LOCATION

1. The Importance of Fuel and Power Costs

The relative importance of fuel and power in the total cost of producing flat glass is seen from Table 17. The lowest costs shown, between

[6] Federal Power Commission Natural Gas Investigation, "Problems of Long-Distance Transportation . . . ," *op. cit., passim.*

[7] "The Glass Industry," prepared by *The American Glass Review* for *The Development of American Industries,* J. G. Glover and W. B. Cornell (eds.), New York, Prentice-Hall, 1946, p. 476.

[8] Cf. Chapter II, B.3.

5 and 10 cents per 1,000 cu. ft., have in the past been found only in flush gas-producing regions. At these rates the fuel and power costs are not an important part of total production expenses. Their importance grows, of course, with increasing fuel costs, and at 25 cents per 1,000 cu. ft., which was the approximate average cost to flat glass plants in 1935, fuel and power costs constitute about 7% of total production costs.

The figures in Table 17 indicate that fuel and power costs are not relatively as important in flat glass production as they are in most of the other manufacturing industries we have considered. They are, however, much greater than their average relative importance in industry as a whole. We must remember, too, that some of the other industries in which fuel and power costs are of greater relative importance, e.g. cement and brick, are not located primarily with respect to fuel, while flat glass factories ordinarily have been located near natural gas. As a result, an additional cost of fuel orientation to be considered later is the cost of transporting flat glass to important markets.

TABLE 17. *Flat glass: fuel and power costs (as a percentage of factory value of product) at varying natural gas prices.*[a]

If natural gas price is (cents per 1,000 cu. ft.)	Fuel and power costs will be (percentage of factory value of product)[b]
5	2.3
10	3.6
15	4.8
20	6.0
25	7.2
30	8.3
35	9.5
40	10.5

[a] Tabulation based on data from *Census of Manufactures, 1939, op. cit.,* for the flat glass industry. Costs other than fuel and power entering into the factory value of products of the industry were $96.5 million; the industry consumed the equivalent of 23.4 billion cu. ft. of natural gas in its furnace operations, and about 254.2 million KWH of electricity.

We have estimated the cost of electric power consumed from the cost of natural gas on the basis of the factors specified in Chapter II, A.1.a, on the assumption that the electric facilities are used at 50% of capacity. The calculation assumes that no electricity is produced from waste furnace heat.

[b] Fuel and power costs are compared with the factory value of product, rather than the costs of production, because the census shows only the former figure. Some data on production costs shown in Tariff Commission, *Flat Glass . . . , op. cit.,* p. 66, suggest that the percentages would not be increased by more than about 1 to 2 percentage points if fuel and power costs were compared with cost of production. (In using the Tariff Commission data the reader should be aware that their fuel and power costs include departmental charges for labor, repair and maintenance, and a share of general plant and office overhead. Cf. U. S. Tariff Commission, *Window Glass,* Washington, G.P.O., 1929, p. 22.)

2. *Comparative Costs of Natural Gas and Electricity in Glass Production*

Electric furnaces have not, so far as we can determine, been used commercially in the American flat glass industry. But they are technically feasible, and have been used abroad in a few instances. Since the glass so far produced in such furnaces is of poorer quality than that produced in gas-fired furnaces, improvements will be required before electric furnaces are accepted in the American glass industry. Nevertheless, their eventual suitability for producing glass of accepted commercial quality is not unlikely.[9] The comparative energy inputs for glass furnaces using natural gas and electricity cannot be specified with a high degree of accuracy since energy requirements vary widely with different conditions, but it is believed that, per ton of merchandisable glass, it is fair to compare 20,000 cu. ft. of natural gas with 1,000 KWH of electricity for glass melting. Equal fuel and power costs for the two methods will, on this basis, obtain at the comparative costs of natural gas and electricity shown in Table 18.

TABLE 18. *Flat glass: rates for natural gas and electricity which equalize fuel and power costs in natural gas and in electrically-fired processes.*[a]

Natural gas (cents per 1,000 cu. ft.)	Electricity (mills per KWH)
5	1.6
10	2.5
15	3.4
20	4.3
25	5.2
30	6.2
35	7.1
40	8.0

[a] The energy requirements for the two types of furnaces have been derived from figures supplied by the Toledo Engineering Co. For melting purposes they are: 20,000 cu. ft. of natural gas in the gas-fired furnace, and 1,000 KWH of electricity in the electric furnace. The tabulation also allows for 200 KWH of electricity per ton in both methods for mechanical, lighting, and other purposes. This has been estimated from data in the *Census of Manufactures, 1939, op. cit.* All data are in terms of merchandisable glass, which has been taken at 60% of glass melted.

Costs other than fuel in these alternative production processes are not available to us. Experts are of the opinion that the cost of an electric furnace might be about ⅔ the cost of a comparable gas-fired furnace, and that maintenance labor required in the electric operation might be ¼ that required in the gas method.[10] This suggests that the exclusion of other

[9] U. S. Tariff Commission, *Flat Glass . . . , op. cit.,* p. 20. Also letters to the author from the Technical Editor of *Ceramic Industry,* and particularly from the Toledo Engineering Co. which specializes in the construction of glass plants.

[10] Letter from Toledo Engineering Co.

costs of production from the above comparisons probably biases the computations in favor of the gas-fired furnaces.

3. Possible Effects of Atomic Power

a. PRODUCTION COSTS IN PRESENT LOCATIONS.   Table 19 sets forth data which are useful in determining whether atomic power might reduce glass production costs in present locations.   For each of the states where there are important flat glass factories, this table shows (1) the average value at the point of consumption of natural gas used for industrial purposes, (2) the KWH cost of electricity which would yield the same fuel and power costs, and (3) the cost at which electricity could be generated in the region (prewar cost of fuel) on the basis of the cheapest fuel available for that purpose.

The data shown in Table 19 describe the economic setting in which atomic power will operate with respect to the glass industry in its present locations.   According to the estimates of Chapter I, atomic electricity might be generated for as little as 2.7 to 3.2 mills per KWH at 80% of capacity. These are at the lower end of the range of estimated atomic costs.   In the medium range the estimated atomic generating costs at 80% of capacity are between 4.6 and 5.0 mills per KWH.   The tentative conclusions about the

TABLE 19.   *Major flat glass production centers: average price of natural gas; estimated electricity cost to yield equal fuel and power costs; and estimated cost of electricity from cheapest fuel in region.*

Location[a]	Average price of natural gas for industrial purposes[b] (cents per 1,000 cu. ft.)	Electricity price to equal fuel and power costs based on gas[c] (mills per KWH)	Cost of generating electricity from cheapest available fuel in modern plant operating at 80% of capacity[d] (mills per KWH)
Central West Virginia	23.2	4.8	3.3–3.5 (coal)
North Central Ohio	39.5	8.0	4.0–4.2 (coal)
Western Pennsylvania	32.4	6.6	3.5–4.0 (coal)
North Central Illinois	23.0	4.8	3.9–4.2 (coal)
Southeastern Missouri	19.6	4.3	3.9–4.1 (coal)
Northwestern Louisiana	9.3	2.4	3.4 (gas)
Central Oklahoma	9.9	2.5	3.3–3.5 (gas)

[a] U. S. Tariff Commission, *Flat Glass . . . , op. cit.,* pp. 24, 41, 93, 235.

[b] *Minerals Yearbook, 1945, op. cit.,* pp. 1175–1176.   Prices shown are averages for the entire state for the industrial consumption of natural gas, and are not necessarily the costs to glass plants.

[c] Derived from Table 18.

[d] The fuel indicated is the one on which electricity production is based in the region.   The data on fuel costs are derived from sources indicated in Chapter II.   The cost of generating electricity is estimated for a plant such as that assumed in Chapter II.   Operations at 80% of capacity are assumed because the production process is continuous, and can therefore provide a steady load with the possible exception of the electricity required outside the furnace.

possible importance of atomic power in present locations suggested by the data are:

(1) Atomic power will not reduce production costs in glass manufacture in the Southwestern natural gas regions.

(2) Compared with gas-fired furnaces, atomic power might reduce production costs in glass manufacture in all other major producing regions. But it is probable that in none of these regions would it reduce costs in existing installations, since these cannot be made to operate on electricity except through the complete replacement of the gas-fired furnace by an electric-fired unit. It is likely, then, that electric operation would not be introduced unless (a) existing facilities are worn out and must be replaced, or (b) production capacity in the region is enlarged.

(3) Electricity generated from coal probably would be a cheaper source of fuel and power in glass manufacture than natural gas in all but the two lowest-cost gas regions. The failure to use electric furnaces today in those regions in which their operation appears less costly than gas-fired units is probably due to the newness of the process, and to the as yet inferior quality of glass produced in electric furnaces.

(4) The cost of generating electricity from coal on the basis of prewar coal prices in the regions favorable to electric furnaces was in all cases below the medium range of estimated atomic costs. This is probably no longer true with the increases in the cost of coal mining and transportation of recent years. Still it is probable that atomic electricity will be introduced in these regions only if its costs fall below the medium cost estimates derived in Chapter I.

(5) Even if electric furnaces based on atomic power should be used in expanding glass production in these regions, the savings in costs would be only moderate. If we assume an atomic electricity rate of about 3.5 mills per KWH the savings involved in glass production in North Central Ohio (the highest-cost gas region) would be about 6% as compared with manufacture based on gas; in West Virginia the savings would be about 2%. As compared with manufacture based on coal-based electricity, the potential savings are probably smaller.

We cannot stress too strongly how conjectural the preceding statements are. Their dominant theme is that atomic power may promise some cost reductions in many present sites compared to those which would occur if natural gas or electricity based on coal should be used in glass manufacture. As we shall see, however, it is doubtful that new glass manufacturing capacity will be introduced into these regions, except to serve nearby markets.

b. NEW PRODUCTION SITES.   Earlier in this chapter we indicated that the widespread availability of the raw materials from which glass is made probably would encourage manufacturers to locate close to markets, were it not for certain unusual characteristics (in the past transportation difficulties, in the future a possible inbalance between supply and demand) of natural gas.   It appears possible that improvements in the transportation of natural gas, coupled with the potential use of electric furnaces in the industry, may bring about a higher degree of market orientation for factories.   Would the use of atomic power significantly affect this possible future trend toward moving glass manufacture closer to consuming markets?   In our entire discussion it should be understood that the degree of market orientation exhibited by an industry such as brick, for example, can never be attained in glass manufacture, since glass is produced in very large factories.   We have in mind, therefore, major market areas, the only regions which could support economically-sized glass factories.

Table 20 brings together a set of estimates which are useful in answering this question.   Two major markets for flat glass products, New York and Chicago, are used in this table to illustrate tentatively certain highly relevant relationships.   The data appear to support the following conclusions:

(1) On the basis of today's costs for the pipeline transportation of gas, it is cheaper to pipe gas for glass production in glass-consuming regions than to ship glass from factories located in gas regions. This conclusion has already been stated earlier in the chapter; we must emphasize, however, that it is based on the estimated delivered cost of gas in the consuming region, which might differ considerably from the price at which gas is sold in the region.   In terms of electricity rates, charges of from 11 to 15 mills per KWH would mean that there would be equal costs for glass in the consuming markets if the glass is shipped from the gas regions, while it would require rates of 5 to 8 mills per KWH to meet costs based on piped gas.

(2) If glass production in the New York or Chicago market area based on piped gas is not undertaken owing to uncertainty as to the future cost and availability of gas in these regions, the costs of glass probably can be reduced through the use of electric furnaces using electricity generated from coal.   At 80% load the cost of generating electricity from coal (prewar fuel prices) might be around 4.6 mills per KWH in New York and about 4.3 mills in Chicago. Even with substantial increases in coal prices, electric glass furnaces using electricity generated from coal probably will be

competitive in these regions with glass shipped from gas-bearing locations.

(3) It appears likely, therefore, that in the absence of atomic power glass production based either on piped gas or coal-generated electricity will move closer to markets. The further savings in glass cost which might result from the substitution of atomic electricity at a rate of, for instance, 3.5 mills per KWH, are not expected to alter significantly the economics of plant location.

(4) Under these circumstances the benefits of atomic power would be limited mainly to reductions in production costs such as those which were discussed in the preceding section. If, for example, atomic electricity costing 3.5 mills were to replace coal-generated electricity costing 5.0 mills in the New York region, the savings would amount to about 2% of the cost of producing glass.

TABLE 20. *Estimated electricity cost in New York and Chicago which would equalize delivered costs of flat glass in these markets under varying assumptions.*

Assumptions	Electricity cost in the consuming region which will equalize the delivered cost of glass (in mills per KWH)
I. Glass is produced in West Virginia using natural gas costing 23 cents per 1,000 cu. ft. and is shipped by rail to the New York market.[a]	12
II. Glass is produced in West Virginia, as above, and shipped by rail to the Chicago market.[a]	12
III. Glass is produced in Oklahoma or Louisiana using natural gas costing 10 cents per 1,000 cu. ft. and shipped by rail to the New York market.[a]	15
IV. Glass is produced in Oklahoma or Louisiana, as above, and shipped by rail to the Chicago market.[a]	11
V. Natural gas is piped to New York from the Gulf Coast region and glass is produced in or near New York.[b]	8
VI. Natural gas is piped to Chicago from the Mid-Continent (North Central Texas and Kansas) field and glass is produced in or near Chicago.[b]	5

[a] For gas prices see Table 19. Transportation charges estimated from data in U. S. Tariff Commission, *Flat Glass, . . . , op. cit.,* pp. 71, 109, 274, 277.

[b] The cost of gas in the producing region assumed to be 5 cents per 1,000 cu. ft. The cost of transportation estimated from data in Federal Power Commission Natural Gas Investigation, "Problems of Long-Distance Transportation . . . ," *op. cit.,* pp. 13, 55.

C. GENERAL CONCLUSIONS

Several hypothetical examples based on data about the United States appear to support the conclusion that in the future glass production will be increasingly located near major markets for glass. Therefore, in glass manufacture the importance of atomic power probably will be the same as in other industries with market-oriented production. Since market orientation implies varying fuel and power costs, an energy source which probably will be characterized by geographic uniformity in costs should bring cost reductions in some regions. The possible extent of the reductions is limited for the glass industry since, as noted earlier, fuel and power are not as important a part of total costs as in most of the other manufacturing industries we discuss in these chapters.

So far as foreign countries are concerned, those in which reasonably-priced fuel is available probably will not benefit greatly from the use of atomic electricity in glass production. For those countries in which fuels are scarce and/or expensive, atomic power probably would be important in permitting the increased production of glass—a vital building material—and at costs comparing favorably with those in the United States.

CHAPTER X

Iron and Steel

FUEL is of critical importance in the production of iron and steel for the following reasons: the cost of fuel and power is a large element in the total cost of producing iron and steel, the presence of coal is an important factor in the location of iron and steel works, and the production of steel in large amounts requires very large quantities of fuel. The effects of atomic power on the cost of steel, the location and integration of steel plants, and the expansion of steel production are the major topics we will consider here. Could the use of atomic power in steel production reduce production costs in present locations? Are production locations likely to be changed through the use of weightless atomic fuel as a substitute for bulky conventional fuels? Could the use of atomic power permit substantial increases in the production of steel in countries handicapped by the limited supply of conventional fuels?

The chapter begins with a brief discussion of the process of producing iron and steel, the necessary raw materials, and the factors affecting the location and integration of iron and steel works. This is followed by a long section analyzing the possible economic effects of atomic power under each of several assumptions. A final section states some general conclusions for the future of the iron and steel industry in the United States and other countries.

A. PRODUCTION PROCESS, RAW MATERIALS, AND PLANT LOCATION

1. Production Process

The iron and steel industries of the United States, Western Europe, the U.S.S.R., and all other industrialized countries manufacture iron and steel in three stages: (a) smelting of iron ore to pig iron in a blast furnace; (b) conversion of pig iron to steel in a steel furnace like the open-hearth, Bessemer, or electric furnace; (c) the working of steel in a solid state primarily by rolling. The rolling of solid steel produces the basic forms of modern steel manufacture such as blooms, billets, slabs, sheets, plates, bars, pipe, rails, and structural shapes. These forms may be further worked by additional rolling, forging, drawing, and other working. Casting of molten steel (other than ingots, the initial form for rolling) is used for more complicated shapes.

156

2. Plant Location and Integration

a. INTEGRATION AND SCALE OF PRODUCTION. The smelting of iron ore has two purposes: to reduce ore to metal and to eliminate impurities. The iron in naturally occurring ore is generally a chemical compound of iron and oxygen, plus miscellaneous elements largely consisting of silicon. To convert the iron oxide in the ore to metal requires the elimination of oxygen. Coal, in the form of coke, is used in modern blast furnaces both as a source of heat and to eliminate the oxygen from the ore. This is the largest use of fuel in the entire process of converting iron ore to steel shapes.

Fuel and power are also needed in the later stages of production, in steel furnaces, heat treating furnaces, rolling mills, and electro-plating processes.

The need for large amounts of heat at successive stages of production has been an important factor in effecting the contiguous location of coke ovens, blast furnaces, and steel plants in order to minimize fuel consumption. Integration of the different stages of production saves fuel by allowing the transfer of pig iron in molten condition to the Bessemer or open-hearth furnace, and by permitting the use of coke oven and blast furnace gases in subsequent melting and preheating processes. On the basis of data in the U. S. Census of Manufactures for 1939 it appears that the use of these by-product gases reduces the cost of producing steel products by about 2% to 4%, entirely aside from the considerable savings (which we are not able to estimate) involved in using molten pig iron.

While fuel economies are most important in tieing the blast furnace to the open-hearth and Bessemer furnaces, savings in transportation are most important in explaining why rolling and finishing mills are joined with the steel works. Contiguous location of these stages obviates the transportation of ingots to the rolling mills and of scrap, which is generated in rolling mill operations (approximately ¼ the weight of ingots), to the open-hearth and electric furnaces. Fuel economy is also involved, for by-product gases are used in reheating ingots to rolling temperatures and in many other ways in the rolling plant. Thus fuel and transportation economies have led to the integration of the three stages of iron and steel production within the confines of single gigantic plants.

b. LOCATION OF PRODUCTION. The location of iron and steel works is strongly influenced by the relative amounts of various raw materials required in steel making and the different freight rates for these materials and the finished steel products. The most important raw materials by weight (excluding air and water) are iron ore, scrap, fuel, and limestone. Limestone is unimportant in location because the amounts needed are relatively small and because limestone deposits are quite widely distributed. The exact proportions in which the other materials are required vary with such factors as the grade of iron ore, the relative cost and availability of iron ore and scrap, and the type of steel-making equipment.

Because finished steel shapes are the end product of integrated iron and steel works, it is useful to calculate the weight of materials required per unit of finished steel. The proportions in which the major economic raw materials were used in the production of finished steel products by the American iron and steel industry in 1939 were approximately as follows (in pounds per 100 pounds of finished steel products) : [1]

Fuel and power (in coal equivalents)	177 lbs.
Coal, as such	140
Other fuels and purchased electric energy (in coal equivalents)	37
Iron ore	138
Purchased scrap iron and steel	36
Limestone	31

The "weight of the market" comparable to the weights specified for coal and iron ore can be derived by adding together the weight of finished steel products and of purchased scrap (since it is generated in large amounts at steel markets). Therefore, the weights of the important factors in the location of an integrated steel plant based on 1939 practice are (in pounds per 100 pounds of finished steel products) :

Coal	140 lbs.
Iron ore	138
"Market"	136

From these weights we see that, aside from differences in transportation rates on coal, iron ore, scrap, and finished steel (which would tend, because of generally higher rates on finished steel, to increase the relative attraction of the market), the three main locational forces are about equal. Clearly, then, the coincidence of any two of these forces generally would be strong enough to attract steel plants. In the past, the coincidence of forces usually has been of coal and market; this is because the presence of coal in a region has provided the basis for the development of diversified industries which constitute the major markets for steel. Iron ore has therefore usually moved to coal (and market). This explains the location of steel works in Pittsburgh and the Ruhr.[2]

Although the coincidence of coal and market has influenced the major steel locations, other factors in location have also been very important. The following are examples of such locations:

[1] Based on data in *Census of Manufactures, 1939, op. cit.*, and American Iron and Steel Institute, *Annual Statistical Report, 1940*, New York, 1941.
[2] Cf. Richard Hartshorne, "Location Factors in the Iron and Steel Industry," *Economic Geography*, Vol. 4, July 1928.

(a) Coincidence of coal, ore, and market   Birmingham, Ala.
(b) Coincidence of ore and market   Fontana (Los Angeles), Calif.
(c) Noncoincidence (market lying between ore and coal)   Chicago

The actual amount of capacity installed at any location will usually be determined by the size of its tributary market area. Thus Pittsburgh and Chicago with the largest tributary markets are the largest steel areas. The effects of pricing systems and freight rates are also quite important in determining actual plant locations, although we cannot examine them in detail here.

The questions of location, scale, and integration of steel production will figure prominently in our subsequent analysis. At this point two facts touched upon in the preceding paragraphs should be stressed. The first is that economy in fuel utilization is one of the keystones of the present integrated structure of steel production at single locations in very large plants. Important changes in the industry's fuel and technological base might eventually cause this structure to collapse. The second is that, as we have noted, coal requirements have had an important influence on the location of steel plants. If coal were replaced by an energy source of greatly reduced weight, so that the weight of fuel and ore were no longer roughly equal, the relative influence of the iron ore deposits and the market on plant location would increase. The scope and direction of the possible changes are considered in the remainder of this chapter

B. POSSIBLE EFFECTS OF ATOMIC POWER ON PRODUCTION COSTS, PLANT LOCATION, AND INTEGRATION

1. Assumptions of the Analysis

Several assumptions important to an understanding of our analysis should be made clear. There is, first, the assumption that atomic energy will be utilized in iron and steel production through the generation of electric power rather than through heat obtained directly from the nuclear reactor. This does not mean that direct nuclear heat will never be used in this industry. Rather, our assumption reflects the current belief that the generation and transmission of heat at a sufficiently high temperature to melt iron ore is less likely than the use of nuclear reactors to generate electricity. Since important possibilities can be delineated for atomic electricity we chose not to burden the analysis with the additional comparisons which the consideration of nuclear heat would have required.[3]

[3] Some of the technical problems arising out of the operation of nuclear reactors to

Secondly, we will discuss only the direct effects of atomic energy on the iron and steel industry.   Since the steel industry is of such basic importance in any industrialized economy, it will be influenced by the impact of atomic power on industries which constitute important markets for steel, and on industries which are competitive with steel, as well as by the effects of nuclear power on overall economic activity.   For example, as a result of the commercial use of atomic power steel demand might decrease because the production of railroad equipment, a very important steel use, might be reduced.[4]   But the problem of indirect effects cannot be treated satisfactorily until we have a better understanding both of the overall economic effects of atomic energy and the direct or first order effects on specific industries.   It therefore remains untouched in our discussion.

In addition to our two general assumptions the effects of atomic power on the steel industry are discussed under three specific assumptions: (a) There are no basic changes in technology; the industry continues to use electricity in the same proportions as today and in the same production locations.   (b) There are no basic changes in technology, but, in response to reduced electricity costs, electric steel furnaces using scrap iron and steel as raw materials for producing steel might produce a larger share of the total steel output.   (c) There are basic changes in iron and steel technology under which electricity will be used as a metallurgical fuel for reducing iron ore to pig iron.

2. Effect on Costs, Assuming No Locational Changes, No Changes in Technology, and No Relative Increase in Electricity Consumption

The energy consumed by the United States iron and steel industry in 1939, when the industry operated at approximately $\frac{2}{3}$ of its capacity, came to about 69 million tons of bituminous coal equivalent, of which 55 million tons were bituminous coal, and the remainder chiefly fuel oil, natural gas, and electricity.   The fuel consumed as bituminous coal accounted for about 14% of the total American bituminous output in 1939, while the total fuel consumption amounted to about 8% of all mineral fuels produced.   The amount of coal consumed as such is especially significant, since in large part this represents coal converted to coke for blast furnace purposes.   For this

---

produce heat at high temperatures are discussed in Chapter I.   The decision to consider nuclear heat in the cement analysis in Chapter VII does not reflect any greater confidence in its development for that purpose since the technical problems are about the same.   The analysis of direct heat was used merely as an example to illustrate the way in which atomic energy might find economic use (if technically feasible) in an industry where operations based on electricity appear quite costly.

[4] The drop in the need for railroad equipment might follow from the decline in railroad transportation due to fewer coal shipments or from shorter hauls resulting from the location of certain industries closer to markets or raw materials.

use there is today no effective substitute for bituminous coal of coking quality.

The cost of energy is an important element in the cost of making steel. In 1939 energy cost accounted for about 11.5% of the value of the steel industry's production.[5] A further breakdown of the energy cost figure reveals that the cost of fuel accounted for 10% of the value of production, and the cost of purchased electricity for only about 1.5%.[6]

When operating at capacity the steel industry is the largest user of electric power among American manufacturing industries, but the amount and cost of electricity used per dollar of product is quite low. When the electricity generated within the steel industry for its own use is assigned a value and added to the purchased electricity, the total cost of electricity accounts for no more than 3% of the value of the product. We conclude, therefore, that unless the cost of electricity to the steel industry goes much beyond the cost in 1939, the mere substitution of (possibly) cheaper atomic electricity in the industry in present locations and with present-day technology will have negligible effects on the cost of steel products.

3. Effects of Atomic Power Allowing for the Possible Increase in the Relative Importance of Electric Furnace Steel Production

a. COMPARATIVE COSTS OF STEEL PRODUCTION BY OPEN-HEARTH AND ELECTRIC FURNACES. The smelting of pig iron to steel today takes place chiefly in furnaces fired by fuel rather than electricity. In the United States in 1946 somewhat less than 4% by weight of all steel ingots were produced in electric furnaces; the remainder came almost wholly from open-hearth furnaces. The hypothesis of the preceding section did not allow substitution of electric furnaces for other furnaces in steel production. We now broaden the conditions to allow increased utilization of electric furnaces in the production of steel. The immediate problem, then, is to determine the relative prices of electricity and fuel (chiefly fuel oil) at which the electric furnace will be substituted for the open-hearth furnace in the production of steel, and the effects of such substitution.

On the assumption that costs other than fuel in the two types of furnaces

---

[5] Both the value of production and the cost of fuel and power were estimated from data in the *Census of Manufactures, 1939, op. cit.,* for the "blast furnaces" and "steel works and rolling mills" industries. The value of products reported in the Census represents the selling value f.o.b. factory of the products manufactured, not their production costs.

[6] There is another cost attributable to energy in steel production: the additional transportation cost incorporated in the final product as a result of the choice of production sites well-situated with respect to coal. Were it not for the relatively large weight of the coal requirements per ton of steel products, locations either closer to iron ore or to market might be chosen, thus reducing the costs of transporting ore or finished steel, respectively. These costs will be considered in the subsequent discussion of relocation possibilities based on atomic power.

162 IRON AND STEEL [x.

are about the same, an assumption which is approximately correct, the comparative costs of open-hearth fuel and electricity which will equalize total costs in the two processes may be determined from the data in Table 21.[7]

These comparisons suggest that the electric furnace using atomic power might range from a slightly more expensive melting cost to a slightly cheaper melting cost per ton than the open-hearth. At 6 to 7 cents per gallon for fuel oil (the approximate 1946–1947 price), the equalizing atomic power price would be roughly at the borderline of realizable atomic costs. Above the borderline costs, atomic power at 4 mills would be equivalent to 8 cent fuel oil, and at 5 mills to 10 cent fuel oil. In any event, differences in melting costs between the two methods would be trivial, especially when compared to current finished steel prices of over $80 per ton. Nevertheless, the electric furnace might account for increasingly greater proportions of steel ingot production because of its advantages as a quality steel producer and because of the flexibility of location it would have if based on atomic power.

TABLE 21. *Comparative prices of fuel oil and electricity which will equalize costs of melting tonnage steel in electric and open-hearth furnaces.*[a]

(1) If price of fuel oil is (cents per gallon)	(2) The energy cost in the open-hearth furnace will be[b] ($ per ton of steel)	(3) Costs in electric furnace will equal those in col. (2) if electricity price is[c] (mills per KWH)
3	$ .78	1.5
4	1.04	2.0
5	1.30	2.5
6	1.56	3.0
7	1.82	3.5
8	2.08	4.0
9	2.34	4.5
10	2.60	5.0
11	2.86	5.5
12	3.12	6.0

[a] Assumes costs other than fuel in the two processes are equal.
[b] Assumed input of 26 gallons per ton of steel melted applies to open-hearth furnace benefiting from fuel economy practice in an integrated steel works.
[c] Assumed input of 525 KWH per ton of steel melted applies to cold scrap charge.

b. LOCATION CHANGES RESULTING FROM ATOMIC POWER-BASED ELECTRIC FURNACES. As we have seen, it is possible that the price relationships

[7] This assumption and data on comparative energy requirements in the two types of furnace are derived from J. M. Camp and C. B. Francis, *Making, Shaping and Treating of Steel,* 5th ed., Pittsburgh, Carnegie-Illinois Steel Corp., 1940, pp. 384, 504. Though the electric furnace has an additional expense of electrode consumption (equal to about $1.00 per ton), it has offsetting advantages compared to the open-hearth of better quality steels, economy in the use of alloy metals, greater yield of metal per unit of charge, and virtually complete control over the phosphorus, sulfur, and oxygen content.

of atomic power and other fuels will be such that costs in the electric furnace and open-hearth steel reduction processes will be virtually equal. Under these circumstances electric furnaces might account for increasingly large proportions of tonnage grade steel production because of locational advantages.

In the United States electric furnaces produce steel almost wholly from steel scrap; out of 3.2 million tons of iron charged into electric furnaces, 3.0 million tons were scrap.[8] Electric furnace plants based on atomic power and scrap steel would have two major location determinants: the supply of scrap in a given area, and the size of the consuming market. Such plants should therefore tend to locate at substantial steel-consuming centers, for these areas automatically generate some scrap through obsolescence and the fabrication of steel products, and the amounts might be large enough to make installations practical. By locating electric furnaces at steel markets, the cost of transporting (a) finished steel from present locations to market, and (b) scrap from market areas to steel production centers would be saved. The size of the transportation saving which might result from such relocation is shown by the average freight per ton of steel products shipped, which was $5.23 in 1944 and $4.97 in 1945.[9] On the other hand, relocation of steel furnaces would probably involve the loss of some benefits of integrated operation to which a dollar value should be assigned, so that transportation economies would not be a net saving.

How much decentralization in steel production could be expected if cheap atomic power is brought to steel-consuming centers which generate large amounts of scrap? Two major factors must be considered, one pertaining to the overall limitation on this sort of development, and the other to the kinds of localities which could support facilities of the proper size.

Turning first to the overall limits, it is obvious that the ceiling on electric furnace expansion would be determined by the supply of scrap. Since it is unlikely that purchased scrap (i.e. scrap exclusive of scrap generated in steel and pig iron production) will exceed 30 to 40 million tons for decades to come, it is clear that iron ore and coal would still be the major source of American steel production. Furthermore, integrated works based on iron ore and coal would continue to compete for scrap and might bid up scrap

[8] American Iron and Steel Institute, *Annual Statistical Report, 1945*, New York, 1946, p. 26. Electric furnaces can also produce steel from pig iron and iron ore, but this is an uncommon practice.

[9] Computed from figures of transportation cost of outbound iron and steel products in "Steel Facts," American Iron and Steel Institute, New York, April 1947, pp. 1–2. The size of the possible transportation savings strongly suggests that even without atomic power there might be cost savings in producing electric steel from scrap in consuming regions compared with conventional methods in existing locations. Atomic power might, in this case, strengthen the factors favoring such a change.

prices so that the development of decentralized electric steel plants would be made very difficult. It seems reasonable to conclude, therefore, that at best only a fraction of the available scrap would reappear as tonnage steel produced in decentralized electric furnaces.

The kinds of localities which could support electric steel plants based on scrap will be determined partly by the success of the efforts which are now being made to perfect small-scale steel-finishing techniques. The trend of development in rolling mills has emphasized large units both in the semi-finishing and finishing stages. The continuous mills for wide flat products (sheets, tinplate, plates), for example, are capable of rolling up to a million tons per year. The smallest continuous mill in the United States, the Allegheny-Ludlow mill at Brackenridge, Pennsylvania, has an annual capacity of 308,000 net tons.

There are today efficient small-scale mills for the rolling of narrow steel, such as the McLouth mill, which can approach the efficiency of the larger mills. The output of that type of mill ranges from 125,000 to 250,000 tons per year. There are also efficient types of small mills rolling the smaller sectional material. Continuous casting of the basic steel sections (bars, small structures, etc.) has already been accomplished on an experimental basis. This process of converting steel into useful forms is a small-scale process. Smaller and more flexible rolling mills for many types of rolled steel (with the possible exception of large structural sections) also may be developed to a state of efficiency comparable to the large continuous mills. If cheap atomic power were accompanied by efficient small-scale rolling or other finishing processes, the number of localities in which steel could be produced and manufactured into finished products would be greater than on the basis of present techniques.

Large cities and industrial areas consuming 1 million tons or more of steel per year in the manufacture of products made of steel, with a current steel production much less than this, or even non-existent, would be natural locations for electric steel plants and rolling mills. Consumption of 1 million tons of steel in manufacturing operations would normally generate about 200,000 tons of recoverable scrap. This together with the old scrap available in such regions would be sufficient to set up a 500,000 ton ingot plant providing about 300 to 350 thousand tons of finished steel or about 30% or 35% of the steel requirements of the region. The eastern coastal region from Philadelphia and New York City to Boston, the Los Angeles, San Francisco, and Seattle areas, the Gulf Coast cities, and large inland cities such as Detroit and St. Louis are places where such development might be feasible.[10]

[10] Some of these places already produce steel, but not enough to meet their requirements.

4. Effects of Atomic Power with Fundamental Changes in the Technology of Iron Ore Reduction, i.e. the Substitution of Electricity for Coking Coal

If iron and steel technology remains unchanged, as we assumed in the preceding sections, the major effect of atomic power on the steel industry of the United States under the most favorable assumptions would be to bring about some decentralization in steel production based on the use of electric steel furnaces in producing tonnage-grade steel. But the bulk of production would continue to be based upon coal rather than atomic power and would presumably continue along established lines and in established locations, except for changes brought on by shifts in the industry's sources of iron ore or coal.

We will now examine the effects of atomic power, assuming that iron and steel-making technology will change so as to make the smelting of iron ore independent of metallurgical coke. Two types of processes are known by which iron ore can be smelted without the use of coking coal:

a. The electric smelting of iron ores to pig iron or directly to steel. The Electro-Metal furnace in Sweden is an example of furnaces which produce pig iron from iron ore using electric power as the source of heat and carbon as the reducing agent.

b. The sponge iron process, or low temperature reduction process, using carbonaceous gases, hydrocarbon gases, or pure hydrogen as the reducing agent and various carbon fuels or electric power as the source of heat.

The first type of process, though not requiring coking coal, does require some form of carbonaceous material. Electric smelting furnaces are used or have been used commercially in Sweden, Norway, Finland, Italy, and Japan. Some sponge iron is produced in Sweden and a sponge iron plant using hydrocarbon gas was built during the war at Warren, Ohio, by the Defense Plant Corporation, a government agency. The pure hydrogen process, though discovered by Stuart, a Canadian metallurgist, at the close of World War I, has only recently undergone systematic experimental treatment, and no actual furnaces on a commercial scale have been constructed.

a. THE ELECTRIC SMELTING OF IRON ORE. We shall confine ourselves in this section to the reduction of iron ore to pig iron. Even though it is also technically feasible to reduce the ore directly to steel in an electric furnace, on the basis of the data available to us the former process appears to be much less costly.[11]

(1) Comparative Costs of Smelting in Electric Furnaces and Blast Furnaces. Table 22 summarizes the comparative prices of coking coal and

[11] Cf. Alfred Stansfield, *The Electric Furnace for Iron and Steel*, New York, McGraw-Hill Book Co., 1923, p. 432.

electricity which would result in equal costs for the production of pig iron in the coke blast and electric shaft furnaces. The tables assumes iron ore of about 50% iron content, and ordinary conditions of operation. Equal cost conditions per short ton of steel grade pig iron are approximately specified by the equation: [12]

> electric shaft furnace (2,268 KWH + 635 lbs. charcoal, or gas coke + 10 lbs. of electrodes + 3.6 man-hours) = coke blast furnace (1,850 lbs. of coke + 1.0 man-hours).

Capital costs may be cheaper for the electric furnace per ton of capacity because of the simplicity of its design compared to the coke blast furnace, but since the difference in fixed charges per ton is not likely to exceed $0.50, we assume them the same.

Charcoal is usually as expensive or more expensive than metallurgical coke, but gas coke and other carbonaceous materials such as lignite and peat are cheaper. We assume, for simplicity, that the price of all carbonaceous materials will be the same. We assume, further, that a man-hour of labor costs $1.00, and a lb. of electrodes 8 cents. By substituting these prices in the above equation we derive the equation:

2,268 KWH × (mills per KWH) + $3.40 =
1,725 lbs. of coking coal[13] × (price of coking coal).

TABLE 22. *Comparative prices of coking coal and electricity which will equalize costs of producing pig iron in coke blast furnaces and electric shaft furnaces.*

If the price of coking coal is ($ per short ton)	The electricity rate which will equalize production costs is (mills per KWH)
4	0.0
6	0.8
8	1.5
10	2.3
12	3.1
14	3.8
16	4.6
18	5.3
20	6.1

a Assumed inputs per short ton of pig iron, where they differ, are: for the electric furnace, 2,268 KWH of electricity, 635 lbs. of charcoal or gas coke, 10 lbs. of electrodes and 3.6 man-hours of labor; for the coke blast furnace, 1,850 lbs. of metallurgical coke and 1.0 man-hour of labor. Labor is taken at $1.00 per man-hour, and electrodes at 8 cents per lb.

(2) Cost-Reducing Possibilities of Atomic Power in Major Steel Centers. The comparisons in Table 22 indicate clearly that the electric furnace reduc-

[12] *Ibid.,* p. 47, and U. S. Department of Commerce, *Foreign Commerce Weekly,* Vol. 19, May 12, 1945, p. 47.
[13] On the assumption that 1.42 lbs. of coal are required per lb. of coke.

tion of iron ore to pig iron based on atomic electricity cannot gain an important foothold in the major steel-producing regions of the United States unless the future price of coking coal is much above the prices of the past. In 1937 the average price of coal used for coke manufacture in the United States was $3.74 per short ton; even zero-cost atomic electricity could not compete with coking coal at this price.   By 1946 the average price of coal used for coking in this country had risen to $5.77 per short ton, but to compete with coal even at this price atomic electricity would have to be less than 0.8 mills per KWH.   Our estimates indicate that it is most unlikely that atomic electricity will ever be produced so cheaply by production techniques now envisaged.

Electric iron ore reduction based on atomic electricity will not come into the same general area of costs as coke blast furnaces, according to the data in Table 22, until the price of coking coal is at least $12 per short ton.   The precise point at which costs will be equalized cannot be known with certainty because there are so many variables to consider, but a range of coking coal prices from $12 to $15 per short ton may be accepted as a reasonably good benchmark for our purposes.   Even with today's inflated prices such levels have not been attained in the United States.   It seems reasonable to conclude, therefore, that atomic electricity does not offer cost-reducing possibilities for the major steel-producing centers in this country if used to reduce iron ore in electric shaft furnaces.   The reduction of iron ore probably would continue essentially along the well-known lines of present technology.

(3) The Possibility of Iron Smelting in New Locations.   Present iron-smelting regions are, of course, usually favorably located with respect to coking coal.   Particular locations have become important iron-producing centers because their combined costs of material assembly, in which coal is today an important element, and of finished product transportation are low compared with other locations.   But the ore-reducing technique now being discussed would, if adopted, probably have different locational determinants because the amount of coal required in smelting is only about $\frac{1}{3}$ to $\frac{1}{2}$ that required in the coke blast furnace.   Therefore the relative attraction of iron ore and the market on plant location would be greater.   But whatever the location, the electric furnace method would not be competitive (Table 22) unless coal costs are at least $12 to $15 per ton.   In the United States, coal prices for large consumers would generally be below this level even in regions remote from coal mines.   We therefore conclude that electric smelting of iron ore in an electric shaft furnace, such as the Electro-Metal furnace, based on atomic power probably will not open the way to the relocation of iron smelting in this country.

b. LOW TEMPERATURE PROCESSES (SPONGE IRON).   Sponge iron is an iron oxide in a porous form ranging from about 85% to 97% iron obtained by reducing iron ore at temperatures below that of fusion.   The process has interested metallurgists for generations since it requires less energy per unit of output, and since such diverse fuels as charred vegetable waste matter (e.g. sawdust and coffee husks), charred or naturally occurring bituminous coal, wood, peat, and lignite, as well as gaseous fuels such as natural gas, coal gas, oil gas, water gas, or the flame of pulverized coal can be used as the reducing agent and source of heat. Electric power can also be used as a fuel together with some carbonaceous or hydrogen reducing agent. In addition, iron ore in almost any physical condition is suitable for the process, although prior treatment may be required to put it in more suitable chemical form.

Several factors, technical and economic, have kept sponge iron processes from being commercially exploited. The principal difficulties in sponge iron furnaces when built to commercial scale appear to be the stickiness of the iron, which adheres to the walls of the furnace, and carbon deposition, both of which eventually clog the furnace. Although the elimination of such technical imperfections raises considerable engineering problems, these do not appear to be different in magnitude from the difficulties which usually appear at the initial development of any new process. In view of the theoretical soundness of the chemistry and metallurgy of the sponge iron processes and the commercial production of sponge iron, though on a very small scale, in Sweden, it is reasonable to assume that a satisfactory commercial process may eventually be found.[14] So far relatively little effort has been devoted to the solution of the engineering problems involved.

Although a complete consideration of the potentialities inherent in the techniques of sponge iron production requires the analysis of numerous methods and their possible application in specific regions, we have chosen to concentrate on only one of many possibilities. This is the technique involving the use of hydrogen as a reducing agent. The reason for selecting this technique is that hydrogen can be derived through the electrolysis of water, and this method thereby opens the possibility of completely substi·tuting electricity for carbon fuels in the smelting of iron ore. It should open the way, if successful, to more important changes than any of the other possible methods, when considered in connection with atomic energy as a source of electricity.[15]

[14] For a description of sponge iron processes, see U. S. Department of the Interior, Bureau of Mines, Bulletin 270, *Production of Sponge Iron,* by C. E. Williams, E. P. Barrett, and B. M. Larsen, Washington, G.P.O., 1927; and U. S. Department of the Interior, Bureau of Mines, R. I. 4096, *Use of Sponge Iron in Steel Production,* by R. C. Buehl, M. B. Royer, and J. P. Riott, Washington, July 1947.

[15] There are other methods of reducing iron ore without the use of carbon. For example, Prof. C. S. Smith, Director of the Institute for the Study of Metals of the University of Chicago, has brought to our attention the possibility of a direct electrolytic process recovering iron either from a fused chloride bath, the chlorine being used directly to extract iron from the ore, or an aqueous electrolysis in which ores are leached with acid regenerated electrolytically. We do not discuss other methods because even fewer data are available for them than for the hydrogen method.

During the war two processes using hydrogen were tested. The Madaras Steel Corporation at Long View, Texas, built a furnace based upon the reduction of preheated iron ore with hot hydrogen (contained in reformed natural gas) under pulsating pressure (the Madaras process). The other attempt was by the Bureau of Mines, at the Mines Experiment Station, Minneapolis, Minnesota. Difficulties were encountered in both attempts. Continued research by the Bureau of Mines Station in Minneapolis has led to the conclusion that sponge iron can be produced most successfully by the hydrogen reduction method if iron ore is used in the form of pellets rather than finely ground as in earlier attempts. Sponge iron made at the Bureau of Mines by means of the pellets and hydrogen was found to be excellent in quality, and has been converted to sound steel of varying composition. E. P. Barrett, under whose direction the Mines Bureau experimental work was carried through, believes that pure hydrogen reduction using iron ore pellets is a perfectly feasible method and that the critical problem commercially would be the cost of hydrogen.

(1) Comparative Costs of Electrolytic Hydrogen Sponge Iron and Coke Blast Furnace Pig Iron.   Although the probable costs of a hydrogen sponge iron process cannot now be estimated with any degree of exactness, the rough order of magnitude of the varying cost components can be given. The estimates here presented draw heavily on an article by C. F. Ramseyer, then Metallurgical Engineer of the H. A. Brassert Company,[16] and on the work of E. P. Barrett and his associates at the Bureau of Mines Experiment Station in Minneapolis.[17]   The only categories of cost considered are those assumed to be affected by the difference in production process.

(a) Fixed Charges.   Ramseyer estimates that the capital cost per ton in the hydrogen process (including the electrolytic hydrogen plant) would be about 45% more than in a coke blast furnace plant (including the coke ovens).   Fixed charges on the blast furnace investment (calculated on the approximate investment cost in 1946) are about $2.60 per ton of iron; estimated cost in the hydrogen process (45% more) would be about $3.75 per ton.

(b) Operating Costs.   Labor requirements in the two processes are considered equal despite Ramseyer's view that they would be lower in the hydrogen process.   Maintenance charges in the hydrogen process are estimated at ⅔ the blast furnace plant charges because of the low temperature of the hydrogen process and the comparative simplicity of the equipment.   These charges, approximately $1.25 per ton in the blast furnace, are estimated at $0.85 per ton for the hydrogen plant.

[16] "Sponge Iron—Its Possibilities and Limitations," *Iron and Steel Engineer,* Vol. 21, July 1944. This article was supplemented by correspondence in which Mr. Ramseyer kindly provided more detailed information on the estimated comparative investment costs in coke blast furnace and electrolytic hydrogen sponge iron plants.

[17] Described in a conversation with Mr. Barrett in Minneapolis on December 31, 1947.

(c) Energy Requirements. We estimate that the production of one ton of electrolytic sponge iron would require 2,440 KWH of electricity. Of this total, 2,240 KWH would be consumed in electrolysis to yield 16,000 cu. ft. of pure hydrogen required to reduce the iron ore, and 200 KWH would be consumed for heat and other uses.[18] The production of 1 ton of pig iron is estimated to consume 2,627 lbs. of coking coal.

(d) By-product Credits. The coke blast furnace and the coke oven yield by-products which are useful in later stages of steel manufacture, or which can be sold. The electrolytic production of sponge iron will yield oxygen, a by-product of possible commercial significance. We estimate the value of the by-products of the coke blast furnace at $3 per ton of pig iron, their approximate value in recent years. An electrolytic plant such as we have assumed will produce approximately ⅓ of a ton of oxygen per ton of sponge iron produced. Oxygen has become a valuable agent of steel making. The commercial value of oxygen produced in such quantities may be set at about $3 per ton. The by-product credit accruing to the sponge plant is therefore equivalent to about $1.00 per ton of sponge iron.

These factors are brought together in Table 23, which summarizes the comparative prices of coking coal and electric energy which will equalize the costs of producing iron by the two methods. This table is basic to the analysis in the next two sections.[19]

(2) Competitive Possibilities of Atomic Power in Major Steel Centers. The cost comparisons summarized in Table 23 are based on estimates which, for the hydrogen process, assume a production technique which has yet to be perfected. Clearly, therefore, the figures can be used only to gain a general idea of the possible economic significance of atomic electricity to steel production, when used in conjunction with the hydrogen reduction process. Our purpose in the following two sections is to evaluate the likelihood and explore the nature of changes in steel industry operations in the United States, which might result from the development of atomic power. Obviously this analysis gives no basis for indicating that the changes which appear to be possible will materialize.

According to Table 23, atomic power at its lowest conceivable cost would not be used in the reduction of iron ore except where the price of coking coal is above $8 per ton. Operations based on hydrogen reduction using electricity priced at our estimated borderline of realizable atomic costs

---

[18] Hydrogen requirements from Ramseyer, op. cit. Electricity required to produce hydrogen estimated from Columbia River and Minor Tributaries, op. cit., p. 151.

[19] An assumption underlying the analytic use of Table 23 should be noted. The coke required per ton of iron is based on current practice even though future advances in blast furnace fuel efficiency are anticipated. We assume that this increase in efficiency may be ignored because it probably will be balanced by a rise in coke prices in the future, quite apart from changes in the general price level, as a result of the depletion of coking coal resources.

would, at this coal price, be on the competitive threshold of the coke blast furnace.

TABLE 23.   *Comparative prices of coking coal and electricity which will equalize costs of producing iron in coke blast furnaces and electrolytic hydrogen sponge iron plants.*[a]

If the price of coking coal is[b] ($ per short ton)	The electricity rate which will equalize production costs is[b] (mills per KWH)
4	1.0
6	2.1
8	3.2
10	4.3
12	5.3
14	6.4
16	7.5
18	8.6
20	9.6

[a] Assumed inputs per short ton of iron (because of the roughness of this calculation, we ignore, for simplicity, the small difference in iron content between sponge iron and pig iron), where they differ, are: for the coke blast furnace, 2,627 lbs. of coking coal, fixed charges at $2.60, and repair and maintenance at $1.25; for the electrolytic sponge iron plant, 2,440 KWH of electricity, fixed charges at $3.75, and repair and maintenance at $0.85.

[b] By-product credit is $3.00 per short ton of iron in the blast furnace, $1.00 in the electrolytic sponge iron process. For simplicity we ignore the changing value of by-products which would undoubtedly accompany the change in the price of coking coal.

How does the price of $8 per short ton compare with the price of coking coal at the blast furnace in major steel production centers in the United States?   Before the war the price of coking coal was between $2.50 and $5.00 per ton at various steel centers, well below the $8 price required to bring atomic electric-powered hydrogen plants into competition with coke blast furnaces.   However, costs of mining and transporting coal have risen substantially since that time.   Because our estimates of atomic electricity costs reflect 1946 prices, the most reasonable coal prices for this comparison are probably those which prevailed in 1946.   On this basis, hydrogen plants using atomic electricity probably still could not compete with coke blast furnaces in important steel-producing centers, since the price of coking coal ranged from about $4 to $7 per ton.

Should coking coal prices almost reach $7 per ton—for example in the Chicago-Gary area in 1946—atomic electricity would fall just short of the competitive threshold.   Of course, coal prices have risen even further since 1946; during 1947 they were about $8 per ton in Chicago-Gary and other steel production centers.   But atomic costs would presumably also be affected by rising prices during this period, and it is not clear what significance to attach to the rise in coal costs between the two years.

(3) The Possibility of Iron Smelting in New Locations. A cost comparison between the hydrogen method and the coke blast furnace in existing steel-producing centers is not really sufficient to determine the relative economic worth of the two methods. As we indicated earlier, the weight of coal required in iron smelting influences the location of blast furnaces, and it is to be expected that they will be more favorably situated with respect to coal than other possible locations. Furthermore, in the coke blast furnace process, fuel economies are achieved by utilizing by-product blast furnace and coke oven gas in later stages of production, and by utilizing the pig iron in molten condition in the steel furnace. These economies are a basic physical factor leading to the location of blast-furnaces, coke plants, steel furnaces, and rolling mills within the confines of one plant.

The hydrogen process would undoubtedly have different locational determinants. If it is based on atomic electricity, the weight of fuel would no longer influence the location of reduction works. So far as raw materials are concerned, iron ore would therefore become the major locational factor. Since the hydrogen process would not result in a great volume of hot waste gases, the iron furnace and the steel furnace in this process could be separated without significant loss of economy. Whether these two stages would be contiguously located would be primarily a matter of comparative transportation costs on sponge iron and steel ingots. Thus, unlike modern iron and steel plants, a steel industry based on atomic-powered hydrogen reduction plants as the basic source of metal could be separated, with some of the process located at the ore, and the remaining stages located at the market.[20]

These later stages would include the steel furnace and the rolling mill. Integration of these two stages is due to transportation economies, internal economies of operation, and cheaper utilization of rolling mill scrap. Melting schedules can be more efficiently determined if rolling schedules are known. Thus the integration of steel works and rolling mills within one ownership group makes it unnecessary for the rolling mill and the steel works to carry large stocks of ingots. Locational integration also eliminates the cost of shipping rolling mill scrap to the steel works.

Let us therefore examine the competitive position of atomic electricity in steel production on the assumption that the hydrogen process is used at the ore site, iron being shipped elsewhere for conversion to steel and for roll-

[20] Another possibility is location at cheap hydroelectric centers, but this does not seem likely unless the ore or market is close by, since the power requirements per unit of output are not large enough to justify much extra transportation to take advantage of low-cost power. For example, a power cost advantage of, say, 1 mill per KWH amounts to $2.44 per ton of iron. At transportation rates of 1 cent per ton mile, 2 tons of iron ore could be shipped an additional distance of no more than 120 miles to take advantage of the lower cost electricity.

ing.   To illustrate, we take the Northern Minnesota iron ore fields which supply the bulk of the iron ore consumed in the American steel industry, and compare the costs of producing iron there with costs in the Chicago-Gary and Pittsburgh regions.   The essential points in this comparison follow.

(a) According to Table 23, the hydrogen process based on atomic electricity will be a cheaper source of iron at any production site only if the cost of coking coal is about $8 per ton or more.   We estimate that the cost of coking coal in the iron ranges of Northern Minnesota was about $8 per ton in 1946 (the Duluth price plus estimated railroad transportation costs to the iron mines).[21]   On this basis we conclude that atomic electricity might be on the competitive borderline with coke blast furnaces in this region. Furthermore, since we are here assuming that the region would not produce beyond the iron reduction stage, the coke blast furnace in Northern Minnesota would not compare as favorably with the hydrogen process as Table 23 indicates, because of its inability to take full advantage of the by-product economies made possible through integration with later stages of steel manufacture.

(b) How would the delivered cost of iron produced by atomic electricity and the hydrogen process in Northern Minnesota compare with the cost of producing iron in Chicago-Gary or Pittsburgh?   The cost of coking coal in 1946 was about $4.80 per ton in Pittsburgh and about $6.76 in Chicago-Gary.[22]   But to make a comparison between Northern Minnesota and costs in these two locations, the costs of transporting iron ore from Minnesota mines must be included.   The cost of ore transportation to Pittsburgh was about $6.00, and the cost to Chicago-Gary about $3.50 per ton of iron.[23] If the ore were smelted in Minnesota, transportation requirements would be less, since about 1.85 tons of ore were required per ton of pig iron.   If, furthermore, the same transportation rates applied on sponge iron as on iron ore, Northern Minnesota would have had a transportation cost advantage of $1.60 per ton of iron in Chicago-Gary, and $2.75 in Pittsburgh.   This could be expressed in the following way: As compared with the hydrogen process in Northern Minnesota, the equivalent cost of coking coal in

[21] The price of coking coal in Duluth in 1946 was $6.86 per ton (*Minerals Yearbook, 1946*, U. S. Department of the Interior, Bureau of Mines, Washington, G.P.O., 1948, p. 421.   See also following footnote).   Cost of transporting the coal by railroad to the iron-producing regions would be about $1.00 per ton, on the assumption that coal took the same rate as iron ore now traveling in the reverse direction.   (I.C.C. Ex Parte No. 162, No. 148, from *I.C.C. Reports*, Vol. 266, section on "Increased Railway Rates, Fares, and Charges," Washington, G.P.O., 1946, for iron ore rates.)

[22] Based on state-wide averages of the cost of coking coal at the coke ovens for Pennsylvania, Illinois, and Indiana (*Minerals Yearbook, 1946, op. cit.,* p. 421).   We assume that the figures relate primarily to the major steel centers in each state.

[23] "The Western Steel Industry," *Utah Economic and Business Review,* Vol. 3, June 1944, p. 19.   Costs in 1946 would be above these figures.

Chicago-Gary would be $6.75 (1946 price of coking coal) plus $1.60 (transportation cost disadvantage on iron), or $8.35 per ton, and in Pittsburgh the equivalent cost would be $4.80 (1946 price of coking coal) plus $2.75 (transportation cost disadvantage), or $7.55 per ton.[24]

These calculations would support the view that iron-reducing plants located in Northern Minnesota which used the hydrogen process and atomic electricity might be a lower-cost source of iron supply for the Chicago-Gary market than blast furnaces located in this area, while in comparison with Pittsburgh, reduction plants in Northern Minnesota might be a slightly higher-cost source of supply. However, the precise cost comparisons are of little importance. The really significant consideration is that the location of iron smelting in Northern Minnesota might be brought within reach of economic feasibility by the combination of two innovations, atomic electricity and hydrogen reduction, each providing support for the adoption of both in steel production. Pittsburgh and Chicago-Gary have been used in these calculations because they provide convenient benchmarks of comparison with the present-day steel industry. Actually, if iron reduction took place in Minnesota, there would be no compelling reason for steel production and rolling to be located in Pittsburgh or Chicago, except to satisfy markets for steel in those regions. For more distant markets, say New York or Boston, it would probably be more economical to ship sponge iron directly to these points for conversion than to ship steel products. The intermediate location for producing steel products is justified today only as one of several stages in an integrated steel operation consuming huge amounts of coal.

The assumptions on which our tentative conclusion about relocation rests must not be forgotten. First is the assumption that atomic electricity might be made available at as little as 3.2 mills per KWH (1946 prices). We consider this to be a possible but by no means certain realizable cost at some time in the future. Second is the assumption that sponge iron could be transported at the same cost as iron ore. The transportation involved is mainly Great Lakes shipment, and there has been developed an elaborate system for handling and moving iron ore at low cost over this route. There is no basis for predicting that the costs of sponge iron transportation and handling would be just as low. On the other hand, if they should be about twice as much as the iron ore rate, the transportation cost advantage of reducing iron ore in Minnesota would be completely offset, and the cost comparisons we have shown would obviously be invalid.

(4) Some General Observations on Locational Factors in Hydrogen Reduction Using Atomic Power. The influence of relative freight rates on

---

[24] This does not mean that coke blast furnaces in Northern Minnesota using coking coal priced below these equivalent costs would be competitive with Chicago-Gary or Pittsburgh, for if they should ship pig iron they would lose some of the by-product advantage of integration.

iron ore, pig iron, ingots, and finished steel on production sites in a steel industry based on atomic power and hydrogen reduction would be of the utmost importance. If freight rates on products at each stage of production were identical, then obviously the minimum ton mile cost location would be at the iron ore site. However, in the United States transportation rates usually increase with the value of the product. Thus the iron ore rate is usually lower than the pig iron rate which in turn is usually lower than the steel ingot or semi-finished steel rate, which is lower again than the rolled steel rates. The minimum cost location is therefore a function of rates and the weight advantage of shipping iron at the various stages. In this country the situation is also complicated by the great use of both circulating and purchased scrap in the production of steel. As a result, less iron ore and pig iron per ton of steel need be shipped than in an industry which is based solely on virgin iron and circulating scrap.

We have not explored the question of freight rates in enough detail to know with certainty what locational pattern might develop for steel production based on atomic electricity and hydrogen reduction. We believe the most reasonable possibility is that relative transportation rates would favor location of the rolling mill at the market, and that the ingot plant would therefore also be located at the market because of the scheduling advantages and the economy of scrap movement. This would leave the reduction plant either as a single unit located in the iron mining regions, or combined with the steel works and rolling mill at the market, depending on the relative costs of shipping iron ore and sponge iron. In either case there would be a shift from present locations.

More likely than a shift in location either to raw materials or to markets is the possibility that both types of sites will see the development of iron and steel production if atomic-powered hydrogen reduction is commercially successful. For the size of the individual hydrogen reduction furnaces can be small—100 tons per day as compared to blast furnaces of 1,000 to 1,500 tons per day. The size of the electric furnaces to convert the iron to steel can also be small without a significant loss of efficiency. It is therefore possible that the production of iron and steel under the low temperature processes may be carried on by large numbers of relatively small producers.

Many deposits of iron ore of not more than a few million tons, for example, could be profitably exploited by means of hydrogen reduction and made the basis for a complete small-scale electrically operated steel industry. The existence of small-scale primary metal production would undoubtedly lend stimulus to the development of small-scale rolling, casting, and other operations.[25] The method of continuous casting of steel forms, already close

[25] On the question of small-scale finishing equipment, see Section B.3.b of this chapter.

to commercial operation, is an indication of what can be done to transform the steel industry from a converter of raw materials measured in the millions of tons by means of gigantic furnaces and rolling mills, and dependent upon a few gigantic deposits of coal and iron ore, to an industry in which relatively small apparatus based on widely scattered and small deposits can be efficiently operated for the production of some of the basic shapes.

Thus there is implicit in successful sponge iron processes the possibility of relatively small-scale local production of sponge iron, supplying local foundries and electric steel furnaces which would be competitive at least in local markets. Northern Minnesota, with its large ore deposits, and other ore-producing regions might on this basis produce iron and steel to supply demands within a moderate radius; on the other hand, market areas in some parts of the country might find it advantageous to bring in iron ore and establish local iron and steel production.[26] This opens up on a considerable scale the possibility of long-run decentralization of iron and steel production in this country.

C. THE USE OF NEW IRON AND STEEL TECHNOLOGY IN THE UNITED STATES AND OTHER COUNTRIES

The preceding discussion revealed the possibility of two major effects of atomic power on the steel industry of the United States. The first is that electric furnace production of steel from scrap might become important in certain major steel-consuming centers which today produce far less steel than they consume, or often produce no steel whatsoever, as a result of the availability of low cost atomic electricity. The second is that the use of atomic power in conjunction with the hydrogen process for reducing iron ore might move iron reduction to the iron ore site. Both effects are important mainly because they imply the use of smaller-scale units and a much wider degree of geographic decentralization in steel production than are common in industrialized nations today. In neither case do large cost reductions appear likely as a result of the change. Instead, the new technology and the new locations were estimated, on the most favorable assumptions about atomic costs, to be at a virtual cost equality with present-day methods and locations.

1. The Possible Importance of Electric Steel Furnace Operations

The electric steel furnace used in the manner assumed in our analysis appears to be of limited importance because of its dependence on scrap supplies. Overall limitations on scrap supply, and competition with established steel producers for this limited supply, should place a fairly low ceil-

---

[26] At coastal points the ore might very well be imported from foreign suppliers, particularly in South and Central America. This is a very important possibility in light of the threatened depletion of the high-grade iron ore in the Mesabi range.

ing on steel production based wholly on scrap, as compared with total steel production in the United States.   So far as other countries are concerned, this method could be important only to those which are already quite highly industrialized and generate large amounts of scrap annually.   Countries like India, Brazil, China, etc., where industrialization has made slight headway, obviously would not be able to develop a steel industry on this basis; and some countries, like Japan and Italy, might suffer from the decreased availability of American scrap for export.

2. The Possible Importance of Hydrogen Reduction

a. IN THE UNITED STATES.   The adoption of the hydrogen method in the United States will depend first on the development of a practical hydrogen furnace.   The first steps toward this have already been taken.   However, given the great investment in existing steel-making property and the relatively small cost saving possible by the new method, it appears unlikely that there will be a sufficient spur in the short run to the research and development necessary to the introduction of commercial methods.   But as capacity is enlarged, or as old capacity is replaced, the combination of the higher cost of coking coal and the exhaustion of large high-grade iron ore reserves should strengthen the attraction of the hydrogen process.

The increase in the price of coal, especially in the coking grades, in recent years is due to gradual exhaustion of the more productive mines and to improvements (especially in the last decade) in miners' wages and welfare. These factors appear certain to continue and probably will further increase the relative cost of coal as compared to atomic power.[27]   Another important factor is the forseeable exhaustion of the high-grade ore in the great Mesabi fields in the next two or three decades if steel production continues close to recent levels. One consequence of this may be to increase the importance of relatively small deposits of iron ore in the various regions of this country which have not been exploited owing both to their smallness and to their distance from coking coal.

The hydrogen method is ideally suited to the exploitation of these small ore deposits because production units based on hydrogen could be only $\frac{1}{10}$ (or less) the size of coke blast furnaces.   In addition, small-scale electric furnaces for the conversion of iron to steel are quite feasible.   Under these circumstances the individual rolling mill, if it maintains its present size, would represent a far greater capital outlay than the iron and steel plants and would probably be supplied by many individual iron and steel plants. This type of organization, not uncommon in the very early days of the steel industry, is parallel to the organization of large assembly operations such as in aircraft and autos, where the final manufacturer is often very much larger

[27] Cf. Chapter II, B.2 and 3.

than his suppliers.   On the other hand, the recent successful experimentation with the continuous casting process gives promise that small-scale rolling
and finishing processes will be forthcoming in the next decades, although
there is serious doubt as to the feasibility of producing large structural sections in small mills.   It is possible that with low cost commercial atomic
power the trend in steel towards larger equipment, plants, and ownership
organizations might be halted and perhaps reversed.   The feasibility of
exploiting relatively small-scale ore deposits scattered throughout this country would tend to lead to greater geographical diffusion of steel production
than is now possible.   Since operations would be on a smaller scale, a smaller
market area could absorb the output of a single steel works.   As a result, the
market for many steel products and its iron ore base may be drawn closer
together, and integrated operations on a considerably smaller scale than we
have today may be the dominant characteristic of future steel industry
operations.

b. IN OTHER COUNTRIES.   We have not had the time to attempt a
systematic analysis of the possible importance of hydrogen reduction using
atomic power to the iron and steel industries of other countries.   Yet this
is a question of the utmost importance since the production of steel, vital to
modern industrialization, requires energy in such large amounts.   It is
particularly important to know whether countries in which steel production
might otherwise be severely limited because of the lack of conventional fuels
(mainly coking coal) would be able to break through these limits by using
atomic power.   The analysis of the comparative costs of producing iron by
hydrogen reduction based on atomic power and by the coke blast furnace in
the United States suggests that atomic power might indeed be one of the
keys to expanded steel output where coking coal resources are limited.

Which countries would benefit from the use of hydrogen reduction?
Although countries with large established steel industries might find
early change unprofitable for the same reasons as in the United States, the
upward trend of coking coal costs might eventually encourage the introduction of the hydrogen process using atomic power.   Some countries such as
Great Britain, in fact, might apply atomic power to steel-making prior to the
American industry because their coking coal reserves are more limited.

But in the long run the most revolutionary effects of hydrogen reduction
and atomic power might very well be experienced by those countries which
today produce steel in small amounts or not at all.   The development of
these countries will require steel in large amounts for the building of factories,
machines, transportation facilities, and so forth.   Without attempting to
answer definitely the question of how great the increases in world steel production would have to be to achieve a minimum acceptable standard of living in all countries, we may adopt certain goals expressed in terms of pounds

of steel consumed per capita per year which we know to be roughly corre-
lated with various levels of industrialization.

It is apparent from Table 24 that steel consumption per capita varies
enormously from country to country.[28] The United States, for example,
consumes about 7 times as much steel per person as the world average and
about 150 times as much per person as India. The goals we use are solely
for the purpose of determining the approximate order of magnitude of world
steel requirements at various levels of industrialization. The consumption
targets are as follows (expressed in pounds of steel ingot consumption per
year per person):

Lbs. per capita	Approximate level of industrialization
100	Spain, Hungary
200	Austria, prewar U.S.S.R., Japan
300	Prewar Czechoslavakia
400	Sweden, France
600	United Kingdom, prewar Germany
1,200	U.S.A. prosperity level

The steel requirements shown in Table 24 are calculated for the first four
levels of consumption, on the basis of 1940 population figures and on the
assumption that no country produces below the assumed level. The average, therefore, for world consumption will always be greater than the minimum level goal, since each country exceeding the minimum is taken at its
peak production figure.

Just two cases will illustrate the implications of Table 24—India and
Brazil. Both countries have vast iron ore reserves (actual and potential).
India's reserves are estimated at 14 billion tons and Brazil's at 15 billion tons,
compared with approximately 10 billion tons each of low-grade ore in Great
Britain and France. However, peak steel production in Brazil was only
about 500,000 tons and in India about $1\frac{1}{2}$ million tons, compared with
peak production of about 15 million tons in Great Britain and 10 million
tons in France.

If India and Brazil should attempt to raise their per capita level of steel
output, would their fuel resources be adequate? For India to produce 100
lbs. of steel per year per person [the approximate prewar level of steel production in Spain and Hungary, but a very low standard by comparison
with consumption in Great Britain (600 lbs.), France (400 lbs.), or the
United States (1,200 lbs.)] would require annual steel production of about
20 million tons, or about 13 times its peak production. Yet in 1937 India's

[28] For purposes of simplicity the steel production per capita is substituted for steel
consumption per capita. For Europe as a whole, the steel production per capita is
about $\frac{1}{5}$ to $\frac{1}{4}$ higher than the true steel consumption, since Europe is a great steel exporting region. The steel production of the Benelux countries in particular is primarily for export. The U. S. production figures are about 10% above the consumption
figures. For the great importing areas (Asia, Africa, and South America) the steel
production per capita is lower than the steel consumption per capita.

TABLE 24. *Estimated world steel requirements compared with peak production.*

Continent	Total population (millions)[a]	Peak annual steel production (million short tons)[b]	Peak production per capita (lbs.)	Total steel requirements, no country consuming less than:			
				100 lbs. per capita	200 lbs. per capita (million short tons)	300 lbs. per capita	400 lbs. per capita
Asia	1,154	12.2	21.1	62.7	116.7	173.2	230.8
China[c]	450	1.6	7.1	22.5	45.0	67.5	90.0
India	382	1.7	8.9	19.1[d]	38.2	57.3	76.4
Japan	73	8.6	235.6	8.6	8.6	11.0	14.6
All others	249	0.3	2.4	12.5	24.9	37.4	49.8
Oceania	11	1.9	345.4	1.9	1.9	1.9	2.2
Africa	158	0.7	8.9	7.9	15.8	23.7	31.6
South America	88	0.5	11.4	4.4	8.8	13.2	17.6
Brazil	41	0.5	24.4	2.0	4.1	6.2	8.2
All others	47	neg.	neg.	2.4	4.7	7.0	9.4
North America	190	93.2	981.1	95.0	97.1	99.2	101.4
U.S.A.	136	89.6	1,317.6	89.6	89.6	89.6	89.6
Canada	11	3.2	581.8	3.2	3.2	3.2	3.2
Mexico	20	0.4	40.0	1.0	2.0	3.0	4.0
All others	23	neg.	neg.	1.2	2.3	3.4	4.6
Europe (excluding U.S.S.R.)[e]	380	69.9	367.9	69.9	69.9	69.9	76.0
U.S.S.R.	194[f]	22.6	233.0	22.6	22.6	29.1	38.8
WORLD TOTALS							
Population	2,175						
Steel production (million short tons)		201		264.4	332.8	410.2	498.4
Steel production per capita (lbs.)			185	243	306	377	458

[a] Population as of December 31, 1939 except for the United States figures which refer to 1943 (*Statistical Yearbook of the League of Nations, 1942–44*, Geneva, 1945, pp. 12–23).

[b] Crude steel production, including both ingots and direct castings. Figures are from *Statistical Yearbook*, United Nations, Lake Success, New York, 1949, pp. 238–239, and American Iron and Steel Institute, *Annual Statistical Report, 1947*, New York, 1948, pp. 186–187.

[c] The data for China include Manchuria.

[d] The chief of the Indian Supply Mission to the U. S., Mr. A. R. Palit, estimated that if India were to raise her standard of living to one tenth of the U. S. level, India's annual steel requirements would be 18 million tons (*Wall Street Journal*, October 20, 1948, p. 1, col. 6). This estimate agrees very closely with ours when adjusted for population difference. The Indian government, excluding Pakistan, has 350,000,000 people now, according to Mr. Palit. Adjusted for comparability, our figure would be 17.5 million tons compared to Mr. Palit's estimate of 18 million tons. The significance of the agreement of the figures lies in the fact that the leaders of the great undeveloped regions themselves look upon the relation of steel production to standards of living as being a linear one, at least in the lower ranges of this function.

[e] Some European countries have a per capita production of less than 100 lbs., but this has been disregarded in our calculations, which treat Europe as a unit.

[f] Population as of December 1939, including the population of territories and countries incorporated in the U.S.S.R., in accordance with various Soviet laws, between September 1939 and August 1941.

reserves of coking coal were estimated to have a life of 60 years at the then prevailing rate of consumption.   If this estimate is correct, India's coking coal reserves could not support the increased level of steel output.   In Brazil the case is even more extreme; so far as is known, that country has very little coking coal, all of poor quality.

Of course, the availability of adequate amounts of fuel would not by itself guarantee expanded steel production in these or other economically under-developed countries, but certainly the absence of adequate amounts of fuel could serve to limit expansion if other factors were favorable.   Atomic power, together with hydrogen reduction, might in this case be the answer to the problem posed by the shortage of coking coal.

But atomic power is not the sole answer.   Countries with limited do-mestic coal supplies can buy coal from abroad, although there might be a natural reluctance to base so vital an industry as steel production on external coal supplies.   We believe that the importation of coal in the amounts re-quired to maintain a reasonable level of steel output in India, Brazil, and other underdeveloped countries might at best find these countries buying coal on extremely poor terms of trade, or at worst they might encounter absolute limits on the amount of coal which would be forthcoming from coal-exporting countries.   This is certainly a debatable point.

On the other hand, coking coal is not the only form in which carbon might be used to reduce iron ore.   Such widely scattered fuels as peat, lignite, semibituminous coal, and vegetable waste could be used in carbon-aceous sponge iron processes.   Such processes appear to have higher labor costs than the hydrogen process, but this should not be important to countries in which labor is relatively cheap.

Because there are other possibilities for increasing steel production in countries where coking coal is scarce, but where iron ore is plentiful, we cannot state with any assurance that atomic power will be decisive in permit-ting steel production at higher levels or at lower costs than otherwise.   We can state that a metallurgy independent of naturally occurring coking coal will allow the attainment of the minimum goals set forth.   Considering the favorable costs of the hydrogen method using atomic power, it is reasonable to believe that the new metallurgy opened up by atomic power might provide one of the most promising alternatives for expanding steel output in undeveloped areas of the world.

CHAPTER XI

Railroad Transportation

THE interdependence of the railroad and fuel industries in modern economic life means that railroad transportation may be vitally affected by the development of commercial atomic power. This interdependence is strikingly revealed in the statistics for the two industries in the United States.[1] In 1945 the railroads consumed the equivalent of 159 million tons of bituminous coal for motive power (of which 115 million tons were in the form of coal, and the remainder in the form of other fuels and electricity), compared with a total production in this country of approximately 575 million tons of bituminous coal.[2] Furthermore, the importance of the railroads as consumers of coal and other fuels is no more marked than the relative importance of coal transportation as a source of railroad revenue. In 1945 revenue from coal and coke traffic came to approximately 1 billion dollars out of a total freight revenue of about 7 billion dollars, while the tonnage of coal and coke carried constituted about ⅓ of all the freight originating on American railways.[3]

Atomic power might affect railroads significantly (1) by providing a new and cheaper source of motive power and/or (2) by taking the place of coal in some industries and thereby reducing the revenues from this important kind of freight. This chapter deals almost wholly with the first of the two possibilities. The second question cannot be dealt with satisfactorily until a great deal more is known about the possible substitution of atomic energy for coal. The future of coal transportation might also be seriously affected by potentially important changes in the technology of coal utilization whereby coal would be changed to liquid and gaseous fuels, both of which

[1] The railroad data are for Class I railways, i.e. those with annual revenues above $1,000,000. The operating revenues of Class I railways in 1945 constituted 99% of total operating revenues of steam railways (including their electrified and dieselized sections) in the United States (*Statistics of Railways in the U. S., 1945,* Interstate Commerce Commission, Washington, G.P.O., 1947, p. 1).

[2] *Ibid.,* p. 66, for fuel consumption by railways. Coal equivalents of other energy sources apparently were calculated on the basis of the amount of coal which would be required in a coal-burning steam locomotive to do the same amount of work. Since the steam locomotive has a comparatively low thermal efficiency, the amount so calculated is generally greater than the amount of coal required in central stations to generate electricity to power the electric locomotives included in the figures, or the coal equivalent in terms of the energy contained in the oil used in diesel locomotives.

[3] *Ibid.,* pp. 43–44.

would probably employ transport media other than railways. (Cf. Chapter II, B.3.) The nature of these changes can be seen now only in outline. So this chapter refers to the second question only where it impinges on the analysis of atomic energy as a source of motive power, i.e. in connection with how the density of railroad traffic would influence the use of atomic electricity in railroad electrification.

Atomic energy as a source of railroad motive power theoretically could develop in two directions. One way would be through the perfection of a small nuclear reactor which could be fitted into a locomotive frame to generate electricity for use in the electric motor propelling the locomotive. This would resemble the diesel-electric locomotive or the projected coal-burning gas turbine locomotive in that the locomotive would house a complete electric generating plant plus electric motors for moving the train. The other way would be through the production of electricity in stationary atomic power plants for delivery to electrified railways in the same manner as on today's electrified roads. For reasons indicated in an appendix to this chapter, the feasibility of the atomic locomotive is questionable. Our analysis of the comparative costs of different types of motive power, therefore, considers atomic power only in connection with conventional railroad electrification.

The purpose of this chapter, then, is to consider the possible impact of atomic electricity generated in stationary power plants on railroad electrification in the United States and other countries. Obviously we can do no more than to sketch some of the important factors which will determine the outcome of the possible competition between atomic electricity and other sources of motive power. To this end we have assembled for analysis some illustrative materials on the economics of different types of railroad motive power.

Since we were not in a position to undertake for ourselves an intensive study of the costs of different types of railroad motive power, we have had to rely on analytic work done by others, and especially on the studies (both published and unpublished) prepared for the Bonneville Power Administration by T. M. C. Martin. But while our quantitative analysis in the next section draws heavily on materials prepared by Mr. Martin, we must emphasize that at several points we have modified his assumptions and that our conclusions are not necessarily in precise agreement with his views.

Altogether our task was considerably lightened by the availability of the Bonneville studies. The reader should be aware, however, that since these studies are based on a hypothetical railroad line operating over a terrain similar to that found in parts of the Pacific Northwest, they have only limited applicability. This difficulty, in some form, is unavoidable. The subject can be brought within a reasonable compass only by constructing a hypothetical situation which will permit us to focus on some of the important variables influencing the economics of alternative sources of motive power.

It is well, therefore, to state at the outset what conclusions we believe should and should not be drawn from our analysis. We do not believe that a definitive case can be made for the electrification of any particular stretch of railroad line on the basis of the figures we show. That question can be settled only by a detailed engineering study. We do believe, however, that in the pages which follow we discuss the important variables affecting the economics of railroad motive power fully enough to block out broadly the types of regions in which electrification based upon atomic power might be economically justified, and the types of regions for which a reasonable case for electrification cannot be made. That is the task we have set ourselves. The important variables in the economics of railroad motive power are considered in the next section. A concluding section deals with the implications of atomic power for railroad electrification in the future.

A. COMPARATIVE COSTS OF DIFFERENT FORMS OF RAILROAD MOTIVE POWER

1. The Cost of Railroad Motive Power

The amounts spent by railroads in the United States for fuel and electric power constitute approximately 8% of their total operating expenses.[4] These costs, while not relatively as large as in some of the other industries we have discussed, are considerably more important than in most industrial activities.

In a discussion of different forms of motive power the cost of fuel and power is but one factor. Actually, many other items of cost are affected by the form of motive power; it is often variations in these costs rather than in fuel costs which determine the choice among competing energy sources. Thus, it has often been pointed out that the growing preference for diesel-electrics over steam locomotives is because such items of expense as maintenance, repairs, fixed charges, cost of train crews, etc., may be less in the diesel.[5]

In the future choice among diesel-electrics, coal-burning gas turbines, and electric locomotives certain elements which have influenced the comparisons between steam and electric locomotives, or steam and diesels, may disappear. Locomotives of each type probably will not differ greatly in respect to haulage capacity, speed, availability for service, etc. It is possible

---

[4] The data, which are for 1939 and 1945, relate to Class I Railways, and include the cost of lubricants (*Statistics of Railways, 1945, op. cit.*, pp. 82–83; *Statistics of Railways, 1939, op. cit.*, pp. 83–84).

[5] See, for example, "Diesel Locomotives," *The Lamp*, Vol. 29, June 1947, pp. 5–6; "The Coal-Burning Gas Turbine Locomotive," by John I. Yellott, *The Analysts Journal*, Vol. 3, No. 1, First Quarter, 1947, p. 24.

therefore that the choice in the future will be determined by a more limited number of factors among which comparative energy costs and the first costs (and fixed charges) of the particular forms of motive power will be of major importance.   These are the main items which figure in our subsequent comparisons.

2. Recent Tendencies in the Use of Railroad Motive Power

Today, as in the past, steam locomotives are the predominant type of motive power on American railways.   This is clearly seen in the following figures: [6]

Number of revenue-service locomotives owned by Class I Line-Haul Railways in the U.S. on August 31, 1945

Steam	39,315
Diesel-electric	3,021
Electric	784
TOTAL	43,120

But the continued predominance of steam locomotives is less significant for our purposes than the growing popularity of the diesel-electric.   Recent changes in the relative importance of the two forms of motive power are summarized in Table 25.   Between 1941 and 1947, the figures show a marked growth in the relative importance of the diesel-electric locomotive at the expense of the coal-burning steam locomotive.   Since the total volume of freight traffic increased between the two years, the absolute volume of freight carried by the coal-burning steam locomotive also increased a little, but clearly its relative position was undermined by the diesel.

TABLE 25.   *Freight service: gross ton-miles of cars, contents, and cabooses; calendar years 1941, 1946, 1947.*

Kind of locomotive in train[a]	1941		1946		1947	
	Number (millions)	Per cent of total	Number (millions)	Per cent of total	Number (millions)	Per cent of total
Coal-burning steam	951,600	79.66	949,238	69.55	990,366	66.98
Oil-burning steam	212,344	17.78	253,853	18.60	274,886	18.59
Diesel-electric	2,654[b]	.22	132,863	9.73	182,971	12.37
Electric	28,010	2.34	28,826	2.11	30,292	2.05
Other	[b]	—	113	.01	83	.01
TOTAL	1,194,608	100.00	1,364,893	100.00	1,478,598	100.00

Source: Interstate Commerce Commission, "Monthly Comment on Transportation Statistics," March 11, 1948.

[a] Principal locomotive.

[b] Diesel and other combined in reports for 1941.

[6] Interstate Commerce Commission, "Postwar Capital Expenditures of the Railroads," Washington, G.P.O., March 1947, p. 17.

The trend in favor of diesel-electric locomotives is revealed most strikingly in the figures on locomotives ordered for domestic use.   For the years 1944–1946 locomotive orders were distributed among the various types as follows:[7]

Year	Steam	Diesel	Electric	Total
1944	74	680	3	757
1945	148	691	6	845
1946	55	856	8	919

The overwhelming preference of American railway managers for diesel power over steam power is unmistakable.

Two factors of importance emerge from the foregoing figures.   The first is that electrified railroads account for an extremely small percentage of the traffic carried on American railways, with no indication that their relative importance will change appreciably.   The second is that, in considering whether atomic power will encourage the growth of railroad electrification, it is more realistic to compare the probable costs of electric operation with diesel power than with steam power.   Therefore our subsequent cost comparisons are between diesel and electric locomotives.

Several assumptions which entered into our decision to compare only diesel power with electrification should be clarified.   The exclusion of steam power should not be interpreted to mean that steam locomotives will disappear completely over the next 25 to 50 years.   Rather it reflects our belief that for the type of service with which we are concerned, i.e. main line locomotives averaging about 200,000 miles per year, diesels will generally prove more economical than steam locomotives so long as relative fuel prices do not change radically.

There is the possibility, however, that the price of diesel oil will rise sharply if domestic crude oil production fails to keep pace with the growth in demand for oil.   Also, other countries which have coal but little oil may decide not to introduce diesels because of the high cost of diesel fuel to them, or because of a desire to avoid dependence on foreign sources for railway fuel (partly perhaps to conserve foreign exchange, and partly for security reasons).   We have been influenced by two major considerations in evaluating these possibilities: (1) the possibility that sometime in the future the cost of synthesizing diesel fuel from natural gas and coal or from oil shales will not be much more than present costs of deriving it from crude oil; (2) we believe that the projected coal-burning gas turbine locomotive holds sufficient promise of commercial success to justify the view that coal-burning locomotives in the future will be of this type, rather than steam engines, at least for the kind of service dealt with in our analysis.

These two considerations lead us to believe that diesel operating costs

[7] "Locomotives Ordered and Built in 1946," by Fred C. Miles, *Railway Age,* Vol. 122, January 4, 1947, p. 110.   Data for 1944 exclude United States government purchases.

based on prevailing fuel prices are to be preferred to steam power costs as a standard for comparison with railroad electrification costs despite the force of the possibilities noted above.   This involves the assumption, supported by preliminary cost estimates, that the operating costs of the coal-burning gas turbine locomotive, based on current coal prices, will approximate those of the diesel.[8]

Excluding steam power from the comparisons in the following pages considerably simplifies the analysis, but we must emphasize that the justification for this omission rests on the assumptions stated above.   Should events show that steam power is preferred to diesel or to coal-burning gas turbines, then our conclusions about railroad electrification based on atomic power would be affected, and a stronger case for electrification could probably be made than that presented here.   For electrification may not be economically justified in comparison with diesel power based on oil costing 5 to 7 cents per gallon—the 1946 price range, which is used in our analysis for comparability with atomic power costs also estimated in 1946 prices.   It may, however, be justified in certain cases where diesel oil prices reach 14 cents a gallon.   In the latter case there might otherwise be a return to steam power (if, for example, the coal-burning gas turbine is unsuccessful) and the electrification would need to be compared with steam power operating costs (including fuel and other charges) greater than the diesel costs we have used in our analysis.   For countries in which oil scarcity is already a problem, and where steam locomotives would be used, the comparisons we show would be biased against electrification from the outset (again, in the event the coal-burning gas turbine is unsuccessful).

3. Comparative Capital Requirements and Operating Costs: Diesel and Electric Motive Power

Some illustrative material on the comparative costs of diesel and electric railway motive power is presented in Tables 26 to 27.   As we noted earlier, a good deal of the data included in these tables are derived from studies made by T. M. C. Martin for the Bonneville Power Administration.

Martin's figures are based on a hypothetical single-track line of 400 route miles over mountain terrain.   This may bias the results somewhat in favor of electrification.   For example, the relationship between maximum load and speed which obtains on the particular mountain profile which Martin has chosen would not prevail on a stretch of level track.   As a result, a smaller number of locomotives might be required for the tonnage carried; for the electrification calculations (cf. Table 26), this would mean distributing the fixed investment in electric supplying facilities over a smaller number of locomotives.   On the other hand, we compensate to a certain extent for

[8] Yellott, *op. cit.;* "Coal: The Fuel Revolution," *op. cit.,* p. 253.

this element of bias by assuming that the diesel and electric locomotives will cover the same annual mileage, although actually the electric locomotive probably would be available for use a greater part of the time than the diesel and is capable of higher speed.

While the figures shown in these tables are only approximate, they reveal certain important characteristics of the two types of motive power, and of the conditions necessary for the costs of electric operation to approximate diesel costs.   In all cases the data cover only those items of cost which are not the same for the two types of motive power.

a. CAPITAL REQUIREMENTS.   A factor of primary importance in the comparative economics of diesel and electric motive power revealed in Table 26 is that the investment required for railroad electrification is much greater than for diesel power.   This is true because the electric locomotive is not an independent unit, but rather one which can operate only if electric supply facilities are provided.   These include transmission lines and other facilities whose costs vary in general with the mileage which must be covered rather than with the density of railway traffic.   The table assumes complete electrification in one case and complete dieselization in the other.   Diesel has an additional advantage, of course, in that diesel units can be introduced piecemeal into a railway operating on steam power as the steam locomotives become obsolete.   The introduction of a single electric locomotive would require an additional heavy investment for the overhead transmission system.

In Table 26 the burden of the investment in electric supply facilities is assigned to the individual electric locomotive on a pro-rata basis in order to compare the total investment required per electric locomotive with that per diesel.   As the density of traffic on the railway line increases and more locomotives are required, obviously the burden of the investment in electric supply facilities per locomotive declines.   This is shown by the figures in column (4) of Table 26.   It is, however, significant that for the varying traffic densities assumed the investment in electrification always exceeds that in diesel power, despite the fact that the electric locomotive itself is less costly than the diesel.   The size of the additional investment required by electrification as compared with diesel power is shown, per locomotive, in column (6) of Table 26.   At the highest traffic density assumed, the excess almost equals the cost of an electric locomotive of the type referred to in the table. At the lowest density the extra cost per locomotive is equal to the cost of 8 electric locomotives or 5½ diesel locomotives.

Since the data in Table 26 relate only to the investment in motive power, a comparison of these figures with costs covering the total investment required in railroads will help us evaluate the relative importance of the extra investment required for electrification.   To do this we select the extra cost involved at an intermediate traffic density—10 million gross ton miles per route mile—and compare the excess electrification investment with the average investment in American railways.[9]   Per route mile, the extra investment at this density comes to about $20,000 in our example, while the average total investment per route mile in Class I railways in the United States in 1945 was $125,000.[10]   This comparison indicates that the extra cost of electrification would at least be an appreciable part of the total investment in constructing a new railway line.

b. OPERATING COSTS.   The degree to which the initial handicap involved in the extra investment required for electrification may be overcome

TABLE 26.   *Diesel and electric railroad motive power compared: capital expenditures not common to both systems, according to varying traffic density of the system (data expressed in terms of a single locomotive of each type running 200,000 miles per year).[a]*

| Traffic density of the system per year (million gross ton miles of cars, contents, and cabooses per route mile) | First cost of diesel locomotive[b] | First cost of electrified system | | | Excess of electric system investment over diesel investment |
		Cost of locomotive[b]	Pro-rata cost of electric facilities to bring power to locomotive[c]	Total cost	
(1)	(2)	(3)	(4)	(5) = (3) + (4)	(6) = (5) − (2)
2.5	$600,000	$400,000	$3,500,000	$3,900,000	$3,300,000
5.0	600,000	400,000	2,000,000	2,400,000	1,800,000
7.5	600,000	400,000	1,500,000	1,900,000	1,300,000
10.0	600,000	400,000	1,000,000	1,500,000	900,000
15.0	600,000	400,000	750,000	1,150,000	550,000
20.0	600,000	400,000	550,000	950,000	350,000

Source: Adapted from published and unpublished data by T. M. C. Martin.   Published source is "Railway Motive Power for the Pacific Northwest," U. S. Department of the Interior, Bonneville Power Administration, August 1946.

[a] Varying traffic density affects only the investment in the electrical supply facilities since it is assumed that each locomotive runs 200,000 miles per year regardless of the traffic density of the system (cf. footnote c).

[b] Assumes locomotives of 5,000 rail horsepower (RHP) at a cost of $120/RHP for the diesel and $80/RHP for the electric locomotive.

[c] Cost of electric facilities is for a section of 400 single track route miles, and is estimated at $10 million.   The pro-rata share of this cost assigned to the individual electric locomotive varies with the number of locomotives required to handle the traffic of the system, and this, in turn, is determined by the traffic density assumed in col. (1).   Cols. (4) and (5) show that as density increases the investment in electric supply (which in general does not vary with traffic) is smaller in relation to the investment in locomotives (which in general does vary with traffic).

The number of locomotives required at different traffic densities can be derived by dividing $10 million by the figures in col. (4).   This calculation yields numbers of locomotives not exactly proportional to the densities specified in col. (1).   The discrepancy results from an arbitrary adjustment, when the exact figure on number of locomotives required at a particular density included a fraction of a locomotive.   In such a case we decided sometimes that the fraction of a locomotive should be considered an extra locomotive, and sometimes that the fractional requirement should be completely disregarded.

[9] The average density of traffic on U. S. Class I railways in 1945 was 8.2 million gross ton miles of cars, contents, and cabooses per route mile (computed from *Statistics of Railways, 1945, op. cit.*).

[10] *Ibid.*, p. 124.

TABLE 27. *Diesel and electric railroad motive power compared; annual operating costs (maintenance and fixed charges) other than fuel, not common to both systems, according to varying traffic density of the system (data expressed in terms of a single locomotive of each type running 200,000 miles per year).*

| | Annual traffic density: million gross ton miles per route mile of cars, contents, and cabooses | | | | | | | | | | | |
| | 2.5 | | 5.0 | | 7.5 | | 10.0 | | 15.0 | | 20.0 | |
	Electric	Diesel	Electric	Diesel	Electric	Diesel	Electric	Diesel	Electric	Diesel	Electric	Diesel
Maintenance of locomotives ($0.55 per mile diesel; $0.30 per mile electric)[a]	$60,000	$110,000	$60,000	$110,000	$60,000	$110,000	$60,000	$110,000	$60,000	$110,000	$60,000	$110,000
Maintenance of electric power supply facilities ($0.07 per mile)[a]	14,000	—	14,000	—	14,000	—	14,000	—	14,000	—	14,000	—
Fixed charges on locomotives (depreciation at 6% for diesel, 5% for electric; interest at 5% for both)[b]	40,000	66,000	40,000	66,000	40,000	66,000	40,000	66,000	40,000	66,000	40,000	66,000
Fixed charges on electric power supply facilities (depreciation at 3%; interest at 5%)[c]	280,000	—	160,000	—	120,000	—	80,000	—	60,000	—	44,000	—
TOTAL	$394,000	$176,000	$274,000	$176,000	$234,000	$176,000	$194,000	$176,000	$174,000	$176,000	$158,000	$176,000

Source: Adapted from published and unpublished data by T. M. C. Martin. Published source is "Railway Motive Power for the Pacific Northwest," *op. cit.*
[a] Based on 200,000 miles per locomotive per year.
[b] Based on costs of locomotives shown in Table 26.
[c] Based on pro-rata costs of electric supply facilities shown in Table 26.

through lower operating costs for electric power as compared with diesels is indicated in Tables 27 and 28.   For varying traffic densities Table 27 shows operating costs (maintenance and fixed charges) other than fuel, not common to both forms of motive power; and Table 28 completes the picture by indicating the comparative diesel oil and electricity prices which will equalize total operating costs.   It should be noted that, following Martin's analysis, the costs include a return of 5% on the capital invested in road and equipment.

TABLE 28.   *Diesel and electric railroad motive power compared: prices of diesel oil and electricity which equalize total operating expenses for the two types of motive power, according to varying traffic density of the system.*[a]

If price of diesel oil is (cents per gallon)	Annual traffic density: million gross ton miles per route mile of cars, contents, and cabooses					
	2.5	5.0	7.5	10.0	15.0	20.0
	Total operating costs will be equal if price of electricity is (mills per KWH)					
4	*	*	0.1	1.9	2.8	3.5
6	*	*	1.4	3.2	4.1	4.8
8	*	1.0	2.8	4.6	5.4	6.2
10	*	2.3	4.1	5.9	6.8	7.5
12	*	3.6	5.4	7.3	8.1	8.8

* Even at zero cost electricity, total operating expenses for the electric system would exceed total diesel operating expenses.

[a] Operating charges other than the cost of energy not common to both systems are shown in Table 27.   Energy costs have been added to the other charges to determine the comparative diesel oil and electricity costs which would equalize total operating expenses.   The assumed equivalent energy requirements per locomotive mile are 6.8 gallons of diesel oil for the diesel locomotive, and 111.7 KWH for the electric locomotive (derived from data in Martin, *op. cit.*).   In addition, 10% has been added to the energy cost for the diesel to provide for lubrication costs.   (Cf. Martin, *op. cit.*, and Yellott, *op. cit.*; these experts state that lubrication will cost from 10% to 20% of the energy cost.)

Examination of Table 27 reveals that the disadvantage of higher first costs in electrification is mitigated to some extent by lower costs for electric locomotive maintenance and depreciation.   But fixed charges on the electric supply facilities bulk very large, and costs for electrified operations other than fuel do not fall in the same general range as those for diesel power until a density of 15 million gross ton miles per route mile is reached.   Beyond this density, electrified costs, other than fuel, are below those for diesel power.

How cheap must atomic electricity be under these circumstances to compete with diesel power?   This question is handled in the comparisons shown in Table 28.   It should be clear from the qualifications surrounding the data that the figures give us a useful general impression rather than a precise answer.   In general terms, then, we find that at the lowest traffic density assumed even free electricity would not yield operating costs as low as those possible with diesel power.   As traffic density increases, electric operations are able to compete with diesel power, and the precise point of equalization is affected significantly by the price of diesel fuel.   On the assumption that the electricity demand of electrified railways will have a 50% load factor (this question is considered in the next section), atomic electricity costs might fall between 4 and 10 mills per KWH.   At 4 mills electrification appears to be economically feasible for a good percentage of the cases covered in Table 28; at 10 mills electrification could not be economically justified in any case covered in Table 28; while at an intermediate rate of 6.5 mills electrification could be justified only if diesel fuel prices are around 10 cents per gallon or higher and quite high traffic densities prevail.

B. IMPLICATIONS OF ATOMIC POWER
FOR RAILROAD ELECTRIFICATION IN THE
UNITED STATES AND OTHER COUNTRIES

In this section we will make some general observations on the possible impact of atomic power on railroad electrification.   The United States exemplifies those nations where supplies of oil and coal are sufficient to support a large (and if necessary expanding) railroad industry, and where atomic electricity will be used only if it can demonstrate its superiority in a highly competitive fuel market.   "Other countries" are meant to include those where alternative sources of railway fuel are limited, and where electrification based on atomic power might be introduced even at a somewhat higher apparent money cost as a means of reducing fuel import requirements.

1. The United States

Data useful for an appraisal of the possible importance of atomic power to railroad electrification in the United States are shown in Table 29.   Here we have figures showing average traffic density on railway lines and average prices paid for diesel fuel by major traffic regions.   These data enable us to estimate in column (3) the average electricity cost necessary to equal diesel operating costs in the region.   Since the railway lines within the region are characterized by varying degrees of traffic density, only the average being shown, this tabulation is obviously useful only for giving a very general impression of the possible role of atomic power in railroad electrification in the region.

In determining what cost of atomic electricity to assume for electrified railways, it is essential to know something of the load factor exhibited for this type of demand.   We must bear in mind the warning from a Federal Power Commission report on railway electrification.[11]   "Since no two rail-

[11] Federal Power Commission, *The Use of Electric Power In Transportation*, Power Series No. 4, Washington, G.P.O., 1936, p. 33.

TABLE 29.   *Traffic density, cost of diesel oil, and estimated electricity rates to equal diesel operating costs for Class I railways in the U. S., by regions.*

I.C.C. region	(1) Gross ton miles of cars, contents, and cabooses per route mile, 1945[a] (in millions)	(2) Average cost of diesel oil in 1946[b] (cents per gallon)	(3) Approximate price of electricity to equal diesel operating costs based on 1946 diesel oil prices[c] (mills per KWH)
New England	6.4	6.8	1.2
Great Lakes	11.3	6.4	3.8
Central Eastern	13.3	6.1	3.8
Pocahontas	17.5	7.1	5.2
Southern	6.8	6.1	1.0
Northwestern	4.6	5.8	*
Central Western	8.5	5.0	1.6
Southwestern	6.2	4.8	*

* Even at zero cost electricity, total operating expenses for the electric system would exceed total diesel operating expenses.

[a] Derived from data in *Statistics of Railways, 1945, op. cit.*   These are region-wide averages; main-line densities would run higher.

[b] Interstate Commerce Commission, *Comparative Statement of Railway Operating Statistics, 1946 and 1945*, Washington, G.P.O., 1947, p. 41.

[c] Estimated from data in Table 28.   Rates shown are very rough approximations.

roads have the same type of service or operating conditions, it is obvious that their power requirements and load curves will vary widely.   While each electrification will produce a load curve typical for its operation, there cannot be a load curve typical for electrified railroads in general."   The same report indicates that a fairly high prewar load factor for electric power consumed by railroads with heavy through passenger and freight service was about 50%, while in some major electrifications (e.g. the Pennsylvania Railroad between New York and Washington, D.C.) load factors of about 60% had been realized.[12]   We shall assume that demand conditions on the railways covered in our example are best characterized by an average load factor of 50%.[13]

At 50% load the estimated costs of generating atomic electricity shown in Chapter I are approximately as follows (in mills per KWH) :

minimum estimate	4.0–4.5
intermediate estimate	6.5–7.0
maximum estimate	about 10.0

It is clear immediately that for most of the regions shown in Table 29 atomic electricity, even at the minimum estimate of costs, will be too expensive to be generally adopted.   This does not mean, however, that there may not be

[12] *Ibid.*, p. 33.
[13] This is the approximate load factor assumed by Martin in an unpublished report for the Bonneville Power Administration.

individual, possibly important, stretches of railway lines within these regions that might benefit from electrification; here we can only indicate how the region as a whole may be affected.   Three regions show some promise according to this analysis: Great Lakes, Central Eastern, and Pocahontas,[14] all with traffic densities considerably above the average for the United States. The Great Lakes and Central Eastern regions might in general derive benefits from electrification at costs falling at the absolute minimum estimate of the costs of generating atomic electricity; the Pocahontas region could probably benefit generally with electricity costs somewhere between the minimum and intermediate range of estimated atomic generating costs.

There are additional important considerations influencing whether atomic power will encourage electrification in these regions.   It is necessary to note first that the costs shown are for generating atomic power, and that additional expenses would usually be incurred in delivering the electricity to the railroads.   Depending upon the distance from the railroad at which the atomic power station is located, transmission expenses might add from 1½ to 2 mills to the delivered cost of electricity.[15]   In this case the general economic feasibility of electrification even in the Pocahontas region might depend on generating costs which are near (and possibly below) the estimated minimum atomic costs.

Assuming that a general trend toward electrification in these regions will depend on the possibility of generating atomic electricity at these minimum costs, another complication arises.   For, according to the analysis elsewhere in this report, atomic electricity at such low costs would in general be less expensive than electricity generated from ordinary fuels, and this would mean that atomic electricity would gradually replace coal used for electric generation, and possibly also for other uses.   It is then at least conceivable that widespread electrification as a result of atomic power would not take place.   For since coal is such an important item of railroad traffic,[16] the traffic densities necessary to justify electrification might not be realized under these circumstances, so that diesel oil and/or coal would continue as the most important sources of railway motive power.

We can tentatively conclude that the average density of traffic in most regions of the United States is not sufficiently high to justify general electrification even at the minimum estimated costs of generating atomic power. For those regions in which traffic is sufficiently heavy possibly to justify a fair measure of electrification based on atomic power, this appears to require generating costs in the neighborhood of the minimum cost estimate for atomic power.   Since at such low costs atomic power might replace coal in

[14] Chiefly covering coal mining regions in Virginia and West Virginia.

[15] Philip Sporn, "Cost of Generation of Electric Energy," *Proceedings of the American Society of Civil Engineers,* Vol. 63, December 1937, p. 1930.

[16] Cf. p. 1 of this chapter.

many of its current uses there is at least a serious question whether, as a result of the decline in coal shipments, the density of railway traffic would not tend to fall below the minimum necessary to justify electrification. In general, then, it appears doubtful that railroad electrification will be significantly increased in this country as a result of the use of atomic power.

2. Other Countries

In countries with meager domestic fuel supplies, where the development of industries consuming large amounts of fuel may be subject to severe limitations owing to exchange or other difficulties in importing fuel, could electrification based on atomic power be a reasonably economic source of motive power?[17] Although there are no criteria for providing a definitive answer to this question, it is at least possible to indicate the magnitude of the additional money cost involved in using atomic power in preference to imported fuels. Whether the additional cost is less than the extra cost involved in buying coal or oil in exchange for goods which might have to be sold at unfavorable terms of trade can be determined only for specific cases.

The pertinence of this question may be judged from the following facts. Railroad transportation consumes an enormous amount of fuel. As indicated earlier, the Interstate Commerce Commission estimates an equivalent coal consumption of 159 million tons on American railways in 1945. Some countries with poor fuel resources but with abundant water power resources have, in the past, reduced fuel imports by using hydroelectricity to satisfy part of their railway motive power requirements. Outstanding examples are Switzerland, where electrified lines carried more than 80% of the total freight and passenger traffic during the 1930's;[18] Sweden, where approximately 40% of the route mileage is electrified;[19] and Italy, where about 10% of the total mileage was electrified in the mid-1930's, with numerous extensions under way.[20] Since hydroelectric power is not necessarily abundant in all countries with poor fuel resources, and since even for those countries which have water power there may be questions concerning the extent of undeveloped resources and their suitability for railway motive power owing to their fixed locations, it is important to consider the cost at which atomic power could perform a similar function.

Data covering the operating costs of Class I railways in the United States have been used in conjunction with the comparisons made in Tables

[17] We assume, for reasons indicated in Part I, that atomic fuel will be more readily available throughout the world than conventional fuel supplies.

[18] Federal Power Commission, *Use of Electric Power . . . , op. cit.,* p. 51.

[19] T. Thelander, "Some Results of Electric Traction on the Swedish State Railways," Fuel Economy Conference of the World Power Conference, The Hague, 1947 (Conference Proof Copy).

[20] Federal Power Commission, *Use of Electric Power . . . , op. cit.,* p. 51.

26 to 28 to provide an approximate answer to this question. We find the following: If atomic electricity costing the railways 7.5 mills per KWH were used as a source of motive power in place of diesel oil costing 10 cents per gallon, the total operating costs for electrified lines would be approximately 5% greater than if diesel power were used at a density of 7.5 million gross ton miles per route mile and approximately 10% greater at a density of 5.0 million gross ton miles per route mile.[21]   It is obviously difficult to know the significance of these figures for a particular country without additional information on the country's foreign trade position. It seems reasonably clear that, because of the importance of adequate transportation to economic development, a cost disadvantage of this magnitude would be worth sustaining if the country's ability to import diesel fuel (or coal) were severely limited.   However, if imports could be managed, albeit at trade terms which were considered unfavorable, it would require a specific comparison to determine whether electrification based on atomic power were preferable.   In any case, atomic power might be very significant in the development of electrified railways in many countries, even though a strict money cost comparison would indicate that alternative fuels would be cheaper and therefore preferable if their availability were not limited.   Since countries which feel the pinch of short fuel supplies might be expected to use the fuel available to them where the substitution of atomic electricity would involve the greatest disadvantage, their railroad system might be so arranged that atomic electricity would be used on the heavily-travelled main lines, while fuel would be reserved for the more sparsely-used lines.   In this way the cost disadvantage of atomic power, if any, could be minimized while fuel imports were kept to a lower level than would otherwise be necessary.

APPENDIX: ON THE FEASIBILITY OF USING NUCLEAR POWER PLANTS IN RAILROAD LOCOMOTIVES

Because of the limited clearance for railroad locomotives at various points along railroad lines and within terminals, an atomic power plant could not be used in locomotives unless the reactor with the necessary shielding could be confined within a very small space.   John I. Yellott, who has pioneered in the design of the coal-burning gas turbine locomotive, indicates that the problem of the size of power plant has been of critical importance in developing this unit.   He states in this connection: "A locomotive power plant must be almost unbelievably compact.   It must be able to pass through a door which is 10 feet wide and 10 feet high.

[21] Gross ton miles is defined here as weight of cars, contents, and cabooses. The average density of traffic on U. S. railways on this basis was 8.2 million per route mile in 1945, and 4.7 million in 1939.

Consequently, designs which are entirely satisfactory for stationary or marine plants might be entirely useless on locomotives."[22]

The published information on nuclear reactors does not provide a definitive answer to the question of whether a nuclear power plant can be made small enough to be carried in a locomotive. A major problem in meeting the size specifications cited by Yellott would be to provide a shield which did not itself almost equal, or exceed, the stated limits.[23] One informed observer has ventured the guess that a reactor small enough to be used in a locomotive might be barely possible, but added that the prospects for this development are not bright.[24]

Although the question of technical feasibility must remain unanswered for the present, the problem of physical size suggests another requirement which may prove a very serious limitation on the nuclear-powered locomotive until long after the development of commercial atomic power. If the reactor is to be made small, it will be necessary that it contain highly enriched fissionable material. In other words, the amount of pure plutonium or uranium 233 in such a reactor will be disproportionately high in terms of the reactor's power rating, as compared with reactors in which compactness is not of critical importance. Furthermore, because of the small size, a larger percentage of the neutrons generated in fission will escape and therefore not produce fresh fissionable material than in the case of reactors with larger dimensions.[25] For reasons explained in Chapter I, pure fissionable substance will be needed as part of the initial fuel stock of all nuclear power plants, and the ability to produce such materials may be a factor limiting the growth of the atomic power industry. The nuclear-powered locomotive, which will require a disproportionately large amount of such material and which will be wasteful in the production of fresh supplies either to replenish its own stock or to feed new reactors, might therefore be considered a particularly poor application of nuclear power.[26]

[22] Yellott, *op. cit.*, p. 27.

[23] See Chapter I, A.1.

[24] Wheeler, *op. cit.*, p. 403. Note that Wheeler implies that minimum railroad clearance is encountered in tunnels, while actually limiting clearances are more often found on "close track spacing, overhead bridges, stations and other structures, center and end excess on curves, etc." (Letter to the author from specialists in the I.C.C.)

[25] The reader is referred to Chapter I for more detailed discussion of some of the ideas expressed here.

[26] The text discussion is in terms of administrative decisions, but the same result would follow from the operation of a price mechanism which may, in fact, be the instrument for effecting administrative decisions. The scarcity of pure fissionable material would be reflected in a very high cost per unit of energy as compared with natural uranium. In this case the costs of building a reactor which required such material in disproportionate amounts would be relatively high. The operating costs would likewise be higher than in stationary power plants of larger dimensions because of the heavier burden of fixed charges, and also because of the need periodically to replenish the supply of pure fissionable material by outside purchases.

Another economic disadvantage of nuclear power plants for locomotives results not from the need for physical compactness, but from the fact that the rated power capacity of such plants is quite low compared with stationary power plants. The cost estimates shown in Chapter I are presumed to relate to power plants in the range of 75,000 to 100,000 kilowatts of rated capacity. Main line locomotives, such as those discussed in this chapter would, on the other hand, have a rated capacity of about 8,000 kilowatts. So small a power plant may be well below the optimum size for nuclear reactors and the investment per kilowatt well above the costs shown in Chapter I. It is unwise to attempt even a general approximation of the increased cost which might be encountered in reactors of less than optimum size, let alone to try to approximate the particular cost which might prevail at 8,000 kilowatts. Nonetheless, it is important to realize that for this reason alone, the figures on investment per kilowatt shown in Chapter I may be completely inapplicable to nuclear reactors for locomotives.

This discussion should be sufficient to indicate that the development of the nuclear-powered locomotive is highly questionable. Limitations on physical size alone may make the design of such a locomotive impossible. In addition, several economic considerations, which seem to indicate that the cost of the locomotive nuclear reactor will be relatively high in comparison with the stationary power plant, argue strongly against its development, at least until long after the commercial atomic power industry is established.

CHAPTER XII

Residential Heating

WHILE all of the other activities considered in these chapters are already performed by existing industries, there is no industry in the United States (with unimportant exceptions) which supplies heat for residential uses. Instead, home heating is generally a service provided by the householder or the landlord. But heat can be produced centrally for entire cities, or neighborhoods within cities, and piped to individual users in a way analogous to the central production of manufactured gas or electricity and its transmission to the consumer. Indeed, there are in this country many such systems for supplying heat to the business districts of large cities, but very few for supplying residential areas. District heating of residential areas is more common in some European cities.

Would the development of commercial atomic power encourage the growth of the district heating industry in the United States at the expense of the fuels now used for domestic heating? This depends essentially on whether centrally-generated nuclear heat could be brought to the home at a cost low enough to compete with the heating methods commonly used today. While the cost figures brought together in this chapter will not enable us to determine whether nuclear-powered district heating might be economically feasible in any particular city, they should provide a general estimate of its possible overall significance.

It is important that we examine this question since a very large part of all the fuel consumed in the United States is for residential heating. In 1945 space heating took virtually our entire output of about 50 million tons of anthracite coal, about 80 million tons of bituminous coal (almost 15% of all the bituminous coal produced in that year), and the equivalent of about 85 million tons of bituminous coal in the form of other fuels, chiefly oil and gas. Altogether, the fuel consumed in space heating accounted for almost ⅕ of the total energy supply from mineral fuels and water power in the United States in 1945.[1] The total amount of energy consumed for this purpose is therefore substantially greater than the amounts consumed by such industrial giants as the railroads or the iron and steel industry.

[1] *Minerals Yearbook, 1945, op. cit.*, pp. 846, 847, 925. The total includes, in addition to residences, the fuel used in heating offices, hotels, schools, hospitals, and probably stores.

199

The first section of this chapter considers the ways in which atomic power might be used in residential heating, and explains why our analysis considers only the central supply of heat. The second part develops estimates of the cost of district heating from atomic sources. A final section explores the extent to which nuclear heating of residential districts might be commercially justified in the United States.

A. THE USE OF ATOMIC POWER IN RESIDENTIAL HEATING

On the basis of currently envisaged techniques for the production of atomic energy in useful form, the heating of residences from this energy source probably would be accomplished by the transmission of energy in some form from a central generating plant. An individual nuclear furnace, even for a very large apartment building, is most unlikely both because of the reactor's space requirements, and because the energy requirements of a single building could be satisfied by a unit well below the probable range of economic reactor sizes.

A central atomic power plant could provide energy either as electricity or as direct heat in the form, say, of hot water or steam. The production of heat in the nuclear reactor at temperatures appropriate for home heating involves technical problems of considerably less complexity than those involved either in producing electricity or in producing high temperature heat for such industrial processes as cement manufacture.[2] Even at the present stage of atomic energy development it is reported that heat produced in the experimental nuclear reactor which began operating at Harwell, England, in July 1948 is already being used to warm buildings there.[3] From a purely technical standpoint the use of the nuclear reactor for residential heating is the only nonmilitary use of nuclear heat so far demonstrated to be feasible.

Electricity from nuclear reactors is nonetheless such a likely development that its use has been assumed in all other parts of this book. From an economic standpoint, is it more, or less, advantageous than direct nuclear heat? Until recently the use of electricity from thermal sources for house heating would not have been seriously considered since the conversion of heat to electricity in electric power plants involves a loss of about ⅔ of the energy content of the fuel, which generally made electric heating more costly than the direct use of fuel in the household furnace. However, with the development of the so-called "heat pump" the situation has changed considerably, because the input of 1 BTU of electric energy in the heat pump can be made to yield from three to five BTUs of heat output in the

[2] We have discussed the latter two earlier in this study; electricity in Chapter I, and high-temperature heat in Chapters I and VII.
[3] "Britain's Atomic Pile," *Discovery*, Vol. 9, August 1948, p. 233.

home. Hence the heat pump could more than redress the unfavorable energy balance formerly involved in the use of electricity for heating.

The heat pump, driven by an electric motor (other types of power may also be used), extracts heat from an outside body such as air, earth, or water and transfers it to the inside of the house. In the summertime the process can be reversed to provide air conditioning. The mechanism is still in an early stage of development, but installations have already been made, mainly in commercial buildings. Units designed chiefly for the home are priced at about $3500 (September 1948); obviously at this price the heat pump cannot serve a mass market. Its appeal today appears to be limited to those wealthy users who are able to afford both summer cooling and winter heating.[4]

Except for the heat pump, with its ability to give three to five units of energy for each unit of electrical energy put in, the nuclear reactor would obviously be used more efficiently as a direct source of heat for residential heating than through the roundabout, energy-losing route of supplying electricity. Since the little that is known about the cost of the heat pump today seems to place it in a luxury class, we will not consider it in our analysis. We shall therefore be dealing only with central atomic plants producing heat for direct residential use. Obviously, if a heat pump should be developed in the same cost class as ordinary home furnaces, we would need to recast our analysis completely.

District heating service is sometimes provided by central thermal power plants whose primary product is electricity. Exhaust heat losses, which would otherwise be wasted, are taken off and used for district heating purposes. When demand conditions permit (as we will see, this would only be in densely settled urban areas) a combined heat and electricity plant would probably produce heat at a lower cost than a straight heat-producing plant. The use of waste heat does not entirely represent a net gain because changes are required in the electric generating equipment to compensate for exhausting heat at a higher temperature than otherwise.[5]

Atomic central stations, when they begin operations, will in certain instances probably produce electricity and useful heat together at a lower average cost than for either one separately. But in this study we have based our cost calculations on the assumption that, with the exception of certain industrial installations, electricity and heat will generally be pro-

[4] On the heat pump see "The Sixth Ingredient—The Heat Pump," by Philip Sporn, *Edison Electric Institute Bulletin,* Vol. 12, August 1944; "The Heat Pump," *Electrical Engineering,* Vol. 67, April 1948; "A Review of Some Heat-Pump Installations," by E. B. Penrod, *Mechanical Engineering,* Vol. 69, August 1947; and "How to Heat a House," *Fortune,* Vol. 38, September 1948.

[5] See *The British Fuel and Power Industries,* A Report by PEP (Political and Economic Planning), London, October 1947, pp. 136 ff.

duced in separate plants. Our chief reason for treating the two separately is that district heating will not be feasible in a large number of places where electricity is commonly used. This primarily reflects a fundamental difference between the two in the conditions determining permissible costs of energy distribution. But we have also not taken the two together even in those cases where district heating might be economically feasible. This is because exhaust heat losses in thermal electricity plants could, at current levels of electricity consumption, satisfy only part of a community's heating needs. This may be seen from the fact that the total amount of fuel consumed by electric utilities in the United States in 1945 was the equivalent of 92.5 million tons of coal, of which between 50% and 70% might be available as exhaust heat, while in the same year space heating took the equivalent of about 215 million tons of coal.

B. THE COST OF ATOMIC ENERGY IN DISTRICT HEATING OF RESIDENCES

District heating involves two major operations: the generation of heat at a central plant, and its delivery to individual users. The second operation is the same regardless of the source of heat, and its cost may be estimated from data on district heating systems already in operation. Apparently this cost would be considerably greater than the cost of producing nuclear heat; we therefore turn to it first. Many of the concepts explained in this discussion will also be useful when we examine the generation of heat.

1. Distribution of Centrally-Produced Heat

The cost of distributing heat from a central plant to individual users is determined mainly by the investment in facilities to carry the heat. Cost will depend, therefore, on the size of the installation needed to satisfy the peak heat demand of a district, and on the annual heat output over which the yearly charges on the installation are distributed. A broad treatment of the costs of district heating, such as ours, must deal with general indicators of peak demand and annual heat requirements, and incorporate these into a hypothetical schedule of costs.

a. PEAK DEMAND (ANNUAL PLANT CAPACITY). The maximum demand for an area depends upon numerous variables such as population, the coldest temperature commonly experienced, the type of buildings and their construction, wind velocity, and other lesser variables. Any attempt to handle all of these items would result in an enormous amount of computation; it would, indeed, mean that we treat each city as a special case. Therefore, we have kept the two most important factors as variables and have considered the others as constants. The values assumed for the constant factors are those typical of cities in the northern half of the United States.

Maximum demand is specified in our calculations by two variables: population density and what is called "design temperature range."

1. Population density is used as the best readily available measure of the volume to be heated.[6] We converted population density to volume to be heated on the basis of the average number of persons per urban dwelling unit in the United States in 1940, and the average size of the dwelling units. The final figure we used is 1,900 cubic feet per person. For a particular city such a derived figure probably will not be exact, but the average provides perhaps the most appropriate figure for our purposes.

2. "Design temperature range" is a commonly used concept in heating design. It is defined as the algebraic difference between 70°F and the lowest temperature at which the heating plant is supposed to be adequate. The minimum temperature for which the heating plant is designed is generally taken as 15° warmer than the coldest temperature experienced in that area. (In some cities 10° is used instead of 15°.) Take, for example, the city of Boston which has a recorded minimum temperature of -18°F. The design temperature range would then be: $70 - (-18 + 15) = 73°$.

The maximum amount of heat required is equal to the volume to be heated (as derived from the population density), times the design temperature range, times the amount of heat required to maintain a temperature difference of 1 degree for a period of 1 day for each cubic foot of volume heated. The latter figure is considered constant, but a typical figure is very difficult to get because the actual values for different buildings vary widely. Several figures were compiled and a "typical" figure selected from these. The figures used here are 1,500 BTU/1,000 cubic feet of volume per degree per day for homes and 800 BTU/1,000 cubic feet of volume per degree per day for apartment buildings.[7]

From the maximum heat demand we have calculated the investment required for the distribution system, as follows:

---

[6] What is needed is a measure of the total volume of all buildings, but this is difficult to estimate. We use population density as the easiest guide to the volume of residential space, the only type of heating use considered in this analysis (cf. Section C of this chapter).

[7] Figures were assembled from the following sources: *Heating and Ventilating's Engineering Data Book,* by Clifford Strock, New York, The Industrial Press, 1948, pp. 5–33. *Proceedings of the National District Heating Association,* Vol. 19, 1928, p. 60; Vol. 30, 1939, p. 100; Vol. 34, 1943, p. 76. *Edison Electric Institute Bulletin,* Vol. 14, July 1946, p. 233; data furnished by the Peoples Gas, Light, and Coke Co. of Chicago.

1. A square area with the nuclear steam plant centrally located was assumed.

2. Several possible distribution layouts were tried and the one involving the lowest investment was selected.

3. "Typical" costs for piping materials were estimated from actual construction costs shown in various sources, and adjusted to the January 1947 price level. It should be understood that for a particular city the costs of piping may depart by as much as 30% from the assumed "typical" costs because of differing soil conditions, variations in the extent of underground obstructions, different types of pavement, etc.[8]

Graph 1 summarizes our estimates of annual plant capacity (daily peak capacity times 365) and related investment as determined by population density and design temperature range. To illustrate the use of the graph we take the city of Boston, for which we earlier found the design temperature range to be 73°. Population density in the city in 1940 was almost 17,000 persons per square mile. For this combination of design temperature range and population density the graph specifies for a 4-square-mile area a steam distribution system with an annual capacity of about 53.5 hundred million lbs. of steam, costing approximately $8 million. Given the appropriate figures on design temperature range and population density for any other city, the graph will similarly specify the capacity and cost of a 4-square-mile steam distribution system.

Graph 1 and all of our later calculations are in terms of a service area of 4 square miles. This is an arbitrary choice. As will become apparent later, the choice of a larger area might have increased the unit costs of delivered steam somewhat because of greater heat losses in transit, but the results would not be significantly altered. For the types of cities in which atomic-powered district heating appears to be economically feasible on the basis of our analysis, a 4-square-mile area would require a nuclear reactor falling within the probable economic size range.

The relationships formulated in Graph 1 will bear much of the burden of the analysis in the final section of this chapter. One point of crucial importance to the conclusions reached there is that the cost of the distribution system increases much less than proportionally with increases in capacity. The existence of decided economies of scale in distribution immediately

[8] The reader with a technical interest should note the following points: (a) steam pressures up to 200 lbs. per sq. in. were considered available; (b) steam is delivered to the consumer at not less than 10 p.s.i.g.; (c) steam pipe sizes were recalculated every half mile; and (d) steam piping costs were treated as continuous even though the discrete sizing of pipes makes this impossible. We believe that the treatment of piping costs as continuous facilitates the analysis without introducing important inaccuracies into the calculation.

suggests the possibility that only areas marked by a high density of demand
will prove suitable for district heating systems.

b. ANNUAL REQUIREMENTS. The factors so far considered determine
the peak heating demand of an area and the investment necessary to build

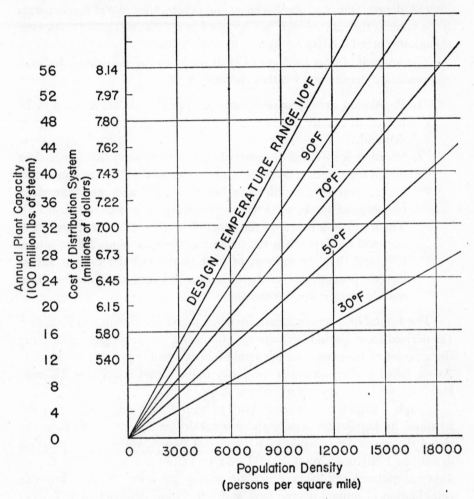

GRAPH 1. Annual plant capacity and investment costs for steam distribution system (excluding cost of generating plant and service connections)
serving four square miles as determined by population density and design
temperature range.

capacity to satisfy peak needs. The cost per unit of heat delivered to the
home will depend on (1) the annual charges on the investment and (2)
the number of heat units over which the annual charges are distributed.
The annual charges consist almost wholly of fixed charges which do not vary
with the number of units produced. Obviously, therefore, costs per unit
of output will be lower as capacity is utilized more fully.

We estimate the annual charges for the entire district heating system, including the generating plant and the distribution facilities, to be 12% of the investment. Of this, 1% represents the cost of labor, maintenance, and supplies, and the remaining 11% comprises fixed-charge items, mainly interest, depreciation and obsolescence, and taxes. The rate of fixed charges is the same as the rate we used in Chapter I in estimating atomic costs, even though it is based on data for district heating systems.

The amount of heat used annually depends, as did maximum demand, on population density and weather conditions.

1. Population density is used here to approximate the volume to be heated in exactly the same way as in the calculation of maximum demand.

2. Weather is specified in terms of "annual degree days" rather than design temperature range. "Annual degree days" is the summation over a year of the difference between 65°F and the mean daily temperature for all days with mean temperatures of less than 65°. The measure is based on the generally observed fact that heating is not used on days when the mean daily outside temperature is over 65°, and that the amount of heat used on colder days is almost directly proportional to the difference between 65° and the mean outside temperature for the day.

The annual heat requirements are equal then to the volume to be heated (as derived from population density), times the annual degree days, times the amount of heat required per cubic foot of volume per degree per day. As we noted earlier, although the latter figure varies widely for different buildings, we have assumed a constant value for it.[9]

Graph 2 summarizes our estimates of annual heat requirements as determined by population density and annual degree days. To illustrate the use of the graph, let us again take the city of Boston, where the population density in 1940 was almost 17,000 persons per square mile. The degree days experienced per average heating season are about 6,000. For this combination of annual degree days and population density Graph 2 specifies for a 4-square-mile area an annual heat requirement of about 12 hundred million lbs. of steam. Given the appropriate figures on the annual degree days and population density for any other city, the graph will similarly specify the average annual heat requirements for a 4-square-mile area.

c. THE COST OF HEAT DISTRIBUTION. Having formulated in Graphs 1 and 2 the basic relationships determining plant capacity, plant cost, and annual requirements, we are now in a position to estimate the cost of delivering steam for any city from data on its population density, design tem-

[9] See Chapter XII, B.l.a.

perature range, and annual degree days.   From the nature of the estimating procedure, it is clear that the costs thus derived may be wide of the mark for any single city since it may depart significantly from some of the "typical" values we have assumed in constructing the graphs.   But we believe that the relationships set forth in the graphs are useful for the general evaluation of the economic feasibility of district heating which we make in the concluding section of this chapter.

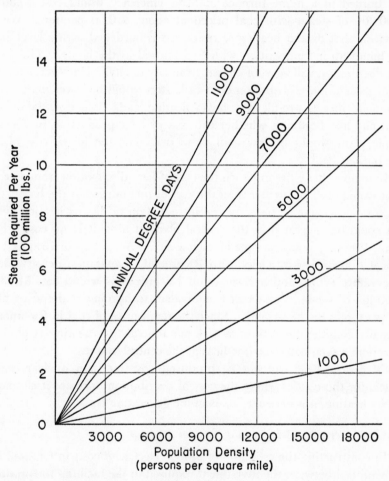

GRAPH 2.   Annual steam requirements for four square miles as determined by population density and annual degree days.

The steps by which the cost of heat distribution is estimated may be illustrated by bringing together the figures we have shown for Boston. Population density and design temperature range specified distribution capacity costing $8 million.   Assuming that 12% of investment represents annual charges on the distribution system, the yearly cost of operation is

$960,000.   Population density and annual degree days specified average annual steam consumption of 12 hundred million lbs.   The cost of distribution in Boston would therefore be $0.80 per 1,000 lbs. of steam.   And, similarly, distribution costs could be estimated for other cities.

The relative costliness of steam distribution becomes clear when the estimated distribution cost in Boston is expressed in terms of the equivalent cost of coal burned in the home.   Heat from an average grade of domestic coal burned in a home furnace at 50% efficiency would cost $0.80 per 1,000 lbs. of steam with coal priced at about $10.50 per ton.   We see, therefore, that district heating of residences is burdened with a large initial cost handicap in competition with fuel burned in the home.   The cost of heat distribution will vary, of course, from city to city.   However, according to our calculation, distribution costs in Boston would be lower than in most American cities as a result of its heavy heating load.   The essential question for a city like Boston is whether the cost of heat production in a central atomic plant would be low enough to overcome the heavy cost of heat distribution.

In many other cities in which the cost of heat distribution from a central plant would itself be greater than the cost of fuel burned in the home, there would be no economic basis for the introduction of district heating even if heat could be generated in the central plant at absolutely no cost.   Richmond, Virginia, is an example of such a city.   According to Graphs 1 and 2, heat distribution in a city with Richmond's population density, design temperature range, and average annual degree days would cost $1.83 per 1,000 lbs. of steam.   This cost is equivalent to burning coal costing about $24 per ton in the household.   Since the retail price of coal in Richmond is normally considerably less than $24 per ton, district heating is obviously uneconomic even with cost-free heat at the central source.

Of course, heat will not be produced cost-free in a nuclear reactor.   Hence, for those cities where the cost of distribution does not itself exclude district heating, it is necessary to examine generating costs.

2. Generation of Heat in a Nuclear Reactor

In constructing the estimates of atomic electricity costs in Chapter I we distinguished between the investment required in the facilities for producing nuclear heat and in the facilities for converting the heat into electricity. The investment required in the heat-producing facilities will now be used to estimate the cost of generating heat in an atomic-powered district heating system.

For our calculations it is most convenient to express the investment in the central heat plant for a district heating system in terms of the dollars of plant investment required, on the average, to produce 1 lb. of steam for 1 hour.   We have, in the other industry chapters, assumed that the borderline of realizable atomic costs might fall somewhere between the minimum and intermediate cost estimates made in Chapter I.   The investment in heat-producing facilities corresponding to this is about $8 per lb. of steam demand per hour.   This figure will serve as the basis for the estimated cost of heat generation in a nuclear reactor.   Although we believe that nuclear energy may be available at this cost, or at even lower cost, it is also possible that costs may run higher.   It should be understood that even the limited importance which atomic-powered district heating might have at this cost would not materialize if costs turn out to be substantially higher.

We may complete the cost calculations begun in the preceding section. Again, we will use Boston as an example.   The capacity required to serve 4 square miles was found to be (on the basis of Graph 1) 53.5 hundred million lbs. of steam per year.   The maximum capacity per hour is, there-fore, 53.5 hundred million lbs. of steam divided by 8,760 (hours per year), or about 610,000 lbs. per hour.   At $8 per lb. of steam capacity per hour the investment in generating capacity would be about $4.9 million (com-pared with $8 million invested in the distribution system), and at 12% the annual charges would be about $590,000.   For Boston's estimated yearly consumption of 12 hundred million lbs. of steam per 4 square miles (on the basis of Graph 2), this is equal to about $0.50 per 1,000 lbs. of steam, compared with $0.80 per 1,000 lbs. as the cost of distribution.   The total cost of atomic-powered district heating in Boston would, therefore, be $1.30 per 1,000 lbs. of delivered steam.

The steps we have followed in estimating the cost of district heating in Boston may be repeated for any area for which the requisite data on popula-tion density and weather conditions are available.   In the final section of this chapter we shall use the same procedure and data to reach some general conclusions on the economic feasibility of atomic district heating in this country.   But before doing this it is necessary to clear up a technical ques-tion to which we have already briefly alluded.   This is the question of heat losses.

3. The Significance of Heat Losses

The literature on district heating usually devotes much space to the subject of heat losses, because such losses increase operating costs, mainly on account of fuel wastage.   Over a 4-square-mile area, for example, heat losses on delivering steam from the central plant to the consumer might amount to about 15% of the annual flow.   Obviously, with conventional fuels heat losses would add considerably to fuel costs.

As we have seen, the cost of fuel probably will be an insignificant item in the cost of producing atomic power.   The major reason for stressing the importance of heat losses in district heating, therefore, is eliminated.   What remains is the effect of heat losses on plant capacity.   Here the relevant measure is not heat losses as a percentage of annual heat flow, but as a percentage of capacity flow.   Steam distribution losses are almost constant in absolute amount regardless of the volume of steam passing through the pipes.   Consequently a loss of 15% in terms of average annual flow would be the equivalent of about 4% loss in terms of peak capacity (i.e. utilization, on the average, is at 25% of capacity).

Plant capacity in a 4-square-mile district heating system should, therefore, be increased by about 4% above peak requirements to compensate for heat losses in distribution.   For an area of this size about 75% of all the losses would occur in distributing the steam within the consumption area and on the consumer's property, and the remaining 25% in the trunk lines bringing the steam to the consumption area.   We have allowed for both types of losses in deriving the capacity estimates shown in Graph 1.

As the size of the distribution area becomes larger, heat losses would increase at a rate of about 2% to 3% of annual flow for each mile of additional trunk line.   Clearly, therefore, where fuel is a large factor in cost, unit costs would tend to increase noticeably as the service area was broadened.   But plant capacity increases on this account would be considerably more moderate—roughly 0.5% to 0.75% for each mile of additional trunk line.   If, therefore, we had assumed, say, a 16-square-mile area instead of an area of 4 square miles, costs would have increased by only a few per cent in those localities in which atomic-powered district heating appears to be economically feasible on the basis of our calculations.

C. THE ECONOMIC FEASIBILITY OF ATOMIC-POWERED DISTRICT HEATING

The materials we have presented on the economics of district heating will now be used in a general appraisal of the possible future importance of nuclear heating of residential districts in the United States.   We believe the data are more useful in such a general appraisal than they might be in making a similar determination for a particular locality.   However, since a limited number of localities will dominate our analysis, there is even here a serious question concerning the correctness of some of the assumed constants for these localities.   Furthermore, we have not examined the data on population density, which constitute one of the keystones in our analytic structure, in sufficient geographic detail for us to do complete justice to the problem.

Our approach to the problem is along these lines.

(1) On the basis of data on retail fuel prices in representative cities in the winter of 1945–1946, we select a range of prices which appears to be typical for reasonably good household fuel.   The range is between $15 and $17 per ton of coal equivalent.

(2) We assume, further, that the average furnace burns fuel at 50% efficiency, i.e. that only one-half of the energy contained in the fuel is made

available as useful heat.   On this basis, 1 ton of coal would yield about
13,250 lbs. of steam.   This estimate is about right for present installations
in this country, and although efficiency could be increased, the cost of the
necessary equipment might make greater thermal efficiency uneconomic.[10]
Given the retail price of coal in a locality, therefore, we can determine the
cost of heat in the home.

(3) By making the appropriate calculations from Graphs 1 and 2,
described in the preceding section, we find out what combinations of population density, design temperature ranges, and annual degree days would
result in atomic district heating costs equal to or below ordinary heating
costs in the locality.

(4) Finally, we select those cities which appear on the basis of step (3)
to provide the necessary conditions for atomic district heating of residential
areas.[11]

(5) In the above comparisons we ignore an additional cost of about $15
per year per dwelling served by district heating for the necessary service
connections.   This is probably no more than the annual depreciation on a
furnace, but if the furnace is already installed, the owner cannot recover its
cost.   In this case we have assumed that the advantages of clean automatic
heat and the saving in basement space resulting from district heating might
balance the additional cost of the service connections.

On the basis of the steps just outlined, we find that the following general
conditions provide a rough guide to the selection of areas in which nuclear
heating might be no more costly than conventional methods.   (1) In a city
such as Duluth, Minnesota, with about the coldest winter conditions experienced in a populous locality in the United States, nuclear district heating
would not be economically feasible, in general, unless the population density
was at least 10,000 persons per square mile.   (Minimum population density
has been calculated in this and other instances mentioned here on the
assumption of 100% subscription of residences to district heating.   Non-
residential steam users are excluded completely.)   But all cities with an
average population density of 10,000 or above are not characterized by such
extreme winters, so that the minimum population density required for economic nuclear heating in the major population centers in the United States
is above 10,000 per square mile.   (2) Weather conditions experienced in
densely populated American cities are such that only with population densities of at least 13,000 per square mile would nuclear district heating be
economically feasible. However, the precise density of population needed
will depend on the design temperature range and the average annual degree

[10] "How to Heat a House," op. cit.

[11] Population density figures based on U. S. Bureau of the Census, Census of 1940,
Population, Number of Inhabitants, Vol. I, Washington, G.P.O., 1942.   Weather conditions from various sources, including Strock, op. cit.; Handbooks of the National District Heating Association; Statistical Abstract of the U. S., etc.

days.   In general, these two variables are so related that only densities above 15,000–16,000 per square mile would justify the use of atomic-powered district heating systems in the North (and then not always), while their use probably would never be justified at densities prevailing in localities experiencing mild winters.

What places within the United States satisfy the conditions required for the commercial success of nuclear district heating?   Obviously, a definitive appraisal of the prospects for residential district heating from atomic sources really cannot be made without examining detailed intra-city data on population density by small districts within all urban centers.   This job probably could be done, but only with the investment of a much greater effort than we could devote to the task.   City-wide averages are therefore the basic population density figures we have used, and we attempt to give a balanced appraisal of their significance.

If we select only those cities in which the city-wide population density is high enough (in conjunction with weather conditions) to make district heating competitive, we are likely to err in two respects: (1) there undoubtedly will be sections within these cities where density is below the average required for district heating, and (2) there undoubtedly will be sections in other cities where population density will exceed the city-wide average so that district heating might be commercially feasible.

The average city-wide population density is high enough, taken together with weather conditions, to bring nuclear district heating into competition with ordinary heating methods in the following cities: Boston, Buffalo, Chicago, Milwaukee, New York, and cities in the Metropolitan New York area including Newark, Paterson, and Jersey City.   The cities in the Metropolitan New York area qualify mainly because of their extreme concentration of population, the others because of a combination of dense population and cold winters.   Milwaukee has the lowest population density and the coldest winters of all the cities listed.

This is a highly tentative listing of cities.   Actually there is no way of knowing with certainty whether conditions in a particular city are favorable for district heating without making a detailed engineering study of the area. We have assumed certain typical conditions for the cost of laying pipe, the nature of the structures to be heated, etc.; naturally the feasibility of district heating in the above-named cities depends on whether the assumed conditions actually apply to them.

If all other questions are resolved satisfactorily, the problem resulting from the use of city-wide averages of population density still remains. Somewhat more than 10% of the population of the United States live in the cities mentioned above (13.5 million persons according to the 1940 census), and their relative importance as consumers of heating fuel is even greater because a considerable percentage of our people live in mild regions where relatively small amounts of heating fuel are used.   Clearly, however, the total heating load in these cities cannot be taken as a measure of the possible extent to which nuclear district heating might be used in the United States because of the inadequacy of our population density figures.

For New York and Chicago, we have examined population density figures for units smaller than the entire city.   The New York figures are for the five counties which make up Greater New York, while the Chicago figures are for "type-of-structure" areas within the city.[12]   Although averages for the smaller units are preferable to city-wide figures, they are not completely satisfactory for determining the percentage distribution of the city's population among areas which do, or do not, have a sufficiently dense population to justify nuclear district heating.   This is particularly true for New York because county-wide averages still encompass large heterogeneous areas.   According to these figures, 1.5 million persons (i.e. the 1940 population of Queens and Richmond) out of New York's 7.5 million people and 366,000 out of Chicago's 3.3 million people live in districts where the population density falls below that required for nuclear district heating. This suggests that if population density figures could be broken down for smaller units in the other cities mentioned, a significant part of their population would be found to live in sections below the required density.

The density figures for smaller areas show that population density varies sharply among different parts of the same city.   In New York, persons per square mile range from 3,000 in Richmond County to 86,000 in New York County; in Chicago the variation is from 8,500 in "single-family structure" areas to 64,000 in "tall apartment" areas.   It is doubtful that many other cities would show as much variation as New York and Chicago, because in few others would "tall apartment" areas be of nearly the same importance as in these two cities.   But density varies greatly even among the types of structures commonly found in other cities, as these Chicago figures indicate:[13]

"Type-of-structure" areas	Density per square mile
One-family structures	8,629
Mixed two- and one-family structures	26,652
Mixed structures, mostly apartments, two- and one-family structures	35,793

Such sharp intra-city variations in population density strongly suggest that the selection of cities on the basis of average population density understates the number of cities in which nuclear district heating might be used, for it is almost certain that those in which the city-wide average is not far

[12] A "type-of-structure" area is an area where a particular kind of residential structure (e.g. 1-family structure, 2-family structure, tall apartment, etc.) or a particular combination of types is characteristic.   (*Master Plan of Residential Land Use of Chicago*, The Chicago Plan Commission, Chicago, 1943, p. 34.)
[13] *Ibid.*

below that required for economical nuclear district heating contain sections where the density would be satisfactory. The cities we have listed might best be designated as cities which are *known* to have districts suitable for nuclear heating (on the basis of our assumptions), while many other cities *probably* have districts which would be suitable.[14]

These results must be qualified in several other respects. Average density of population, as shown in the census, is derived by dividing a city's total population by its total land area. Since in large urban centers like those we are considering a substantial portion of the city's land is devoted to such non-residential uses as office buildings, stores, factories, railroad yards, etc., the averages clearly understate the average density of residential sections. The existence of undeveloped areas within the city limits likewise results in understatement.

We might, of course, have included steam consumed by industrial and commercial establishments in our calculations, but we chose not to do this because of great variations in the pattern of industrial and commercial activity among the cities which figured prominently in our analysis. By excluding these uses we have seriously understated urban steam requirements, since these activities require relatively large amounts of steam.[15] Clearly, therefore, both because (1) our residential population density figures are too low for determining straight residential consumption, and (2) straight residential consumption understates a city's steam requirements, we do not make as strong a case as is warranted for district heating.

Any understatement of the possible economic feasibility of nuclear district heating is balanced to some extent by our assumption that the heat plant is located in the center of the area to be served. Such a location might often prove impossible because of the excessive cost of bringing large quantities of water there. In addition, until more is known about the safety factors involved in operating nuclear reactors, we must consider the possibility that it will be judged hazardous to place them in the center of populated districts.

[14] Since this chapter was written, Mr. William H. Ludlow of the Program in Planning of the University of Chicago has brought to our attention certain figures supporting the belief that there will be districts in a large number of cities in addition to those we list, where population density would be high enough to justify nuclear district heating, on the basis of our cost calculations.

Mr. Ludlow also points out that nuclear district heating might be important in the redevelopment of blighted urban areas. "In Chicago, for example, there are over 20 square miles of blight or near blight that require rebuilding by large scale operations. Densities in such rebuilt areas may approach 30,000 persons per square mile. . . . In any event, nearly all such urban redevelopment in all our large cities will be considerably above the critical ratio [we have derived] for atomic-powered district heating." (Quoted from a letter from Mr. Ludlow, January 31, 1949.)

[15] This point is borne out by comparing the actual steam requirements, including non-residential uses in certain districts in Manhattan served by the New York Steam Corporation, with the requirements we would estimate for the same districts for residential

It seems reasonable to conclude, from the quantitative and qualitative aspects of the analysis set forth above, that nuclear heating might make noticeable inroads into the space-heating market satisfied at present by ordinary fuels and conventional heating equipment. The adoption of nuclear-powered systems for heating city districts would probably be confined to a limited number of northern cities in which the density of population is very high.   On the other hand, the cost reductions which would result from the substitution of nuclear heat for conventional fuels would probably be quite modest.   Actually, the analysis shows nuclear heat costs and ordinary heating costs to be about the same in those places where nuclear heat can compete at all.   We assume that under these circumstances there might be some shifting to nuclear heat either because of slight cost advantages, or because of the convenience involved in subscribing to a central district heating system.   All of this is on the assumption that nuclear heat will be generated at a cost roughly midway between the minimum and intermediate cost estimates derived in Chapter I.   If actual costs should prove to be higher, say in the neighborhood of our intermediate estimated atomic cost, nuclear district heating would be somewhat more expensive than ordinary heating methods at the average population density prevailing in all the cities named above except New York.

---

use only. Our estimate comes to about one-fourth the actual requirements. As a result of the heavier demand, the unit steam distribution costs would be considerably less than those we estimate. (This comparison is based, in part, on figures supplied by Mr. W. F. Davidson of the Consolidated Edison Company of New York.) Of course, the relative importance of non-residential steam demand in Manhattan is unusually high.

Part Three

ATOMIC POWER AND ECONOMIC DEVELOPMENT

CHAPTER XIII

The Effects of Atomic Power
on National or Regional Economies

In Part One of this study we made an analysis of the economic and technological factors involved in the production of electricity through the application of atomic power. In Part Two we examined the effects of the availability of low-cost heat and power upon the productivity, location, and output of specific industries. But the significance of a technological change such as that we are studying is not limited to its direct influence upon power-consuming industries; it has important repercussions on many other parts of the economy. In Part Three we shall attempt to assess the character and magnitude of the direct and indirect cumulative effects that may be initiated by the availability of cheap atomic power.

Our discussions in the next two chapters will fall into four main sections. In the present chapter we shall examine:

(1) The effect of cheap power upon the *real income* (real net national product) of a whole economy. Here we shall consider both the long-run and the short-run effects of the cheapening of power, giving some attention to the effect of technological change upon the degree of employment of labor and other resources.

(2) The effect of ubiquitous atomic power upon the *location* of industry and population, and hence upon the development of particular regions within an economy. Although a technological change may produce only a moderate increase in the productivity of an economy as a whole, it may have much greater effects upon particular regions through the relocation of industry.

The following questions will be discussed in Chapter XIV:

(3) The possible role of atomic power in facilitating the *industrialization of "backward" areas which lack other power sources*. This section will be prefaced by a brief analysis of why and how such industrialization takes place. We will then explore the question of whether industrialization has been hampered more by the scarcity of power or by the scarcity of other factors, especially capital.

(4) The possible role of atomic power in facilitating the *industrialization of "backward" areas by reducing the amount of capital required* for a

219

given scale of industrialization.   In particular, we must examine the possibility of reducing the investment required for the means of transportation, and specifically for fuel-transporting railroads.   This leads naturally to the question of whether industrialization might be regionally decentralized.

Our analysis will be based upon the characteristics of atomic power previously described (Chapters I and II).   In certain areas atomic power may be cheaper than power generated from conventional sources, largely because of its "being available everywhere," its "mobility" or "ubiquity." The cost of transporting the atomic fuel necessary to produce a given amount of electricity in a given place is negligible compared with the cost of transporting coal or oil, or of transmitting hydroelectricity.   Another important characteristic of the cost structure of atomic electricity which distinguishes it from coal-based power (though not from water power) is that the cost of labor and the cost of fuel are relatively unimportant.

The cost estimates already set forth show that atomic power will not be much cheaper, if indeed it will be any cheaper, than old types of power at those sites which will continue to have sufficient supplies of cheap fuel or water power.   It might be anticipated, therefore, that atomic power will be of greatest relative importance (a) to those regions where, owing to natural limitations or even the exhaustion of conventional power resources, the supply cannot keep step with the expanding demand for power; and (b) to those parts of the world which are far from water power or from sources of cheap fuel.   The former areas are often industrially advanced.   The latter ones are, on the contrary, "backward areas"—primarily agricultural, with relatively slight industrial development, and with a low per capita income.

The factors to be considered are so numerous that we can only expect to obtain very general results predicting qualitatively—but hardly quantitatively—the effects of the introduction of atomic power.   We must realize that important changes, including industrialization in many parts of the world, are already taking place quite apart from atomic power; atomic power is only one of a large number of technological, social, and economic factors that will determine the rate of these changes.   Atomic power may, at most, accelerate some processes, overshadowed by such factors as the rate of capital formation in slightly industrialized areas, population changes, the rate of spread of technological knowledge and skills, and government policies in relation to industry and international trade.

It is always a temptation, of course, to underestimate the importance of a radical change in technology.   Few prophets were able to foresee what revolutionary changes would be produced by the internal combustion engine or electricity.   On the other hand, as prophets we have set ourselves a limited task.   In this and the following chapter, we are considering primarily just one of the many applications that may ultimately be made of atomic power—the production of electricity. If really revolutionary results occur they will probably come about through the invention of entirely new applications of atomic energy rather than simply through lessening the cost of electricity. Moreover, our analysis is confined to the effects that may appear during the first 25 or 30 years after the construction of a practicable atomic electric plant, a rather short time for the maturation of radical technological discoveries. Even within these limitations, the task of prophecy is extremely difficult.

A. POSSIBLE EFFECTS ON THE NATIONAL INCOME

We wish to estimate what change might take place in the income (net product) of an economy when atomic power becomes available alongside of traditional energy sources. By income we shall mean the aggregate of goods, services, and leisure that is available for consumption in the economy. The resources released by the introduction of cheaper power could be used to produce additional goods and services (and possibly leisure), that is, according to our terminology, to increase the national income. We may assign a money value to this increase in income by applying the money prices of the original period to the additional goods (and leisure) that could be produced.

1. Estimation of the Increase in Income

Our estimation process can be carried out in a series of steps, to be described briefly as follows:

Step I: We suppose that the resources liberated through the reduction in the cost of energy are applied to bring about an increase in all the goods and services comprising the national income, simply by expanding *proportionately* all the economic activities that are carried on.

Step II: Lowering the cost of energy would actually encourage a greater expansion of those economic activities which are thereby made cheaper (i.e. where the cost of energy is a relatively large part of the cost of final product) than of those economic activities in which the cost of energy is a relatively small part of the cost of product. Hence, as a second approximation, we make allowance for this increased demand for energy.[1]

Step III: Not only would a reduction in the cost of energy cause an expansion of energy-intensive processes already in use, but it would also bring into use new processes (and/or new locations) that had not been previously economical, and cause them to be substituted for present proc-

[1] We will express energy increases in terms of KWH of electric power. When we speak of an increase in energy consumption, we mean a *net* increase in all forms of energy taken together (e.g. waterpower, coal, atomic energy) and not an increase in atomic electricity, say, compensated by a decrease in energy from coal.

esses.   The overall increase in demand for energy would then be composed of the combined increases mentioned in Step II and Step III.

Part Two of this book provides examples:

Chlorine-caustic soda exemplifies a case where atomic energy would not induce changes in either location or process.   If applied, it would, at best, cheapen the product and possibly increase demand.   This corresponds to Steps I and II.

In most of the other industries studied in Part Two, atomic power, if cheap enough, might induce changes in either process and/or location, as follows:

(1) Aluminum: Change in location, but not in process.

(2) Phosphate fertilizer: Change in process, but probably not in location.   Here an energy-intensive process based on electricity would be substituted for a process using only small amounts of energy per ton.

(3) Cement, brick, and flat glass: Changes in process, but probably not in location.   In these industries one energy-intensive process would substitute for another—electrically-fired for gas-fired furnaces in glass, kilns heated by hot reactor gas for kilns fired by conventional fuels in cement and brick.

(4) Residential heating: Change in process.   Energy previously provided by individual home furnaces would be supplied by a central heating system using nuclear reactors.

(5) Iron and steel: Changes in location and process.   Here a process using coke both as a source of heat and as a chemical agent would be replaced by a process using electrolytically-produced hydrogen.   The new process would economize on transportation through shipment of sponge iron rather than iron ore.

(6) Railroads: Change in process.   Here electric locomotives would substitute for diesel locomotives.

The likelihood of changes in each of these industries is analyzed in Part Two.   Changes, where possible, would correspond to Step III above.

We shall now try to gain a very tentative impression of the orders of magnitude of the effects on national income involved in Steps I, II, and III.

STEP I.   Assuming first that all economic activities will expand proportionately, let us estimate the increase in the national income of the United States that might be induced by the availability of cheaper energy.   The national income in 1946 was 180 billion dollars.   This figure may be used as a measure both of the goods and services consumed in that year, and of the resources used in producing them.   In the same year 191 billion KWH of electric energy were purchased by various users, and 46 billion KWH were produced by industrial establishments for their own use—a total consumption of around 240 billion KWH.   Hence, each reduction of one mill per KWH in electric generating costs would save 240 million dollars, or 0.13% of the resources used to produce the 1946 national income.[2]   That is, the goods actually produced in 1946 could have been produced with 240 million dollars less of resources than were actually used.   If the introduction of atomic power reduced average generating costs by 2.5 mills per KWH, the total saving of resources would amount to about 600 million dollars at 1946 prices of labor and other resources.

If the 1946 national income could have been produced by 180.0 − 0.6 = 179.4 billion dollars of resources, and if 180 billion dollars of resources were actually available, then we may estimate that the total income, with cheaper power in the proportionately expanded economy, would have been about 180/179.4 × 180 billion dollars, or 180,600 million dollars.   Hence, the increase in income attributable to the availability of cheaper energy would be 600 million dollars.[3]

STEP II.   We next take account of the fact that energy-intensive processes will tend to expand more rapidly than other processes.   Suppose that we knew that a 2.5 mill per KWH reduction in energy cost would increase the amount of electric energy demanded and produced from 240 billion KWH to 270 billion KWH.   (The question of how we can actually estimate the increased demand for energy will be taken up later.   For the moment we arbitrarily assume a hypothetical figure.)   Note that under the assumptions of Step I, electric energy production would increase from 240 to 240 × (180/179.4), i.e. to 240.8 billion KWH, an increase of only 800 million KWH.   Now suppose that before the cheaper power was introduced, energy cost an average of 1 cent per KWH.   The 270 billion KWH of energy it is now possible to produce at a cost of 2,025 million dollars would have cost 2,700 million dollars at the old price.   The price reduction is therefore equivalent to adding resources worth $(2,700 − 2,025) = 675$ million dollars, or increasing the national income by the same amount.[4]

---

[2] U. S. Bureau of the Census, *Statistical Abstract of the United States, 1948,* Washington, G.P.O., 1948, Tables 306, 539, 534.   Figures on the power consumption and national income of 1948 result in the same percentage.

[3] Let p_1 and p_2 be the cost of energy before and after the introduction of cheap energy, respectively; Y_1 and Y_2 the national income before and afterward; E_1 and E_2 the quantity of energy produced before and afterward; R the quantity of resources actually available (measured in dollars); and R' the quantity that would be required with the cheaper electricity to produce the income Y_1.   Further, suppose that k KWH of energy and l dollars of other resources are consumed per unit of national income.   Then $R = (p_1k + l)Y_1 = (p_2k + l)Y_2$, and $kY_1 = E_1$; and since $R' = (p_2k + l)Y_1$, we have: $R' = R − (p_1 − p_2)kY_1 = R − (p_1 − p_2)E$ and $Y_2 = \frac{R}{R'} Y_1$.

[4] Again let E_2 be the quantity of energy; let $L_2 = l_2Y_2$ be the dollar value of resources used

STEP III.   However, a reduction in energy cost would not only increase the use of pre-existing energy-intensive processes.   It would also bring into use processes that were not previously economical, and would cause geographical shifts from locations that were previously optimal to new locations that are optimal with low energy costs.   Several instances were cited in Part Two where process shifts or locational shifts, or both, would take place if energy were available everywhere at a sufficiently low price.   In the case of process shifts, new processes possibly using larger quantities of electric energy per unit of output, but using smaller quantities of all other factors taken together, replace previous processes.   In the case of locational shifts, sites previously handicapped by high energy costs but favored by low transportation costs (except for fuel) can now compete with sites having low energy costs (e.g. near coal mines or hydroelectric plants) but high transportation costs.

In the case of process change, no resources are released by shifting to the new process until atomic energy costs are reduced below the break-even point between the processes.   The resource saving will be less than proportional to the difference between the new energy price and the old energy price.[5] It is thus smaller than in Step II.

---

for other purposes than energy production, both measured after the introduction of cheap energy.   Let R be the resources actually available (measured in dollars), and p_1 and p_2 the initial and final prices of energy, respectively.

Let $\bar{R}$ be the resources (in dollars) that would have been required to produce E_2 and L_2 without the cheaper method of producing energy.   Then, by definition, $R = p_2E_2 + L_2$ and $\bar{R} = p_1E_2 + L_2$, whence $\bar{R} - R = E_2(p_1 - p_2)$.

[5] Let k_1, l_1 and k_2, l_2 be the amounts of energy (in KWH) and of other resources (in dollars) used per unit of national income with the old and new process, respectively.   The new process is more energy-intensive than the old; that is, $l_1 > l_2$, $k_1 < k_2$.   As before, Y_2 is the new national income.   With the resources R and $\bar{R}$ defined as in the preceding footnote, we have

$$R = Y_2(p_2k_2 + l_2); \qquad \bar{R} = Y_2(p_1k_1 + l_1).$$

Let $\bar{p}$ be the break-even price of energy; that is, $k_1\bar{p} + l_1 = k_2\bar{p} + l_2$; hence $l_1 - l_2 = (k_2 - k_1)\bar{p}$, and

$$\bar{R} - R = Y_2(p_1k_1 - p_2k_2 + l_1 - l_2) = Y_2[p_1k_1 - p_2k_2 + (k_2 - k_1)\bar{p}];$$

(1) $$\bar{R} - R = Y_2[k_1(p_1 - \bar{p}) + k_2(\bar{p} - p_2)];$$

also $\bar{R} - R = Y_2[(k_1 - k_2)(p_1 - \bar{p}) + k_2(p_1 - p_2)] = (k_1 - k_2)(p_1 - \bar{p})Y_2 + E_2(p_1 - p_2)$, where $E_2 = k_2Y_2$ is the amount of energy used after the change.   Since $k_1 - k_2 < 0$ it follows that

$$\bar{R} - R < E_2(p_1 - p_2)$$

as stated in the text.   In the special case where $k_1 = 0$ (i.e. if the old process uses no energy), we get from (1)

$$\bar{R} - R = E_2(\bar{p} - p_2),$$

i.e. in this limiting case the resource-saving is proportional to the difference between the new energy price and the break-even price of energy.

Examples of this limiting case would be provided by the change-over from gas to electrical ranges or by the introduction of the sponge iron process into the steel industry. It is true that in the general case, the resource-saving is larger than $E_2(\bar{p} - p_2)$ though, of course, smaller than $E_2(p_1 - p_2)$.   However, we can break down the reduction of energy price into two parts: from the old price p_1 to the break-even price $\bar{p}$, and from $\bar{p}$ further to the new price p_2.   Equa-

In the case of a locational shift, no resources are released unless energy costs at the new location are reduced below the break-even point of the two locations. In this case the resource saving is proportional to the difference between the new energy price at the new location and the break-even price of energy at that location (i.e. less than proportional to the difference between the new energy price and the old energy price at that location).[6]

Process shifts and locational shifts are illustrated by the hypothetical example given in Table 30. This table shows the unit costs for manufacturing a product by two processes, Process I requiring 1,000 KWH per unit of output, and Process II requiring 8,000 KWH per unit. The costs are also shown for two hypothetical locations, one with high transport costs, the other with low transport costs.

To illustrate process change, assume an initial energy cost of 10 mills at all locations. Under this assumption Process I would be used at the low freight site. The cost would be $67.00 per unit. If new energy costs were reduced to 4 mills everywhere, Process II would be used at the same site, with a cost of $54.00 per unit—a saving of $13.00 per unit. This saving is much smaller than the difference in cost with Process II at the old and new energy prices ($102 − $54 = $48). The break-even price of energy for the two processes is 5 mills.

To illustrate locational change, assume an initial energy cost of 10 mills at the low-freight site, and 5 mills at the high-freight site. If only Process II is known, production will take place at the high-freight site at a cost of

---

tion (1) shows that resource-saving is more strongly influenced by the second part than by the first part of the price reduction, since $k_2 > k_1$. If k_2 is very large compared with k_1 (i.e. if the new process is very electro-intensive compared with the old process), the bulk of the resource-saving will be proportional to the difference between the new energy price and the break-even price.

[6] Let p' be the old price of energy at the old location; let p_1 and p_2 be, respectively, the old and new price at the new location. We assume the production process to be the same but the transportation cost (except possibly for fuel) to be higher at the old site. Hence, the cost of production before the cheapening of energy can be expressed as $p'k + l_1$ for the old site, and $p_1k + l_2$ for the new site, where k is the amount of energy per unit of product and where $l_1 > l_2$ and $p_1 > p'$. If $\bar{p}$ is the break-even price of energy at the new (low-freight) location, $kp' + l_1 = k\bar{p} + l_2$, or

$$(2) \qquad\qquad l_1 - l_2 = k(\bar{p} - p').$$

Thus $\bar{p} > p'$. To produce the new output Y_2 one needs at the new site the resources measured by $R = Y_2(p_2k + l_2)$; at the old site one would need $\bar{R} = Y_2(p'k + l_1)$. The resource-saving is, by (2),

$$\bar{R} - R = Y_2[k(p' - p_2) + k(\bar{p} - p')] = Y_2 \cdot k(\bar{p} - p_2),$$
$$\bar{R} - R = E_2(\bar{p} - p_2), \text{ where } E_2 = kY_2.$$

Hence the resource-saving, if the price of energy at the new site were reduced to $p_2 < \bar{p}$ would be, as stated in the text, $E_2(\bar{p} - p_2)$. Since in general $p_1 > \bar{p} > p'$, we have

$$E_2(p_1 - p_2) > E_2(\bar{p} - p_2) > E_2(p' - p_2).$$

In Step II we use the first of these quantities; this yields a maximum estimate of the effect of relocation on income. On the other hand, $E_2(p' - p_2)$ would give an underestimate, and will even, in the case where $p' < p_2$, yield a negative quantity.

TABLE 30. *Effects of alternative processes and locations: hypothetical example.*

	Units	Process I							Process II					
Energy required	KWH/ton	1,000							8,000					
Energy cost per KWH	mill/KWH	15	10	7	6	5	4	3	10	7	6	5	4	3
Energy cost per ton	$/ton	15	10	7	6	5	4	3	80	56	48	40	32	24
Other cost per ton, except transport	$/ton	45	45	45	45	45	45	45	10	10	10	10	10	10
Total cost per ton, except transport	$/ton	60	55	52	51	50	49	48	90	66	58	50	42	34
Transport cost per ton														
High	$/ton	20	20	20	20	20	20	20	20	20	20	20	20	20
Low	$/ton	12	12	12	12	12	12	12	12	12	12	12	12	12
TOTAL COST														
High	$/ton	80	75	72	71	70	69	68	110	86	78	70	62	54
Low	$/ton	72	67	64	63	62	61	60	102	78	70	62	54	46

$70.00. If energy costs at the low-freight site are now reduced to 5 mills, the production will shift to that site at a cost of $62.00. The break-even price of energy at the low-freight site is 6 mills.

Finally, a reduction in energy costs may produce a simultaneous change in both process and location. Suppose the initial price of energy at the low-freight site is 15 mills, and at the high-freight site, 6 mills. Then Process I at the high-freight site will be optimal at a cost of $71.00. If energy costs at both sites are now reduced to 4 mills, production will be shifted to Process II at the low-freight site, at a cost of $54.00.

Hence, if we employ the method of Step II to estimate the change in income resulting from the introduction of atomic energy, and if part of the increased demand for energy in fact results from changes of the sort discussed in Step III, we will have overestimated the increase in income. In any actual case of a reduction in energy cost, the resulting increase in demand for energy will certainly be due to the kind of changes described both in Step II and Step III. We may conclude that the estimation method derived in Step II gives us an upper limit of increase in income resulting from a decrease in energy cost; the actual increase will be somewhat less.

2. *Estimation of the Increased Demand for Energy*

Our calculations, in Step II and Step III, of the increase in national income resulting from a reduction in energy cost depend upon estimates of how great an increase will take place in the quantity of energy used in the economy. We shall discuss the effects of cost reduction upon the industrial as well as the residential use of energy in two ways: first, by trying to estimate the impact of this effect in the past and to apply it to the future; second, by considering the possibilities of this effect being greater than in the past, owing to some spectacular changes in methods and locations made possible at a lower level of energy prices.

Data on manufacturing industries in 44 states in 1939[7] show that energy purchased per wage-earner increased almost proportionately to the decrease in the average price paid for the energy. That is, in states with an average cost for purchased energy of 12 mills per KWH, the average quantity consumed per worker in manufacturing industries was about 5,400 KWH, while in states with a cost of 6 mills per KWH, the average consumption was about 11,000 KWH, or twice as great. If one disregards the interstate variations in annual wage rates, one can conclude that in manufacturing industries the elasticity of demand for energy is about 1, for a region of the size and type of an American state. This elasticity is the ratio between the percentage decrease in the price of electricity and the consequent percentage increase in the demand for electricity. We assume the price of labor and the quantity of labor employed to be unaffected by the price change in electricity. The figure was obtained without including four mountain states, Arizona, Montana, Nevada, and Wyoming. If they are included, the figure rises to 1.3 owing to the role of ore refining plants in those states. On the other hand, the elasticity of demand thus derived refers to electricity, not to energy as a whole. The elasticity of demand for energy must be somewhat lower than that for electricity; the latter is affected by the substitutability between electrically-driven and steam-driven transmission belts, between electric and thermal processes in chemical industries, etc.

Also, the interstate variation in energy sales to manufacturers may occur largely because those industries which consume considerable quantities of energy were attracted to states where the energy cost (or, more strictly, the ratio of the energy cost to wage-rates) is low, compared with other states. The cheapening of electricity in one state would raise its demand for electricity, while possibly lowering the demand for electricity in some other state. If the cost of electricity were reduced everywhere, the total effect would probably be considerably smaller than that calculated above.

Consequently, the elasticity of demand for energy for the manufacturing industries of the United States as a whole is very probably smaller than 1. We shall nevertheless assume for a moment that it is 1, in order to estimate the maximum effect of electricity cost reduction. American manufacturing industries in 1946 used 122 billion KWH, and extracting industries another

---

[7] Gordon, *op. cit.*, pp. 174–176. Based on the *U. S. Census of Manufactures, 1939, op. cit.*

12 billion KWH; the average cost of electricity to these users was about 1 cent per KWH.[8]   Suppose that the cost of generating electricity is reduced by 2½ mills and that this cost saving is fully transmitted to each user.   The user's cost is then reduced from 10 to 7½ mills, i.e. in the proportion 4:3. Suppose also that wage-rates remain constant.   (More generally, the same effect is obtained whenever the ratio of the user's electricity cost to the wage-rate is reduced in the proportion 4:3.)   Then, if the elasticity of demand for electricity is 1, the industrial demand will increase in the proportion of 3:4 or by about 45 billion KWH.

We now proceed to the non-industrial demand for energy.   A total of 103 billion KWH were consumed in 1946 outside of manufacturing and mining: 39 billion by residential customers, the remainder by "rural" customers (i.e. those who payed distinct rural rates), commercial users, railroads, etc.   These groups paid on the average about 2.5 cents per KWH.[9] Suppose again that a fall of 2½ mills is transmitted to the user, thus reducing the price to 2.25 cents (assuming that distribution costs are unchanged). Suppose we had reason to believe that the usual response to a price fall is to increase the quantity demanded in the same proportion as the price change, i.e. assume the "elasticity" of this demand to be 1.   Then a price fall in the proportion 25:22.5 would cause a demand increase of $(103 \times 25/22.5) - 103 =$ about 11 billion KWH.   Should demand elasticity be 2 instead of 1, demand would increase 1.23-fold or by 25 billion KWH.[10] What can we learn from the past about the actual elasticity of the non-industrial demand for electricity?   It suffices, in Table 31, to consider residential customers only.   For other non-industrial consumers of electricity the relative price fall may be larger but the elasticity of demand is probably smaller than in the case of residential customers.

If we compare the change in consumption per head of population or per household (lines 2 or 3) with the change of price (line 5), we find a demand elasticity of about 4 for the period 1913–1930.   But in the period 1930–1948 the demand elasticity was less than 2; it was, in fact, as low as 1 if one takes account, as one must, of the increase in real income per capita that took place in this period (line 7).   We can be sure that the pattern of 1930–1948 carries more interest for any future estimate than that of 1913–1930, since the lower sensitivity to price changes which is plausible at high income levels is likely to persist unless we expect a catastrophic fall in income.

[8] *Statistical Abstract of the United States, 1948, op. cit.*, Tables 929, 539.   The price paid by "large light and power" buyers we assume to be also the cost per KWH of the 46 million KWH of electricity produced by manufacturing plants for their own use.

[9] Average price paid by customers other than "large light and power" buyers.

[10] By definition, if the demand elasticity is η, then the ratio of the new (y') to the old (y) quantity is $y'/y = (p/p')^\eta$ where p' and p are the new and old prices, respectively.   Hence, when $\eta = 2$, $y'/y = (p/p')^2 = (25/22.5)^2 = 1.23$.

TABLE 31. *Changes in the residential consumption and price of electricity, and in national income, 1913–1930–1948.*

	1913	1930	1948	Proportion of rise or fall	
				1913–30	1930–48
1 Annual sales to residential customers, KWH per customer	264[a]	547[b]	1,563	1:2	1:3
2. Same, KWH per capita of population	9	90	350	1:10	1:4
3. Same, KWH per household	41	367	1,275	1:9	1:3½
4. Average price paid by residential customer, cents per KWH	8.7[a]	6.0	3.0		
5. Same, deflated by cost of living (1930 = 100)	14.7	6.0	2.1	2½:1	2¾:1
6. Annual income per capita of population, $	358[c]	560[c] 610[d]	1,540[d]	1:1	1:1¾
7. Same, deflated by cost of living	606[c]	560[c] 610[d]	1,080[d]		
8. Proportion of residential and rural customers to the number of households	.15	.42	.93		

Sources (in addition to those mentioned in the footnotes below): *Statistical Abstract for the United States, 1948, op. cit.; Survey of Current Business,* U. S. Department of Commerce, Vol. 29, 1949; Edison Electric Institute, *Statistical Bulletin, 1948,* New York, June 1949.

[a] *National Electric Light Association, Statistical Bulletin, 1931,* New York, July 1932, Tables II, IV, farms included.

[b] *Ibid.,* eastern farms included.

[c] Old Series of Department of Commerce, from *The Economic Almanac for 1949,* National Industrial Conference Board, New York, 1948, p. 77.

[d] *Ibid.,* New Series of Department of Commerce.

In spite of the fact that 1948 consumption was probably depressed by shortages of new homes and of electrical applicances in the preceding years, the sales of electricity per domestic customer increased more rapidly from 1942 to 1948 than from 1938 to 1942.   We conclude here by estimating the future elasticity of non-residential demand for electricity at not more than 2.

This estimate is not too moderate.   The very rapid increase in residential use of electricity over the past two generations can be attributed only in small part to the reduction of electricity prices.   More important has been the increase in the number and proportion of electrically wired homes and in the number of electrically operated household devices.   The wiring of American households which previously had no electricity is a process that rapidly approaches its end; this is but imperfectly indicated on line 8 of Table 31.[11]   As to the increase in the number of electrically operated devices, we refer to Table 32.   It presents an estimated distribution of resi-

[11] Customers are identified by bills or electrical meters and are thus not identical with the "household" units of the census.

dential electricity consumption among its various uses in 1942, the last year for which data are available.

TABLE 32.   *Residential use of electricity.*

	Per cent saturation 1/1/43[a]	1/1/47[b]	KWH per year per appliance[a]	Total annual use billion KWH 1942[a]	100% saturation[c]
Lighting and miscellaneous	—	—		9.38	15.00
Refrigerators	71	69.1	300	6.05	8.52
Ranges	12	13.0	950	3.19	26.59
Water heaters	4	4.7	2,500	2.67	66.75
Radios and phonographs	100	90.4	80	2.18	2.18
Flatirons	100	91.5	63	1.74	1.74
Oil burners	8	—	185	.43	5.38
Washing machines	50	60.5	24	.32	.64
Vacuum cleaners	60	58.9	20	.33	.55
Clocks	55	61.7	14	.21	.38
Ironers	6	6.6	100	.18	3.00
All other appliances	—	—		1.15	30.00[d]
TOTAL				27.80	160.73

[a] Edison Electric Institute, *Statistical Bulletin, 1942*, New York, May 1943. "Per cent saturation" is an estimate of the percentage of the total number of households in the United States equipped with the appliance in question.

[b] *Electrical Merchandising*, Vol. 77, January 1947.

[c] Estimated from actual 1942 use and saturation.

[d] Assumed increase is principally electricity used for air conditioning.

In assessing the importance of the price of electricity in increasing its use in homes, we must distinguish two types of electrical devices: (a) those that replace non-electrical devices (e.g. electric refrigerators for gas refrigerators or ice boxes), and (b) those that provide some new service (e.g. radios and washing machines). In the former case, particularly if the capital investment in the electrical and non-electrical devices is approximately the same, the relative cost of operation may be the determining factor, although convenience, cleanliness, and fashion certainly play an important role. In the latter case, the decision to use the device will probably depend more on the capital investment involved than on the annual cost of current. For example, a three-hundred dollar electric ironer consumes perhaps two dollars of current per year.

Hence, the important effect of cheaper residential electricity would probably be the replacement of gas and other non-electrical appliances by electrical appliances. The consequent increase in domestic electricity consumption might be considerable. As the last line in Table 32 shows, if all households had used electrical appliances in 1942 for refrigeration, cooking, heating water, ironing, and the like, their consumption would increase 6-fold. (For a later year the figure would probably be smaller.) Compare this with our previous statement that if the electricity price to home consumers fell from 25 to 22.5 mills, and if the elasticity of their demand were 2, one would expect a 1.23-fold demand increase. This estimate of elasticity at 2 does not seem too conservative in the light of the above estimates of saturation. It is very unlikely that complete saturation of homes by electrical devices, or complete substitution of electricity for gas would occur simply because the electricity price to the consumer fell from 25 to 22.5 mills. Note in particular that the most important individual possibilities for increased demand for residential electricity are electric ranges and electric water heaters. In neither of these cases is electricity at present competitive with gas on a cost basis except in areas of unusually high manufactured gas prices and unusually low electricity prices. Furthermore the electric range involves a substantially higher initial investment than does the gas range of comparable quality. We conclude that, although there is a very large potential market for residential electricity above the present rate of consumption, the tapping of this market on the basis of competitive cost alone would require very large reductions in electricity prices, relative to the price of domestic gas. To estimate the demand rise that would result from a retail price fall from 25 to 22.5 mills, a demand elasticity of 2 would appear not too moderate a figure. (For commercial users this figure is probably too high.)

We can now estimate the probable upper limit of the overall income effect of a 2½ mill reduction in the wholesale price of electricity. We employ the method of Step II, above, under the assumption of a demand elasticity of 1 for industrial uses of electricity, and an elasticity of 2 for non-industrial uses.

We have seen that the total increase in national income attributable to the reduction in the cost of electricity cannot be greater than the amount of reduction (2½ mills per KWH) times the total demand for electricity *after* the cost reduction has taken place.

In 1946 consumption of electricity totaled 240 billion KWH. With the demand elasticities assumed above, we have seen that industrial consumption would be increased by about 45 billion KWH, and non-industrial consumption by about 25 billion KWH, so that total consumption would be about 310 billion KWH. The total increase in national income would then be no more than 310×2.5 million dollars, or approximately 775 million dollars. This is ⅖ of 1% of the national income.

Even this may be an overestimate, not only because the elasticity of demand for electricity is probably smaller than 1, but for two additional reasons. First, the probable decrease in the cost of electricity will be hardly as high as 2½ mills—the figure used in the analysis just made. Second, where the increased demand for electricity results from process or location changes we should, in estimating the resource-saving, use less than the entire reduction of 2½ mills. Hence, the income effect in this case should be expected to be smaller than the price reduction times the quantity of electricity consumed. This point was discussed under Step III of Section A.1.

On the other hand, our estimates may be too conservative for three reasons. First, they neglect relocation; second, they do not allow for the introduction of new unforeseen technological processes that may create an enormous demand for electricity. We shall term the latter possibility the "trigger effects," to be considered in Section A.3. Third, we have not taken into account that technological change may stimulate the employment of resources in the economy. If, prior to the change, resources were underemployed, additional income might be produced through the fuller employment of resources. While in Section A.4 we shall exclude such effects in our discussion of the economy as a whole, we will give them further attention when we come to discuss possible effects, through industrial relocation, upon the economies of particular regions. The underemployment may be due, in the country as a whole, to the slowness with which resources, especially labor, move from one place to another; when technological change creates new employment opportunities in a previously depressed region, the result may be a net gain in the employment of the nation. We shall take this up in Section B of this chapter.

3. "Trigger Effects"

It may be argued that past data on the elasticity of demand for electricity cannot safely be projected into the future because at some lower price of electricity important "trigger effects" may appear. By "trigger effects" we mean very large increases in the demand for electricity produced by small changes in its price and occurring in some new, possibly yet unknown, applications of electricity.

In answering this contention, we need not deny that lower electricity prices might bring about an enormous increase in the consumption of electricity, and that in certain price ranges the elasticity of demand for electricity might be much greater than our estimates from historical data. If electricity were cheap enough, for example, to give the sponge iron process a clear economic advantage over present iron-smelting methods, the consumption of electricity in the sponge iron process in the United States might be some 150 billion KWH annually.[12]

But our present interest is primarily in the effect of such changes upon the national income. Suppose the break-even price of electricity for the sponge process is 1½ mills below the present price. Then, as shown in footnote 5, a 2½ mill reduction in electricity price would lead to a 1 mill saving per

---

[12] See Chapter X.

KWH, or, in the event of a complete change-over of the steel industry to this process, an increase in national income of only $150 million (1 mill per KWH × 150 billion KWH).

In this estimate we have not taken into consideration a possible further increase in electricity consumption resulting from an increase in the consumption of the cheaper steel. The reduction in the price of the product will always be less, percentage-wise, than the reduction in the price of electricity, even after the break-even point between the two processes is reached, for the cost of electricity is only part of the total cost of the product. Thus, if a product is manufactured by an electricity-intensive process at $10 a ton, of which $5 a ton represents the cost of electricity, a 50% reduction in the price of electricity will reduce the cost of the product by only 25% (from $10 to $7.50). If the elasticity of demand for the product is 1, the 50% reduction in electricity cost would result in a 25% increase in demand for the product.

To state the matter more positively, income effects of very large magnitude would be expected only if the cheapening in electricity caused a very large increase in demand for the product (and not merely substitution of the electricity-intensive for the non-intensive process). But the large increase in demand for the product would result only if (1) its elasticity of demand were high, and (2) if, in addition, the cost of electricity were a very large part of its total cost of production.

The empirical studies of Part Two revealed few instances, even in the electro-chemical industries, where the cost of electricity is the major part of the total cost of the product, particularly when electricity is cheap. The applications of electricity to production that we can perceive, even on a fairly distant and vague horizon, would all seem to require substantial inputs of factors other than electricity. The industrial application of heat and electricity requires relatively elaborate structures and equipment—e.g. furnaces or electrolytic cells—that involve a substantial capital investment per unit of output.

Let us consider again the example of the sponge iron process. It is estimated (Chapter X) that the cost for capital (interest, depreciation, and maintenance) per ton of output would be $4.60. Assuming there were no other costs whatsoever, the total cost of sponge iron, with electricity at 4 mills per KWH (2,440 KWH per ton of iron) would be $14.36 per ton, of which $9.76, or about 68%, is the cost of electricity. With electricity at 2 mills, the total cost would be reduced to $9.48 per ton, a reduction of about 35%. Reduction of the price of electricity from 2 mills to 1 mill would reduce the cost of sponge iron from $9.48 to $7.04, about 26%; reduction of the price of electricity from 1 mill to ½ mill per KWH would reduce the cost of sponge iron from \$7.04 to \$5.82, about 17%. We see that each successive halving of the price of electricity brings about a smaller percentage reduction in the price of sponge iron.

The chapters of Part Two illustrated a number of possibilities of "trigger effects," so far as increases in the demand for electricity are concerned. But our present analysis shows that strong "trigger effects," so far as national income is concerned, would be produced only by the discovery of some process that used electricity in very large quantities per unit of output, and the other factors of production in very small quantities per unit of output. Since we have no reason to expect that new applications of electricity will not require a substantial capital investment per unit of output (quite apart from the investment required to produce electricity) the prospect is not at all bright for revolutionary changes in technological process that would completely invalidate our methods of estimating possible income effects.

4. Economic Assumptions Involved in Estimates of Increased Income

Two assumptions underlying the previous analysis require closer examination. First, it is true that the release of resources through the cheapening of power increases the productive potential of the economy. However, whether this higher potential will actually lead to an increase in production depends on how fully the economy's resources are employed before and after the technological change occurs. If the principles of classical economic theory are correct, price flexibility will bring about full employment of the factors of production both before and after the technological change has occurred. However, it is at this very point that the classical theory has been severely criticized in recent years by those who argue that price flexibility, and particularly flexibility in money wages, may not bring about full employment, and that economic equilibrium is quite consistent with underemployment.

In the second place, even if electricity can be produced from nuclear fuel more cheaply than from coal, it will probably not be economical immediately to scrap the existing generating plants and to replace them with atomic plants. Hence, the full effects of this, as of any other technological changes, will appear only over a considerable time. While actual physical depreciation of the existing plant may be slow, extending over several generations, the change-over may be hastened if new facilities must be installed because of an increase in the aggregate demand for power or in the demand of certain geographical areas.

The succeeding 4 sections will be devoted to a further exploration of these assumptions: that resources will be fully employed, and that the new technology will actually replace the old without delay.

a. THE FULL EMPLOYMENT OF RESOURCES. A technological improvement in the power and heat industry—the substitution of nuclear fuel for other sources of energy—permits an increase in the real income of an economy because new resources are released, either in the power and heat industry itself, or in industries that directly or indirectly consume power and heat. But are we safe in assuming that those resources which are released will be used? Is it not conceivable that they will simply be wasted— unemployed?

It has long been noted that a technological advance in a particular industry could produce "technological unemployment" in that industry. In the bituminous coal industry, for instance, the relative increase in coal consumption, even with a growing population, has not kept pace with the displacement of miners through the increased mechanization of mining operations. Similarly, the invention of the tractor resulted in the technological unemployment of farm horses and pasture land.

Unless there is a compensating increase in the quantity of production, a technological advance in an industry will always lead to reduced employment of at least some of the factors of production in that industry. As we pointed out in Chapter I, the direct investment per kilowatt in an atomic power plant probably will be greater than in conventional thermal power plants, but the quantity of labor per KWH will be smaller (since labor is the principal component in coal production costs).

All of this would be relatively inconsequential (although not without importance to the particular occupations affected) if we could be sure that the displaced labor would be absorbed by the increased power consumption or the increased production in other industries. As we mentioned above, if classical economic theory is correct this absorption will actually take place, while if the newer Keynesian theories are correct it need not take place unless the government takes adequate steps.

Since it does not appear possible to estimate the magnitude of unemployment and since this magnitude would be affected by now unknown governmental policies, we must limit ourselves to estimating the potential productivity of the economy, i.e. the national income that *could* be produced if all resources were employed.

However, one cause of the unemployment of national resources is their imperfect mobility. Especially labor is often slow in abandoning depressed areas and moving into areas with job opportunities. If new industries are encouraged in a previously depressed area, owing to cheap heat and electricity, there results a gain in the employment of the resources of the nation as a whole. Further details of the location problem will be discussed below (Section B.3 of this chapter).

In the remainder of our analysis we will proceed with the assumption, except where expressly stated to the contrary, that there is full employment of resources both before and after the introduction of atomic energy. We will not examine whether this result is achieved by the "automatic" operation of economic mechanisms or by government intervention.[13]

b. SUNK CAPITAL COSTS. The fact that at least in certain localities energy can be produced at a lower average cost in an atomic plant than in a coal-burning plant does not mean that existing coal-burning plants will be scrapped in favor of atomic plants. Since the "sunk" costs of constructing existing coal-burning plants have already been incurred and cannot be recovered, these costs must be disregarded in comparing the relative economy of an existing plant with that of a new atomic plant. In the short run, the change-over will be profitable only if the *variable* costs of the coal-burning plant exceed the *total* costs of the atomic plant.

Suppose, for example, that electricity can be produced in the atomic plant at a total cost of 4 mills per KWH (Chapter I, C.2.b) while the total cost in the coal-burning plant is 7 mills—3.3 mills for interest and depreciation of plant investment and other fixed charges, 3.7 mills for fuel, labor, and other variable charges. If a new plant is to be built, an atomic plant will obviously be preferred to a coal-burning plant. It will not be profitable to replace an existing coal-burning plant with an atomic plant, however, for the only additional cost (apart from investment already made) in producing electricity from the coal-burning plant is 3.7 mills per KWH, while a new atomic plant would require expenditures of 4 mills.

In such a situation, will a change-over ever take place? Probably it will, eventually, but there are a number of factors that will determine the rate of replacement. The first of these is physical depreciation. Of course, we have assumed that maintenance and repair costs are included among the variable costs of the coal-burning plant. However, a time may arrive when maintenance costs will rise sharply, due to the age of the equipment, or when replacement of major components of the equipment will be necessary. At that point it may prove profitable to permit the plant to wear out, and to replace it with an atomic plant, rather than to undertake major repairs.

Our analysis thus far has been based upon the distinction between fixed, or "sunk," costs and variable costs. We must note, however, that a cost which is variable to the power plant may be a fixed charge for some other sector of the economy. Fuel costs, for example, are variable costs for the power-generating station, but, in so far as they represent charges on investments in railroads and coal mines, fuel costs are fixed costs from the standpoint of the economy as a whole. Hence the immediate effect of the

---

[13] Although we will therefore ignore unemployment of resources, we will not neglect possible changes in the quantity of resources which are employed, that is, changes in the work force through population growth, child labor, and lengthening or shortening of the work week, and changes in the rate of capital accumulation and hence in the amount of capital available (see Section A.5).

competition from atomic energy might be to reduce the price of coal (re-sulting in a smaller return on railroad and mine investment) and thereby to postpone the change-over to the new method of power production.

No attempt will be made here to estimate quantitatively the period that would be required for atomic plants substantially to replace coal-burning plants, after an atomic plant with a lower average cost than a coal-burning plant has been designed. In an economy expanding at a moderate rate, like that of the United States, most existing coal-burning plants might con-tinue in operation for one or even two generations. In a backward area entering into a period of industrialization, practically all new generating facilities might be expected to use atomic power, should atomic power be producible in that area at a lower average cost than power from conventional sources.

Where quantitative estimates are made of the effects upon an economy of the availability of cheap electricity, these estimates apply to a period when the new technology has supplanted the old. For the interim period, while the change-over is taking place, the effects will gradually increase from zero to their full magnitude, as a larger and larger percentage of the electricity generated is produced in the new plants.

The same considerations apply where the change in technology induces a relocation of industry, and we may expect, for industries with high fixed costs, time lags of the same general magnitude in relocation as in the change from coal-burning to atomic generating stations.

5. Long-term Repercussions

Thus far our discussion has been concerned only with some relatively short-term consequences of the cheapening of power. To be sure, even these effects may gain their full magnitude only over a period of years, due to the lapse that would occur before atomic plants would completely replace present generating stations. In addition, however, the short-term effects may produce equally important repercussions on the economy over a longer period of a generation or more. These possible long-term effects include the following:

    a. Changes in the rate of capital accumulation.
    b. Changes in the population and the labor force.
    c. Changes in technology.

a. CHANGES IN THE RATE OF CAPITAL ACCUMULATION. We have seen that the rate at which power consumption can be increased is limited by the fact that power consumption requires machinery and other capital equip-ment. So long as the quantity of available capital is independent of the cheapness of power, effects on income cannot be expected to be very large.

However, if the resources saved through the cheapening of power permit an increase in an economy's income, this increase in income may lead, in turn, to a more rapid accumulation of capital than would otherwise have taken place. The larger supply of capital will result in a further increase in the productivity of the economy. Since savings generally increase more than proportionately with the increase in income, an increase in the annual income of only 1% may lead to an increase in the annual rate of capital accumulation of several per cent. Moreover, since these effects are cumulative, they may achieve a considerable magnitude over the years, quite comparable to the compounding of interest. Counteracting this will be the tendency of the marginal productivity of capital to decrease as it becomes more plentiful, relative to labor and land.[14]

b. Changes in Population and the Labor Force. The responses of population to changes in per capita income and other economic forces is very imperfectly known. In countries that already have a high per capita income, birth rates are often hardly sufficient to maintain the population level, and a further rise in incomes may hasten the date at which the population becomes stable or begins to decline. Moreover, high productivity may lead to a decline of the work force through the diminution in the number of children, married women, and older persons employed—a substitution of leisure for increased consumption of goods.

In an economy that has been near the subsistence level, the consequences of increased productivity may be almost the opposite. A temporary increase in real incomes may so reduce death rates as to cause, in the absence of birth control, a rapid increase in population and renewed pressure of the population upon productive resources. Moreover, a rise in productivity in such an economy may lead to increased desires for goods, and hence may result in an actual increase in the work force through longer hours, child labor, and the employment of women and old persons.

Historical examples can be cited for all the possibilities mentioned above. The effects of increased productivity upon population is particularly important for subsistence economies, and will be examined at greater length in the next chapter.

c. Long-term Repercussions: Changes in Technology. If necessity is the mother of invention, perhaps opportunity is the father. It is generally thought that the invention of the steam engine, providing a cheap source of power for the first time, greatly stimulated the invention and improvement of power-consuming machinery. Hence, it may be that a

---

[14] The marginal productivity of capital does not *necessarily* decrease with an increase in its amount. When capital is more plentiful, economies may be attained through the use of large-scale machine methods, with consequent increasing returns. This is more likely in an economy in which capital has hitherto been scarce than in one that is already highly mechanized.

further cheapening of power would stimulate the invention of new devices of production or consumption that would employ large quantities of power. Such a result would greatly magnify the effect on the national income which, in the absence of changes of this kind, we have considered to be moderate or small.

It should be pointed out, however, that the cost of power is at present such a small part of the cost of employing power-consuming devices that a further cheapening would not greatly stimulate the use of power. Probably, for instance, an additional reduction in the cost of power will not offer a stimulus to invention of the same magnitude as the original introduction of the steam engine.

Long-term possibilities such as we have been considering are of such a conjectural nature that it would be idle to speculate on their magnitude. In general, they may lead to some gradual increase of the initial resource-saving, which will be of importance only after the lapse of a considerable number of years.

B. POSSIBLE EFFECTS UPON LOCATION

Thus far we have been considering how low-cost power may directly increase the overall production of an economy. We have seen that the availability everywhere of cheap power may, by eliminating existing differentials in power costs, cause some industries to change their locations. Apart from the effect of this locational shift upon national productivity, there may be important effects upon the geographical patterns of population and industry, and upon the economic development of particular regions within the economy.

An area of high power costs, but possessing other locational advantages such as proximity to markets or to raw materials, may attract new enterprises if low-cost power becomes available. Other industries may then remove to this location because they are linked locationally to the first enterprises, or because new local markets have been created.[15] An estimate of the total increase in industrial activity in the new location must take into account such "multiplier" effects. In analyzing these effects we must be careful to distinguish the change in partial equilibrium for a particular location and

[15] Throughout this section, when we refer to "removal" or "relocation" of an industry, this terminology is used for brevity and is not intended to imply that relocation ordinarily takes place by the actual movement of existing plants or their immediate replacement by new plants. Except in very unusual cases existing plants will continue to operate until the investment in them has depreciated, and relocation takes place only as new factories are built to replace depreciated or obsolete ones, or to take care of the expansion of the industry. The old plants at the original sites will be operating at a competitive disadvantage and may become insolvent during periods of poor business (e.g. the New England textile mills during the depression of the 1930's), thus hastening the relocation.

industry from the impact upon the economy as a whole.   We shall see that readjustments may take place throughout the economy that will either reinforce or partially offset the effects of the locational change.

We may trace the chain of effects upon the region and upon the economy in this general fashion: Employment in a particular industry is expanded in some region because of a greater fall in power costs there relative to other regions.   This expansion may be very great, even without any change in the aggregate production of the industry in the economy as a whole.   Unless it is accompanied by a trend toward overall expansion of the industry or of other industries, the increase in one region will be compensated by decreases in the regions where the industry was previously located.   (The shift in cotton manufactures from New England to the South has been an important example of such a transfer in the United States, although in that case changing power costs were not directly responsible for the shift.)   The expansion in one region leads to a series of "multiplier" effects:

1. Industries that use the products of the relocated industry as raw materials may also be attracted to the new region.   We will call this the "production multiplier" effect.

2. Industries that produce production goods for the relocated industries (e.g. textile machinery) may themselves be relocated.   We will call this the "production goods multiplier" effect.

3. These relocations will produce an increase in incomes in the new region, either through a migration of population to the region, a shift of working population from agriculture to industry, or both.   We will call this the "consumption multiplier" effect of the initial locational shift. Where the new work force is obtained largely by a shift from agriculture, a mo.e efficient utilization of the labor resources in the economy as a whole may result, and this will increase the overall productivity of the economy in addition to producing the expansion effect discussed above.   This is an important reason why the increase in production in the new region may more than compensate for any decrease in production in the regions where the industry was previously located.

4. Particularly during the period in which the relocation is taking place, the demand for capital goods in the new region will be greatly stimulated. Houses, factories, roads, utilities, and machinery will be installed.   In an economy that has full employment of resources this will simply mean that investment will take place in the new region which would otherwise have taken place in other regions.   In an economy suffering from "stagnation"—underemployment of resources due to the lack of suitable investment opportunities—the relocation may increase the employment of resources, and hence may bring about a further increase in the production of the economy as a whole.   We will call this the "capital equipment multiplier" effect.

To the extent that the new capital goods are supplied locally, the capital equipment multiplier will further increase economic activity in the new region; to the extent that the capital goods are imported from other regions, production in these other regions will be stimulated.

It is difficult to estimate the magnitude of these effects, particularly those connected with the underemployment of resources. However, it may be of interest to assess in a very rough way their potential importance. This will be followed by a discussion of the sources of the new labor force required for the increase in regional production.

1. The Production Multiplier

The process of transformation of raw materials to consumer goods usually takes place in a succession of steps, one or more of which may involve weight reduction, increase or decrease in bulk, or increase or decrease in perishability. Cotton ginning, spinning, weaving, and garment manufacture is an example of such a chain. Another example is iron mining, smelting, steel-making, rolling, and manufacture of farm machinery. Examination of the geographical distribution of the links in any such chain of industries usually reveals that the first links are located near the sources of raw materials, the last links near the markets, and the connecting links at intermediate locations.

Relocation of any one link in such a chain of industries would tend to bring with it (a) those prior links which were market-oriented, and (b) those subsequent links which were raw-material oriented. It should be noted that where two industries are locationally linked in this manner, the relative advantage of competing locations can only be compared for the two taken together. Removal of the first industry to a more advantageous location may be prevented by the disadvantages (e.g. greater distance to markets) that this location possesses for the second. Moreover, since both components have to move to retain the advantages of their linkage, a locational advantage that is large for the first industry may be unimportant for the two considered together.

2. The Production Goods Multiplier

The location of an industry whose products are used by other industries (e.g. machine tools) will correspond somewhat to the pattern of these other industries. However, the magnitude of the multiplier involved here is also difficult to estimate because many of the industries in question are relatively footloose—the advantage of locating near their markets is not decisive.

The manufacture of textile machinery in the United States, for example, is highly concentrated in New England, the older textile manufacturing area, little of it having yet migrated to the South. The manufacture of agricultural machinery is concentrated in the central industrial states—Illinois, Wisconsin, Indiana, Minnesota, Iowa, and Kentucky—midway between the agricultural areas and the western steel-producing areas. So far as "backward" areas are concerned, lack of technological skills, of capital, and of the basic iron and steel complex continues to require the importation of textile machinery long after a local textile industry is established on a large scale.

On the other hand, the building materials industries and the construction industry are largely market-oriented. These enter, of course, both into the production-goods and the consumption-goods multipliers, and will be discussed more fully in connection with the latter. Moreover, these particular industries would be considerably stimulated during the period of relocation and require special consideration as an important component of the capital-equipment multiplier.

3. The Consumption Multiplier

A reasonably definite estimate can be made of the multiplier for market-oriented consumption goods. Construction, transportation and communication, wholesale and retail trade, personal services, professional services, government, and certain manufacturing industries are all related closely to the distribution of the consuming population. Of the 10,573,000 persons engaged in manufacturing in the United States in 1940, perhaps 1,750,000 to 2,000,000 were employed in market-oriented industries. To these must be added about 24,250,000 persons in the non-manufacturing categories listed above.[16] Thus, of the total employed population in this year (about 45 millions), some 58% (nearly ⅗) were employed in market-oriented industries and occupations. From this we would estimate a consumption multiplier for agriculture and for the other industries (those not market-oriented) of about 2.5—that is, the total employment is 2.5 times the employment in those industries that are not market-oriented.

Using a somewhat different classification, the National Resources Committee classified the total of 41.4 million persons employed in 1935 as follows: 11.5 (28%) in industries located close to resources; 10.2 (25%) in relatively footloose industries (those without strong locational pulls to raw materials or to markets); and 19.7 (48%) in industries located close to the consumer.[17] The consumption multiplier here is about 2. This estimate, however, allocates only 887,000 persons to manufacturing industries located close to the consumer, as compared with our estimate of 1,750,000.

A multiplier estimate using somewhat different methods has been made

[16] These figures are rough estimates, based on an examination of the detailed breakdown of employment by industry in the 1940 census, and data in Gordon, op. cit., indicating which industries are largely market-oriented.

[17] The Structure of the American Economy, Part I, Washington, G.P.O., 1939, p. 36.

by Daly for certain English areas. His multiplier is an aggregate of the production goods and consumption multipliers discussed here.[18] Daly arrives at a multiplier of about 2.3.

It must not be supposed that a multiplier of the same magnitude would apply in areas outside Britain and the United States. Colin Clark has presented statistics on the number of persons engaged in primary, secondary, and tertiary activities in various countries. Primary activities include agriculture, fisheries, and forestries. Secondary activities include manufacturing, building construction, mining, and electric power production. Tertiary activities include all other industries—principally services, transportation, and government.[19] The magnitude of these tertiary activities will give us a rough measure of the consumption multiplier in any country. The estimate will generally be too low since, as we have seen, a considerable fraction of manufacturing, construction, and electric power industry employment is market-oriented. Clark finds a very close relationship between a country's real income per occupied person and its tertiary production per occupied person; tertiary production increases more rapidly than real income. He gives percentages of total working population in tertiary production in the period 1925–1934 as follows: United States, 45.8%; Britain, 43.9%; France, 35.5%; Canada, 45.1%; Eire, 33.9%; Esthonia, 24.1%; Hungary, 21.8%; Roumania, 15.0%; Australia, 44.4%; India, 20.8%; Japan, 30.2%; and New Zealand, 48.7%. If the multiplier of 2 for the United States is approximately correct, then we might expect multipliers as low as 1.5 and 1.2 for Japan and India, respectively.

a. THE INCOME EFFECT AND THE CONSUMPTION MULTIPLIER. Where the relocation of industry increases efficiency by the redistribution of labor, the geographical shift may have associated with it not only a consumption multiplier for the new region of industrialization (and a corresponding decrease in consumption in the area from which the migration took place), but also a consumption multiplier for the economy as a whole due to the higher total income. There would consequently be an overall growth of industry. This point has already received comment in Section A.4.a of this chapter.

The scope of the market-oriented industrialization (including production goods industries as well as consumption goods industries) in the new region will also be very greatly affected by the aggregate population attracted to the area, and the aggregate increase in income produced there. Most industries are subject to economies of large-scale operation, at least up to a certain point. A small and isolated industrial area, or one with low consumer incomes, may continue to import goods which, if the local demand were

[18] M. C. Daly, "An Approximation to a Geographical Multiplier," *Economic Journal,* Vol. 50, June–September 1940, p. 254.

[19] C. Clark, *The Conditions of Economic Progress,* London, Macmillan and Co., 1940. pp. 337–338.

larger, would be produced within the area.   For example, the industrialization of the coastal parts of Brazil, involving several million workers, might
be expected to attract proportionately more market-oriented industries than
the exploitation of a specific mineral resource, employing a few thousand
workers, in Rhodesia.

Other things being equal, high transportation costs to a region will
encourage a greater degree of self-sufficiency in market-oriented goods than
low transportation costs.   Thus industrialization of the interior of China
might result in a larger local consumption multiplier than would result from
a corresponding increase in the coastal areas.   However, improvement of
transportation facilities in that country would be even more favorable to the
consumption multiplier for the entire country, because only in this way could
the economies of large-scale production be realized.

4. The Capital Equipment Multiplier

Finally, we consider the possible magnitude of the capital equipment
multiplier.   This multiplier, while closely related to the production goods
multiplier, deserves special treatment because it is peculiar to areas of expansion or new development, and because some economists have insisted on
its importance for the maintenance of full employment.

The capital equipment multiplier will be greatest in a region in which
capital is used intensively in industrialization, in which transportation and
other utilities have been poorly developed prior to the industrialization, and
in which industrialization produces a large movement of population, even if
this be only from farms to cities.   Where these elements are present, industrialization requires the construction of new houses, stores, and factories,
and the manufacture or importation of machinery.

It has been estimated that the capital plant (buildings and equipment,
including residences and utilities) in any economy is about 3 to 4 times the
annual production.[20]   Hence, we may conclude that when an industrial
population is moved into a new region, capital amounting to the equivalent
of at least 3 man-years production is required for each person employed.
Part of this capital (building construction, for example) must be produced
in the new location, part (e.g. machinery) may be imported from other
areas.[21]

The relation of the capital equipment multiplier to full employment can
only be mentioned here.   As we pointed out earlier in the chapter, tech-

[20] Clark, *The Economics of 1960*, London, Macmillan and Co., 1943, pp. 72–87;
Clark, *Conditions of Economic Progress, op. cit.*, pp. 389–395.

[21] Even if the capital is produced over a period of several years, a multiplier of considerable magnitude must be applied to the entire population of the new area: in the
case of two years, a multiplier of $(3 \div 2) + 1 = 2.5$, or thereabouts; in the case of three
years, a multiplier of approximately $(3 \div 3) + 1 = 2$.

nological change usually implies that a given quantity of goods can be produced with a smaller expenditure of resources, and the resources thus saved may become unemployed, unless government action is taken. In the short run, however, technological change may actually increase the employment of resources not already fully employed. The application to practice of changes in technology requires the construction of new capital equipment, and hence provides new investment opportunities. Some economists (in this country, for example, Hansen and Schumpeter) have considered these investment outlets to be a primary stimulus to economic growth. If this is the case, we should expect the effect to be particularly pronounced when the technological change, by bringing about relocation of industry, requires a new stock of capital equipment to be brought to the areas of new industrialization.

5. Source of the New Labor Force

The labor force in a new industrial region may be obtained from persons previously unemployed, persons previously employed in industry elsewhere and moving into the region, persons previously employed in agriculture, or from the natural increase in the working age population resulting from changes in birth and death rates. Any increase in income when the labor force is obtained from the second source would be approximately balanced by a decrease in income in some other region; but if labor is obtained from the first, third, or fourth sources, there may be a net increase in income produced in the economy.

Of these four possible sources of the labor force, the third needs further discussion. It has often been observed that generally the productivity in agriculture (measured as the net money value of product per person employed) is much lower than in secondary and tertiary activities. Colin Clark has presented data to show that such a relation holds for most countries.[22]

At least three causes cooperate to produce this result, two of which are compatible with economic equilibrium. First, agriculture is usually less capital-intensive than manufacturing. Second, the cost of living is generally lower in rural than in urban areas.[23] Clark's data show that this is by no means the whole explanation. Third, most countries have been experiencing a relative rise in secondary and tertiary occupations; and because of the imperfect mobility of labor, this has increased relative wage rates in these occupations.[24]

[22] Clark, *The Economics of 1960, op. cit.*, p. 38.
[23] Clark, *Conditions of Economic Progress, op. cit.*, p. 342.
[24] Because of the greater income-elasticity of demand for industrial than for agricultural products, any increase in productivity in both agriculture and industry will inevitably produce this result. (See Herbert A. Simon, "Effects of Increased Productivity Upon the Ratio of Urban to Rural Population," *Econometrica*, Vol. 15, January

Because of this third source of disparity in wage rates, any change which increases the mobility of labor will lead to a more efficient distribution of resources and will increase the income of the economy as a whole. The location of new industries in regions that suffer from an agricultural overpopulation in this sense lowers the barriers which the necessity of long-distance migrations presents to the transfer of population from agriculture to industry, and hence increases labor mobility.

We have now examined in detail the various "multiplier" effects, as outlined at the beginning of this section, that accompany the relocation of a particular industry. It can be seen that although only a few industries were definitely identified, in Part Two of this study, whose relocation might come about if atomic power were available, the relocation of even these industries would bring about significant consequences. To the relocation of the industry in question there has to be added: relocation of industries locationally linked to the first; relocation of production goods industries; relocation of market-oriented consumption goods industries; and the immediate demand for new capital equipment in the area of relocation. What effect all this will have on the remainder of the economy will depend on what additional resources are drawn into the productive process by the relocation, either because of the transfer of less productive agricultural labor to industrial employment, or because of increased employment of resources through the stimulus provided by new investment opportunities.

C. SOME GENERAL CONCLUSIONS

In this chapter we have considered the effects of the introduction of cheap atomic power, available anywhere, upon the economy of a nation or a region. Our conclusions as to the effects of relocation of particular industries upon the region of relocation have been summarized in the preceding paragraph. This included also the effects upon the national economy through the fuller employment of the resources of a region. These should be added to the other nation-wide effects which we will now summarize from Section A of this chapter.

1. The principal short-run effect upon an economy like that of the United States would be a moderate increase in productivity, and a consequent increase in income; it does not seem likely that it will be more than 1%.

2. The cheapening of power might stimulate the invention and introduction of new power-consuming devices that have not yet been applied and that, if applied, would use energy in very large quantities not now

---

1947.)   Observers are in general agreement as to the imperfect mobility of labor, particularly in the Far East. Further evidence of the correctness of this explanation is the fact that no disparity in wages exists in Australia, where labor mobility is high and agricultural production has not been declining.

economically practicable.   The likelihood of such an occurrence can only be speculative.

3. Long-run effects of larger magnitude might be produced over a number of years if the increase in income resulted in a more rapid accumulation of capital, thus further increasing the productivity of the economy.   Whether this increased productivity would actually be translated into increased production depends upon the adequacy of general investment opportunities and government policies to maintain full employment of resources.

4. Apart from the effect produced by a more rapid accumulation of capital, or from the inventions mentioned above, the cheapening of power alone cannot be expected to produce any substantial increase in the demand for energy.

CHAPTER XIV

Atomic Power and the Industrialization of Backward Areas

SUPERFICIALLY, it might appear that the effects of atomic power upon the industrialization of backward countries could be treated within the framework of the last chapter.   To some extent this is true, but there are several reasons why a supplementary analysis, proceeding from a different viewpoint, has to be undertaken.

An area like China is neither a self-contained economy, nor simply a region within an economy.   Because of international trade and international capital movements, it cannot be treated as completely isolated.   On the other hand, because of the imperfect mobility of capital and the almost complete immobility of land and labor across national boundaries, it cannot be treated as a segment of an economy that permits free movements of these factors.[1]   The same may be true even of smaller political units.

Furthermore, our analysis thus far has been largely in terms of an economic equilibrium, and the disturbance and reestablishment of that equilibrium after the introduction of atomic power.   It is very doubtful whether the existing international division of labor can be correctly described as an equilibrium situation.   Important dynamic changes are taking place, and the effects of an innovation such as the availability of cheap power everywhere would be superimposed upon these changes.

Finally, in our analysis thus far we have assumed no government policy except that of maintaining full employment.   Yet the national economies of most backward countries are likely to develop under the direct and close control of governments, mostly bent on industrialization.

Consequently, in this chapter we shall investigate industrialization as a dynamic process, carried out under the direction of a government trying to industrialize as rapidly as possible.   We shall be particularly interested in determining whether the availability of cheap atomic power will have any effect upon the rate at which this industrialization can proceed, or upon the lengths to which it can ultimately be carried.

[1] On the relaton of location theory to the theory of international trade, see Bertil Ohlin, *Interregional and International Trade,* Cambridge, Mass., Harvard University Press, 1933, particularly Chapters 1 and 12.

In the first section of the chapter we shall examine the characteristic stages through which an industrializing country passes. We shall then try to determine what factors place an economic or a technological limitation upon the rate and extent of industrialization. In particular, we shall want to know whether there are any important areas in the world where, in the absence of atomic power, the lack of energy resources will prevent or limit industrialization.

A third section deals with the question of whether the availability of atomic power may accelerate industrialization by reducing total capital requirements. A fourth section examines the possibility that atomic power may permit a geographically more decentralized industrialization than would otherwise be possible.

Concrete data for our analysis will be drawn from a number of industrializing areas—particularly China, India, Japan, and South America. These data are intended purely as illustrations, employed because of their availability. No attempt is made at a comprehensive survey of the industrialization problem throughout the world. There are certainly other areas, especially Eastern Europe, the Mediterranean coast, and the Near East, to which the general analysis would apply, but which are not discussed in detail.

A. TYPICAL STAGES OF INDUSTRIALIZATION

"Backward areas" are characterized either by a subsistence economy of agriculture and native handicraft, or by a cash crop agricultural economy depending upon the importation of manufactured goods (and sometimes of food, as well), or by some combination of these. Most non-European portions of the world, when they were brought into contact with western culture and trade, went through certain typical stages of development culminating in gradual industrialization.[2]

Stage 1: The Village Economy

All the non-westernized areas of the world, regardless of the level of civilization they have reached, have had certain economic characteristics in common. First, the economy has been primarily a village economy, the main wants of each village being satisfied in the village itself or by an interchange of goods with other villages within a very small radius. Trade over considerable distance has been restricted chiefly to high-value goods of small bulk. Non-agricultural commodities have been produced within the household, or by handicraft methods. The level of productivity has been low, and the real income of the mass of the population never far above a subsistence level. As a result, production and con-

[2] See G. B. Jathar and S. G. Beri, *Indian Economics,* 3rd Ed., Vol. I, London, Cambridge University Press, 1931, pp. 120–133.

sumption of agricultural products has been the main economic activity. Money has played an insignificant role.

Stage 2: The Single-Crop Economy

When such areas are opened to western trade, the effects are felt first near the ports of entry; only gradually do they spread into the interior. This spread may be much impeded by the lack of adequate means of communication, and as a result, there may be a great lag between the economy in the interior and its development near the ports of entry. Stage 2, the first stage of westernization, is characterized by two important shifts: the area becomes an exporter of agricultural products, and native handicrafts (except in so far as they become articles of export) are gradually replaced by the importation of western manufactured goods, particularly textiles, to satisfy the simple non-agricultural needs of the population. Even though its agricultural activity increases, the area will often specialize in cash crops to such an extent that it becomes a heavy importer of food.

At first the entrepreneurs who open the area to trade merely purchase agricultural commodities and handicrafts from the natives. In those areas that are relatively primitive this method usually proves unsatisfactory to the western entrepreneurs, and there is a gradual development of plantation agriculture which, in turn, brings about cultural changes of the greatest significance. The native population takes its first steps toward "westernized" work habits and consumption patterns.

In Stage 2, then, a backward area becomes a specialized segment in the international division of labor. The new industry, which requires little capital investment, gradually results in the destruction of native handicrafts, except those that are exportable. Money begins to assume a more important role, although subsistence agriculture may continue to exist side by side with the cash crop agriculture.

An important variant of Stage 2 is found in areas that are exploited for their mineral rather than their agricultural resources. Here imported capital assumes a much more important role, the development of transportation to the sites of the mineral deposits becomes essential, and the westernization of that part of the native population which comes into the new sphere of activity is accelerated. Even more than in an agricultural community, the old and the new economies here may continue to exist side by side, particularly when the scarcity of available capital and mineral deposits limit the scale of development. There is often a considerable mobility of native labor back and forth between the two economies.

In still a third variant of Stage 2, a sparsely populated area is settled and developed by western people, with an initial specialization in agriculture and/or mining.

In the whole world today, only the interiors of the continents have not developed at least to Stage 2. The first variant described above has been typical of the densely populated areas of the Far East, including the Indian peninsula, China, Malaya, Burma, and Indonesia; the second has been typical of the west coast of South America, Rhodesia and the Congo, and some Balkan areas; the third is most strikingly seen in Argentina, Australia, and at an earlier period, North America.[3]

Stage 3: Initial Industrialization

There next occurs a recession from the extreme degree of international division of labor reached in Stage 2. The backward area begins to develop domestic industries, utilizing technological methods and capital supplied by the industrial countries. The rate of industrialization is limited by the available supply of capital, and by the rate at which the native population continues to be westernized.

Acculturation to western ways is of fundamental importance in the industrialization process for several reasons. The new industries require a supply of labor used to working regularly and sufficiently reliable and educable to operate the mechanical equipment. The industries also require markets for their products, which implies further growth of transportation, communication, and the money economy, and a willingness of the population to consume factory-made goods in place of the products of native handicraft.

Even in countries of very high population density, like India and China, the new industries have usually been plagued by a scarcity of labor. The "surplus" agricultural population is not easily induced to migrate to the cities; the migration is often temporary and even seasonal. The laborers have not been used to working beyond the point where their basic subsistence needs have been met; illiteracy, complete unfamiliarity with western machinery, and physical debility caused by the always poor diet and low living standards make the labor extremely inefficient. On the positive side, the introduction of rudimentary health and famine relief measures brings about almost everywhere a substantial drop in death rates and a rapid increase in population, increasing the pressure on the land and encouraging the movement to the cities, where acculturation proceeds much more rapidly than in the villages.

[3] In the case of Eastern Europe and the Mediterranean basin, of course, stages 1 and 2 proceeded contemporaneously with the industrial revolution of Western Europe, rather than subsequently. In these areas, which shared the pre-existing European culture, the change did not involve the same kind of cultural replacement as in the Far East. So far as economic developments are concerned, however, the stages described here are applicable to these areas as well as to the areas specifically discussed in the text.

It appears, then, that acculturation is at the same time a consequence of industrialization and a necessary condition for its continued progress.

Industrialization usually occurs first in industries manufacturing low-quality goods consumed in large quantities in the domestic market and requiring only a small amount of capital for the number of workers employed.[4]   Industries are also developed to produce export goods using large quantities of hand labor.   In some cases these areas, using less capital-intensive production methods, produce goods in competition with those of the more industrialized countries.   As the industrialization progresses, the low-quality consumption goods may be exported in quantity to nearby areas that have not reached the same stage of development, and may there replace goods previously supplied by the industrial countries.   The textile, food processing, and shoe industries are typical areas of concentration in this stage of development.

Countries in Stage 3 continue to import from the more industrialized countries most of their capital goods requirements, as well as high-quality consumption goods.   Raw materials for the consumption goods industries are often imported.

The cotton textile industry in the Far East affords an illustration of the economic interrelations of a number of countries that reached Stage 3 at different times, and of their relationships with the industrial countries. Japanese cotton textiles replaced English textiles first in Japan itself, and later in China, Indonesia, and India (where a native textile industry using handicraft methods and exporting to Europe had already been displaced by English textiles).   More recently these three countries have begun quantity production of their own cotton textiles, competing with Japanese goods. Finally, exports of textiles from the Far East have made some inroads on the domestic market for European textiles.

This stage of initial industrialization culminates, then, with the "backward" areas exporting not only agricultural commodities, but also low-quality, labor-intensive manufactured goods, while the industrialized nations come to specialize in capital-intensive, high-quality manufactured goods.

Stage 4: The Introduction of Heavy Industry

In the final stage of industrialization, the "backward" country begins to develop those industries which are most characteristic of western industrialization, such as ferrous and nonferrous metallurgy. Here, the availability of particular resources becomes a specific limiting factor. (Of course it was also a limiting factor in Stage 2 for mining areas.)

[4] But in Japan, because of government controls and subsidies, heavy industry appeared at a very early stage in industrialization.   This pattern may be repeated in other countries now undertaking planned industrialization.

Iron ore, coal, and power are needed.   In this stage, also, the availability
of capital takes on critical importance, since the industries in question
are capital-intensive.   On the other hand, when an area reaches this stage,
it is frequently able to supply a considerable quantity of its own savings,
and it may prove possible for the development to take place without large
borrowing from abroad; the imported capital goods—iron, steel, and
machinery—are paid for by exports of agricultural and industrial products.

B. INDUSTRIALIZATION AND REAL INCOME

The economic effects that follow from industrialization depend in very
large measure upon what happens to real income as the industrialization
proceeds.   The level and amount of real income will set limits upon the
possible rate of domestic capital accumulation and will also determine the
domestic market for various kinds of commodities.   This, in turn, will
affect such matters as the magnitude of imports of various kinds, and
the extent to which the economies of large-scale operations can be realized
in producing for the domestic market.

The passage from the village economy to the single-crop economy has
often been accompanied by an actual lowering in real income for some
part of the population.   The farm owner will usually better his condition,
at least temporarily, through the substitution of cheap foreign goods for
the more expensive products of the local handicraft industries.   Persons
engaged in handicrafts will be impoverished by the fall in price of their
products, and many of them may join the ranks of agricultural labor,
thus driving down the wages of this group.

When farmers seek part-time employment in handicraft industries
during quiet seasons, the results of the transition to the single-crop economy appear almost paradoxical.   The farmers will begin to buy, for their
own consumption, the cheaper foreign textiles and other manufactured
goods.   This will reduce the demand for native handicraft labor and
the wages of such labor, and consequently will cut down the total incomes of farmers.   Hence, their gain through the availability of cheap
goods may largely be offset by the decline in their wages for part-time work.

In general, we might expect the transition between these early stages
to have a generally depressing effect upon those who earn their living entirely as handicraft workers or as farm labor, while it might moderately
increase the real income of farm owners, particularly if these are not
seriously dependent upon part-time employment in industry to supplement their agricultural incomes.[5]

[5] Jathar and Beri, op. cit., pp. 133–138.   J. S. Furnivall, Netherlands India, Cambridge, England, Cambridge University Press, 1944, pp. 106–108.   Hsiao-Tung Fei

We have already noted that as the backward area moves into Stage 3 (of initial industrialization), important population changes take place. Agricultural economies, particularly in backward areas, have always had a rate of population increase that pressed them down to a subsistence level. The very inadequate statistics that are available for such areas suggest that the natural increase in population is periodically removed by wars, famines, and epidemics. With the passage to the more productive single-crop economy, there has almost invariably been a rapid growth of population keeping pace with the increase in production and maintaining incomes at low levels. This population increase has been encouraged by peace and public health measures and by improved communications that reduced the famine problem.

With the advent of industrialization, a large part of the population increase is drawn off to the cities, but at the same time the resistance to migration permits at least urban wage rates to rise above the subsistence level. A rise in incomes is further encouraged by declining birth rates in the cities, and often to some extent in the rural areas as well. As is well known, this latter development has progressed so far in the older industrial countries of Europe that many of these now have reproduction rates too low to replace their populations, and appear to have definitely escaped the dilemma posed by the pressure of population against resources.

Whether the areas of high population density that are now undergoing industrialization will also escape this dilemma is still very much in doubt. Even in the case of Japan, the most advanced of these, trends in vital statistics before the war gave no clear indication as to whether a balance of births and deaths would be reached at higher than subsistence-level incomes.[6] In China, India, Egypt, and other areas with high birth rates, various factors in the social structure constitute serious barriers to the reduction of birth rates.

We have discussed these matters in some detail, because they show how important it is to the backward country that the industrialization proceed quickly. Where the process of industrialization is rapid, and is not exclusively concentrated on investment goods, it can be predicted that there will be a substantial rise in real incomes, at least in the cities, during the initial period. If this is accompanied, as has been usual, by a sufficiently rapid drop in the birth rate, the new higher living standards

and Chih-I Chang, *Earthbound China,* Chicago, University of Chicago Press, 1945, pp. 304–306 and *passim.* J. L. Hammond and Barbara Hammond, *The Rise of Modern Industry,* New York, Harcourt, Brace and Co., 1926, pp. 90–96. Edward P. Cheyney, *An Introduction to the Industrial and Social History of England,* New York, The Macmillan Co., 1910, pp. 220–223.

[6] For illustrations of these generalizations as applied to particular areas, see *Demographic Studies of Selected Areas of Rapid Growth,* New York, Milbank Memorial Fund, 1944.

may become definitely established. Moreover, the higher incomes are likely to increase the rate of capital formation by encouraging larger savings.

Where the industrialization process is slow, because of an initial scarcity of capital or for other reasons, it may take a very long time for the backward country to free itself from its population pressure. An increase in population may swallow up the increased productivity resulting from industrialization, and require the available capital to be spread more thinly. This, in turn, will probably have a depressing effect upon the rate of capital accumulation by making saving impossible for most families. Hence, any technological improvement that will even accelerate slightly the rate at which industrialization can proceed may considerably improve the prospects of the backward areas.

The passage, then, from the single-crop economy (Stage 2) to initial industrialization (Stage 3) is marked by increased dependence upon the local sources of supply of low-quality consumption goods. On the other hand, because of the income effect just described, the aggregate demand for consumption goods, and particularly for high-quality goods, may increase. Moreover, the industrializing country may greatly expand its imports of machinery and other production goods, including raw materials, exporting in return agricultural commodities and low-quality manufactured goods.

So far as real incomes are concerned, the remarks of the preceding paragraphs generally apply to Stage 4 as well as to Stage 3 in the process of industrialization.

C. THE LIMITING FACTORS FOR INDUSTRIALIZATION

We next inquire how far the process of industrialization may be expected to go in backward areas, and to what extent the availability of atomic power might facilitate the process, particularly by removing limitations that would otherwise prevent it or slow it down.

The term "limitations" seems to have two meanings, the one economic, the other technological. A factor may impose an economic limit on industrialization if it is scarce, and if an increased amount is attainable only at a high cost. A factor may impose a technological limit on industrialization if it is absolutely scarce—i.e. if there is no way of increasing the amount available—and if there is no possibility of substituting other factors for it. Of course technological limitations are actually only an extreme example of economic limitations, for in almost all cases a technologically scarce factor can be imported, but at a high cost.[7]

[7] Land would seem to be an exception here, but not entirely so, because water and mineral fertilizers can be imported to increase the productivity of the land. As economists have often noted, it is hard to draw a sharp line between land and capital as factors of production.

In one sense, however, there is still some utility to the distinction between "economic" and "technological" scarcity. In the increasing number of cases where industrialization is controlled and directed by a central government, comparative costs may not be the sole criterion in determining the direction of industrialization. Moreover, the large-scale importation of technologically scarce factors may be out of the question because of a lack of foreign exchange or for military or political reasons. In these cases, economic planning may be more concerned with technological than with economic limitations.

What are the specific limitations upon the scope and speed of industrialization? The most important would be capital, skills and "know-how," energy, and specific minerals.

1. Capital

The capital employed per head of working population is generally somewhat greater in manufacturing than in agriculture. The capital-labor ratio is by no means fixed, however, for it varies greatly from one industry to another, and in a given industry it may vary widely through substitution between capital and labor.

Thus, the amount of capital required for the industrialization of backward areas depends upon the types of industries that develop and the types of manufacturing equipment utilized. With respect to the first point, data from Pennsylvania for 1929 show the following amounts of capital per wage earner in various industries: pig iron, $19,658; electrical machinery, $7,483; brick, $4,211; cotton goods, $4,137; boots and shoes, $1,664; men's clothing, $900; shirts, $563.[8]

Data presented by Hubbard show an investment of about $30 per spindle in the Japanese cotton spinning industry, and about 55 spindles per wage earner, or about $1,650 investment per wage earner.[9] China, with approximately 25 spindles per wage earner, would show an investment of only $750 per worker. The investment in the Indian cotton spinning and weaving mills in 1930 was about $330 per worker (estimating the rupee at 33 cents). The lower capitalization of these countries per wage earner is probably attributable to the less efficient labor, and to the use of less automatic and mechanized equipment.

Another set of estimates gives the following figures for the capital requirements of cotton mills in India: modern mill, large-scale, $395 per

[8] Charles A. Bliss, *The Structure of Manufacturing Production,* New York, National Bureau of Economic Research, 1939, pp. 109–110.

[9] G. E. Hubbard, *Eastern Industrialization and Its Effect on the West,* London, Oxford University Press, 1935, pp. 119, 123, 207, 249.

wage earner; power loom, small-scale, $99; automatic loom, cottage industry, $30; handloom, cottage industry, $12.[10]

In addition to the capital required for factories and machinery, there will be large additional requirements for roads and railroads, power plants, housing, and the like.   On the basis of these data, it is not safe to assume that the industrialization of a backward country, using modern equipment and with moderately efficient labor, can be carried out with the provision of less than $1,000 or more in capital for each worker employed (in dollars of 1940, say).   Even this figure assumes concentration in textiles or other capital-extensive industries.

An investment two or three times as great would be required in the iron and steel industries, even excluding investment in mines and railroads.   To employ 20% of the Chinese working population (say 40 millions out of 200 millions) in industry with even moderate capitalization would require an investment of some 40 to 80 billion dollars.[11]   Spread over 15 years, such a development would require the accumulation of capital at the rate of 2.7 to 5.3 billions per year—large, although perhaps not impossible, goals for a country like China.

The capital needed for industrialization can be paid for in one or both of two ways: out of domestic savings (that is, by consuming less than is currently produced), or by borrowing or securing investments from

[10] S. Kesava Iyengar, "Industrialisation and Agriculture in India Post-War Planning," *Economic Journal*, Vol. 54, June–September 1944, p. 202.   Iyengar's data are expressed in rupees, and the text figures are based on the exchange rate of 1 rupee = 33 cents.

[11] Mordecai Ezekiel estimates capital requirements for new industrial workers at a minimum of $1,000 per head.   He finds that total new capital investment in Russia from the Revolution to 1939 was about $3,500 (at 1 ruble = $.26) for each additional non-agricultural worker.   He also presents a table of "estimated investment need" for the first postwar decade totalling 18 billion dollars for Eastern Europe, 10 billions for China, 12 billions for South and Central America, and 10 billions for India, but these figures for India and China are far too small to accomplish his goal of raising the average non-agricultural employment from 40 to 55% of the total work force.   ("Industrial Possibilities Ahead," in *Towards World Prosperity*, New York, Harper and Bros., 1947, pp. 24–26.)   In the same volume, Kurt Lachmann presents industrialization goals for the Balkans that would require capital of 6.6 billion dollars ("The Balkan Countries," *ibid.*, p. 184).   This is postulated on an increase of 11 billion KWH in electric power production, or about $2,400 new investment per kilowatt-installed electric capacity.   Of this investment, 17% would be for the construction of power generating and distribution facilities, the remainder for industrial plant and rehabilitation.   K. Mandelbaum's plan for Eastern Europe requires an annual investment of 1 billion dollars, on the basis of $1,725 to $2,100 per person added to the non-agricultural work force (*The Industrialisation of Backward Areas*, Oxford, Basil Blackwell, 1945, p. 37).   Colin Clark predicts an increase in the Chinese and Indian non-agricultural work forces to 1960 of about 70 million persons each and a required net increase in capital to this same date at 61 billion dollars (at the 1925–1934 average value) for China, and 144 billions for India.   Although the aggregate goals appear quite unrealizable, the ratio of capital to employment—$870 in the case of China, and $2,030 in the case of India—is fairly comparable to that assumed in the present study, although derived by quite different methods (Clark, *The Economics of 1960, op. cit.*, pp. 71, 80).

other countries. The supply of investments and loans from abroad depends upon the rate of saving in the more mature countries, and the comparative marginal efficiency of capital in the developing country (i.e. the return it will bring to investors), and, perhaps even more important, upon political considerations.[12]   Colin Clark has estimated that several mature countries will have savings in the next 15 years far in excess of domestic capital requirements.   In particular, he gives the excess for the United States as 7 billion dollars during that period; for Great Britain, 58 billions; and for France, 14 billions.   (This neglects, of course, political developments which have made the United States a huge creditor.) These funds, through a combination of private investments, government loans, and loans from international public agencies, could be channeled to the backward countries.

Even apart from political factors, events of the past 3 years make it appear that Clark's estimates for Britain and France are far too optimistic, while he has apparently underestimated the exportable surplus of American savings.   Most recent estimates of American spending under conditions of full employment conclude that there will be an export balance of at least 2 to 3 billion dollars per year.[13]   Even if the 1946 balance of about 5 billion were maintained, this would add up to only 75 billions over 15 years, a considerable part of which is likely to go to Western Europe under the Marshall plan or its possible successors.

Unless all these estimates are too low, it is clear that external borrowing could provide only a modest portion of the funds required for even a moderate industrialization of the major backward areas.

Moreover, unless the political risks involved in investments in most backward countries at present can be greatly reduced—and current events are not too reassuring—the amount of private and other loans available for investment overseas at moderate rates of interest will be severely limited.   But of course without some degree of internal stability industrialization will be impossible, whatever the source of investment funds.

On the demand side, one must not conclude that there is a lack of potential investment opportunities because of a scarcity of mineral resources and fuel.   The specific evidence for this statement will be presented in subsequent sections.   The marginal efficiency of capital does depend on the resources available for exploitation, but probably even more upon a variety of factors, among them: (a) the presence or absence of internal

[12] Clark, *The Economics of 1960, op. cit.*, p. 81.
[13] From an unpublished compilation, by L. R. Klein, of "full-employment" estimates. On the other hand, J. Frederic Dewhurst, in *America's Needs and Resources*, New York, Twentieth Century Fund, 1947, Chapter 21, estimates net capital outflow at 1.2 billion dollars in 1950 and 1.1 billions in 1960.

improvements, particularly railroads, roads, sanitation, and internal security; (b) the existing scale of industrialization, which may lead to economies of large-scale production that make further investment attractive; (c) the knowledge of opportunities; and (d) the size of the domestic market for manufactured products (dependent, in turn, upon real incomes per head and upon the size of population in the industrializing country). The nature of these factors suggests the possibility of a cumulative effect; the process of industrialization may itself render investment possibilities more attractive.

The domestic supply of savings depends upon real incomes, the distribution of incomes, the tastes of the wealthier classes regarding conspicuous expenditure and hoarding of precious objects, and government policies. In post-revolutionary Russia, government policy was of course the principal determinant of the rate of saving. In Japan, the wealthy ruling families provided much of the capital by reinvesting their profits, once they were convinced that their own and their country's future lay in the direction of industrialization. (They were brought to this point of view only after early government investment and continued subsidies had demonstrated the feasibility and the profitability of this course of action.) In India and China, on the other hand, industrialization has been greatly impeded by the preference of many persons to accumulate their savings in the form of land, mansions, jewels, and precious metals.

Today, virtually all the important backward areas are proposing and undertaking planned industrialization (although we must not be too confident as to how effective such planning and its implementation will be). Their governments and their industrial leaders show no inclination to leave the processes of economic and technological change to the forces of laissez faire.

Under a program of planned industrialization, the governments of the backward countries will undoubtedly attempt to secure large funds for capital investment through voluntary and forced domestic savings.[14] This will be difficult at first, while incomes are near the subsistence level, but will become progressively easier if success is achieved in the early stages. At the same time foreign loans will be sought, chiefly as a source of foreign exchange, to finance imports of machinery and other necessary goods and equipment not available at home.

Of the 10,000 crores of rupees required to finance the "Bombay plan," for example, it is proposed that 300 be obtained from hoarded wealth, 1,000 from existing sterling balances, 600 from a favorable balance of trade,

[14] The population may be forced to save by some combination of these measures: taxes, refundable taxes (i.e. forced loans to the government), and an inflationary price rise reducing consumption.

700 from direct foreign borrowing, and 7,400 from voluntary and forced domestic savings (including 3,400 by the printing of money).[15]   Mandelbaum's plan for Eastern Europe proposes that perhaps ½ of the investment funds be created within the area itself.[16]

It is not entirely safe to base predictions of capital accumulation in industrializing areas upon the historical rates of saving in these areas.   One of the major efforts of government in stimulating industrialization will be to provide the necessary capital by domestic loans, currency creation, foreign loans, and other measures.   Although there is some dispute as to the Russian figures, the U.S.S.R. may have accumulated capital, at least in the 1930's, at a rate in excess of 20% of the national income.[17]   Even under less complete government control, or in some cases almost no control, rates in the neighborhood of 15% have been estimated for Great Britain (1860–1869), the United States (1900–1910), and Norway (1913); and rates of about 20% for Germany (1900–1910), Holland (1925–1930), and Japan (1919–1937).[18]

Colin Clark, on the basis of rather optimistic assumptions with respect to income, estimates that savings in India in the period 1945–1960 will aggregate 72 billion dollars (about 7.5% of income)—about 5 billion per year—but in China only 3.4 billion dollars (1% of income).   The "Bombay plan," as we have seen, proposes capital investment for India over 15 years of about 33 billion dollars.   Official plans of four Eastern European countries (Poland, Hungary, Czechoslovakia, and Yugoslavia) together require an annual capital investment of more than 2½ billion dollars.

It appears not too unreasonable to conclude that, with strong support from the central governments, 30 to 60 billion dollars may be invested in the industrialization of China and of India during the next 15 years,[19] and smaller amounts in other backward countries.   Our next problem is to determine whether the resources of these countries are sufficient to support an industrialization on this scale.

2. Skills

One of the very important resources required to operate modern industry is technological knowledge and skill.   It has often been observed

---

[15] P. Thakurdas et al., A Plan of Economic Development for India, New York, Penguin Books, 1945, pp. 51 ff.   1 crore of rupes = 10 million rupees = 3.3 million dollars.

[16] Op. cit., p. 84.   The official postwar plans of Hungary, Czechoslovakia, Yugoslavia, and Bulgaria assume that investment (ranging from 10% to 20% of national income) will be domestically financed, while the Polish plan assumes 80% domestic financing for investment.   United Nations, Department of Economic Affairs, A Survey of the Economic Situation and Prospects of Europe, Geneva, 1948, p. 126.

[17] Clark, Conditions of Economic Progress, op. cit., pp. 300–400.   Clark believes the correct figure is somewhat lower.

[18] Ibid., p. 406.

[19] That is, something in the neighborhood of 10% of income.

that industrialization first proceeds in industries requiring simple skills, but that the level of skill improves rapidly. There is no evidence, however, that the areas of newer industrialization actually overtake the older areas in this respect, because the latter have continued to be the centers of technological improvement.

There is considerable evidence that the emphasis placed upon technological training in Imperial Germany was a major factor, not merely in her rapid industrialization, but particularly in allowing her to overtake Great Britain and the United States in technical competence in such industries as chemicals and electrical equipment. The high level of literacy and education in the United States has often been cited as a partial explanation for her industrial expansion.

These lessons have not been lost on the newer areas. Technological education has always formed an important part of the Russian plans. Almost all the proposals for industrialization of the other backward countries place equal emphasis on the expansion of educational facilities. It is generally recognized, however, that it will be a much easier task to develop the necessary labor skills, and engineering and managerial competence, in light industries than in heavy industries.

3. Power

Although power is the productive factor of central importance to our particular problem, it is difficult to disentangle it from capital requirements in general, and also from requirements for particular mineral materials. The relationship between power and capital is twofold. First, capital in relatively large amounts is required for the production of power. We consider this below in connection with possible capital savings through the use of atomic power. Second, a large proportion of industrial capital is in the form of energy appliances.

In Japan (1929) there were about 2 kilowatts of installed capacity in motors and other energy-using devices for each worker occupied in manufacturing, while in the United States the installation of motors (1939) amounted to 8.5 kilowatts for each person employed in manufacturing.[20] An investment in power-producing facilities is required to supply these motors in addition to the capital directly invested in factories.

There is no fixed ratio, of course, for the combination of power with industrial capital, because this will vary from industry to industry, depending largely on the design of the particular plants and machinery. In the United States, the average number of kilowatts per $1,000 of capital

---

[20] John E. Orchard, *Japan's Economic Position*, New York, Whittlesey House, McGraw-Hill Book Co., 1930; and U. S. Bureau of the Census, *Statistical Abstract of the U. S., 1947*, Washington, G.P.O., 1947, respectively.

(1929) was probably about 0.7,[21] but varying from 3 in the production of pig iron to 1 in the textile industries and 12 in tobacco manufactures. In Japan (1929) the figure appears to be about 1.5 for all factory industry, and is in the neighborhood of 1.5 for cotton weaving and spinning, and 2 for metal refining and the cement industry.[22]

In an industrial economy energy is used (a) in transportation, (b) to operate industrial and agricultural machinery, (c) in chemical and metallurgical processes, and (d) as a consumption good. In several of these uses there is a high substitutability among different energy sources, particularly since an increasing proportion of electricity is being supplied by large central electro-generating stations using a variety of fuels. However, as we have seen, atomic power could not be substituted for coal, with existing technology, in many industrial processes.[23]

While statistical data on fuel resources are notoriously inaccurate and speculative, they may be of some interest in indicating the degree of sufficiency in coal of some areas of potential industrialization.[24] The United States, which consumes more than 4 tons per capita per year, has per capita reserves of over 30,000 tons. Great Britain, with a 5 ton per capita per year consumption, has per capita reserves of about 4,500 tons. The per capita reserves of the U.S.S.R., China, India, and Japan are about 7,000, 2,400 (or 550, according to another estimate), 250, and 120 tons, respectively.

If waterpower were developed to the maximum in these countries, additional power would be available amounting to .03 kilowatt per capita for China, .02 for India (including Pakistan), .07 for Japan, .27 for the U.S.S.R., and .31 for Brazil, as compared with .20 in the United States.[25] With a 50% load factor, 1 kilowatt of generating capacity is equivalent to 4,400 KWH per year, and this, in turn, would require about 1.7 tons of coal.[26] Hence the coal-equivalent of the available water power would be about .05 tons per capita for China, .03 for India, .12 for Japan, .46 for the U.S.S.R., and .53 for Brazil. Forgetting for the moment the capital required for the development of this waterpower and its inaccessibility from

[21] Bliss, *op. cit.*, p. 106.
[22] Orchard, *op. cit.*
[23] See Part Two above, particularly Chapter III, B.
[24] The estimates, in billions of tons, are: United States, 4,231; Great Britain, 209; China, 1,097 (241); India 87; U.S.S.R., 1,200; and Japan, 8. The figures for the United States, Great Britain, China, India, and Japan are from *Coal Resources of the World, op. cit.* The figure in parentheses for China is an estimate (1939) of the Geological Survey of China (*China Handbook, 1937–1943*, Chinese Ministry of Information, New York, The Macmillan Co., 1943, p. 481). The Russian figure is from *Electric Power Development in the U.S.S.R., op. cit.*
[25] Based on Map 2, Chapter II.
[26] At 10,000 BTU per KWH of electric power generated, the thermal efficiency assumed in the analysis of conventional power costs in Chapter II.

the standpoint of location, all of these countries could undertake a considerable industrialization based upon hydroelectric power alone, but not even Brazil and the U.S.S.R. could produce more than about $\frac{1}{20}$ of the present per capita consumption of energy in the United States.[27]

In estimating the power requirements of backward countries in the visible future, it is unrealistic to employ American consumption as the standard. A more sensible, although extremely rough, estimate of probable requirements can be obtained on the basis of the estimates of capital accumulation presented in the last section.

We assume a ratio of about 1.3 kilowatts in power-utilizing equipment per $1,000 of capital invested in manufacturing. American experience indicates that on the average about 2,000 KWH will be consumed annually per kilowatt of installed motors, or 2,600 KWH per $1,000 of capital. If all the power were generated from coal, this would require the production of about 1 ton of coal per year for manufacturing alone per $1,000 of manufacturing capital (in dollars of 1940, say).

Not all the energy used in industry, of course, is consumed in the form of electricity. In the United States, electricity represents about $\frac{1}{4}$ of the total energy consumption in manufacturing.[28] This would increase our coal requirement to about 4 tons per $1,000 of manufacturing capital. Moreover, energy consumed in manufacturing represents only about $\frac{1}{3}$ of the total energy consumption in the United States,[29] although the percentage of energy consumed in non-manufacturing uses might be lower in a country undertaking industrialization. We must realize also that not all the new capital in an industrializing country will go into manufacturing equipment; part will be needed for improving transportation facilities and other utilities and for building construction. Putting all those factors together, we will probably not be far wrong in assuming that for each $1,000 of new capital investment there will have to be an annual energy production equivalent to between 4 and 8 tons of coal.

New capital investment in the amount of 33 billion dollars over a period of years (the amount proposed in the "Bombay plan") would require an increase in coal production over this same period to a rate of 130 to 260 million tons a year in excess of the present rate. The coal resources of neither China nor India would be strained by such a rate of consumption, which amounts to less than 1 ton per capita.

---

[27] Total energy consumed in the United States in 1946 was the equivalent of 10 tons of coal per capita, but this included an equivalent of about $1\frac{1}{2}$ representing gasoline consumed by automobiles.

[28] Estimated from data in the *U. S. Census of Manufactures, 1947*, preliminary report on "Fuels and Electric Energy Consumed," Washington, November 18, 1949.

[29] *Ibid.;* and the 1948 *Minerals Yearbook*, section on "Coal—Bituminous and Lignite," U. S. Department of the Interior, Bureau of Mines, Washington, G.P.O. (preprint).

In the case of some areas like Japan and South America, coal resources are inadequate even for a moderate increase in industrialization. It must be remarked also that the sufficiency of coal for power production does not mean that there is enough coking coal for blast furnace operations, and some concern on this score has been expressed in the case of India.[30]

Though the actual exhaustion of coal reserves is not in question, the substitution of atomic power for coal in backward countries has potential importance in a number of ways: (1) it may lead to a cheapening of power; (2) it may be substituted for coal in certain uses, in countries having a relatively limited supply of that mineral;[31] and (3) under certain circumstances it may permit a reduction in the capital required for industrialization by reducing the investment required in dams, transmission lines, mines, and railroads.

The effects of cheapening energy, which were discussed in the previous chapter, apply to backward areas as well as to those already industrialized. The other two points will receive attention later in this chapter.

4. Mineral Resources

In considering the role of mineral resources in industrialization, we must distinguish between an area's production of these resources and its consumption. To be sure, the areas of heavy industry have in the past been those producing large amounts of coal and iron, but this close relationship between production and consumption may be chiefly a historical accident. If the method for reducing iron ore by electrical means without the use of coal discussed in Chapter X is practicable, this may permit the creation of an iron and steel industry in countries lacking coking coal, or in those where its deposits are far from the iron mines.[32]

Per capita iron ore reserves have been very roughly estimated for various countries as follows: United States, 29 tons; Great Britain, 66 tons; U.S.S.R., 27 tons; China, 1.1 tons; India, 10.7 tons; Japan, 1 ton; Brazil, 98 tons.[33] These particular figures are subject to even more serious reservations

[30] See Chapter X, C.2.b. Cf. also H. F. Bain and T. T. Read, *Ores and Industry in South America,* Council on Foreign Relations, New York, Harper and Bros., 1934, p. 117.

[31] See Chapter II for some examples of countries with severely limited conventional energy resources, and for an analysis of why atomic power might be more easily available to them.

[32] For the possible role of atomic power in facilitating the expansion of world steel output, see Chapter X, C.2.b.

[33] The figures given are those for "actual" reserves. India, Brazil, the United States, the U.S.S.R., and Great Britain have in addition large "potential" reserves of less accessible or lower quality ores. E. C. Roper ("The Union of Soviet Socialist Republics" in *Towards World Prosperity, op. cit.,* p. 201) estimates Russian reserves at 267 billion tons, or 1,600 tons per capita, about 10 times Mikami's "actual" plus "potential" figures. See H. M. Mikami, "World Iron-Ore Map," *Economic Geology,* Vol. 39, January–February, 1944.

than the corresponding figures for coal, for the ores in question are of greatly varying quality, those of Britain and Japan being especially poor, and those of China decidedly mediocre. The United States mines over ½ ton of ore per capita per year, Russia in the neighborhood of $\frac{1}{10}$ ton. Neither China nor Japan has reserves that would support a western-scale iron industry over any period of years, as Japan has long since discovered. On the other hand, the Chinese current scale of production of less than .01 ton per capita could be considerably increased without danger of ore exhaustion for several generations.

It will be instructive, again, to calculate the amount of iron ore production that would be required to sustain a given scale of industrialization. This is less easily estimated even than the coal requirement, for the amount of iron used depends on whether the emphasis in industrialization is placed upon heavy industry or light industry. Because of the great emphasis that Russia has placed upon heavy industry, the experience of that country will give an upper limit of the iron ore requirements.

Russia produces about 1 ton of pig iron for every 10 tons of coal mined (we are considering total coal consumption, not merely coal used in blast furnaces). Approximately 2 tons of high-grade ore need to be mined for each ton of pig iron produced, which with a ton of scrap will make 2 tons of steel. If a capital investment of 33 billion dollars corresponds to coal consumption of about 130 to 260 million tons per year, as previously calculated, we may now assume that this same investment would imply iron ore production at a rate not exceeding 26 to 52 million tons per year.

Iron ore production at this rate would probably exhaust Chinese ores in less than a generation (Mikami's estimate is 500 million tons "actual" reserve, and 700 million tons "potential" reserve), but such production could be sustained indefinitely from Indian reserves. We conclude that in India, iron ore will not be a limiting factor in industrialization, even with concentration on heavy industry. China, on the other hand, will be faced with the alternative of importing part of her iron ore requirements, or reducing these requirements by concentrating her development in industries other than iron and steel. In either case, the difficulty is not one that will be removed by the availability of atomic power.

It is not necessary for us to carry out analyses for other industrial raw materials, since if the fuel and iron-ore bases for industrialization are present other minerals can be imported, and the quantities required are not such as to place a strain upon available foreign exchange.

The increasing substitution that has been taking place during the last decade of the light metals, aluminum and magnesium, for iron and steel may, within the next generation, very greatly change the mineralogical base of heavy industry. With the prospective development of the light metals in mind, the relative poverty of some of the backward countries in either coal or iron reserves appears much less serious than it would have a generation ago.   Little is known regarding the distribution of raw materials for the manufacture of light metals, largely because little is known as to which raw materials can be economically employed.   If recent encouraging developments continue in the manufacture of aluminum from common clays and the extraction of magnesium from sea water, the raw materials, except power, for a light metals industry will be virtually ubiquitous.[34]

D. PROSPECTS OF INDUSTRIALIZATION THROUGH ATOMIC POWER

We have now reviewed the typical stages in the industrialization process, the effects of industrialization upon real income, and the factors that limit the rate and scope of industrialization.   In the light of this discussion, what are the prospects of industrialization in some of the principal backward areas, and to what extent may industrialization be facilitated by the availability of atomic energy?

The present international division of labor cannot be considered an equilibrium situation.   Almost all of the backward areas, once they have been exposed to western trade, have in time entered into a period of industrialization concentrating upon cheap consumers' goods that require little capital to produce.   The principal factors that appear to limit and determine the pace of this industrialization are capital and power.   Capital of not less than $1,000 and energy equivalent to 4 to 8 tons of coal per year must be provided for each factory worker employed.

Our analysis indicates that, for the next generation at least, scarcity of capital is a much more serious limitation upon the rate of industrialization of these countries than a technological scarcity of energy resources, although they are not too bountifully supplied with fuels.   This is particularly true if the industrialization is primarily in the direction of producing consumer goods.   South America is perhaps the major exception, since that continent is almost entirely lacking in coal, and its water power is relatively inaccessible.   Energy resources may also be a limiting factor in Southeastern Europe.

Most of the backward countries, however, are proposing for themselves not merely an industrialization of the type indicated, but an industrialization that would include the heavy industries as well, and that would increase their self-sufficiency in all types of manufactured goods.   These intentions have often been criticized as deriving from a misplaced chauvinism.   It is argued that because of the superior skills, capital supply, and resources of the older industrial areas, the backward countries should concentrate on

[34] See Chapter IV for a discussion of the use of atomic power in aluminum production.

labor-intensive manufactures of relatively low quality, and should continue to import most of their producers' goods.

From the standpoint of the most efficient international division of labor (with the *given* supplies of skills and capitals), this argument is quite plausible, but it leaves out of account several factors which the backward countries feel they cannot ignore. First, they are unwilling to remain dependent upon foreign sources for their basic supplies of military equipment. Secondly, under present world conditions, they are unwilling to depend upon the vagaries of world markets and prices for foreign exchange that will permit the importation of the necessary producers' goods, and they are definitely suspicious of too heavy reliance upon loans from foreign governments to finance these imports. Because of foreign exchange problems, they wish to supply domestically as large a portion as possible of the capital goods, as well as the investment funds, required for industrialization.

India and the Indonesian islands[35] both have excellent prospects for a considerable development of heavy industries using local mineral and fuel supplies, except for coking coal in India. In Brazil the absence of coking coal will be an extremely serious handicap, so that atomic energy would radically change her prospects, especially if iron ore can be electrically reduced. Japan is already pressing upon the limits of her energy resources, and atomic piles would provide a very welcome new source of power, with iron ore or scrap or pig iron imported from abroad. In China, iron ore rather than power is the limiting factor. In almost all the backward countries the substitution of light metals for iron or steel would facilitate industrialization and would be helped by ubiquitous atomic power.

1. Capital Savings Through Atomic Power

Several important possible effects of atomic power upon the rate of industrialization remain to be examined. Among these is the possibility of reducing capital requirements through the use of atomic power in place of fuel-generated or hydroelectric power. If invention goes well, atomic power might reduce the aggregate capital required for central generating stations, for railroad construction, and for coal mines. What would this mean in terms of the total capital requirement for industrialization?

Data have already been presented showing that about 2,600 KWH of electricity will have to be produced per year for each $1,000 of capital in manufactures, requiring .65 kilowatt generating capacity per $1,000 of capital. The ratio for non-manufacturing capital is not likely to be any higher. Since coal generating plants require a direct investment of about $130 per kilowatt capacity, to each $1,000 of capital invested in manufac-

[35] The mineral resources of the islands of Indonesia are imperfectly known, but the islands appear to have very large resources of coal and especially iron, as well as the well-known petroleum reserves.

turing must be added $85 for generation of power. There is no reason at present to believe that the capital costs of an atomic plant would be lower than those of a steam plant; indeed we have estimated them to be in the range of $140 to $315 per kilowatt. However, operation of coal generating plants may involve indirect capital costs for coal mines and railroads.

Some Russian data have been published on the total capital investment needed for steam-generating plants, including investment in mines and railroads. Unfortunately, the direct and indirect investments are not explicitly given separately, and there are some internal inconsistencies in the data that make proper interpretation difficult. These figures indicate that in the Soviet Union the maximum investment per kilowatt of power capacity in mining and transporting coal, i.e. for thermal power stations located at the greatest distance from coal mines, is just about equal to the direct investment per kilowatt in the power plant proper. If, therefore, the investment in a thermal power plant is $130 per kilowatt of capacity, as we have assumed in Chapter II, the total investment, including the indirect investment in coal mines and railroads, would be $260 per kilowatt.[36]

Thus, with the maximum indirect investment included, the total investment in the coal plant is $45 per kilowatt higher than the intermediate estimate for the atomic plant (as given in Chapter I). Using the most optimistic estimate ($140) for the atomic plant, a saving of $120 per kilowatt might be realized as compared with the maximum for the coal plant; while if the highest estimate ($315) for the atomic plant is correct, this will exceed by $55 per kilowatt even the maximum investment in the coal plant, including indirect costs. On the basis of the most optimistic estimate for the atomic plant, a total investment of $1,091 (including power-generating facilities) would be required for each $1,000 of manufacturing capital, whereas the maximum for the coal plant would be $1,170; the atomic plant would conserve about 7.0% of the total capital required.

According to the Russian data at least 80% of the maximum indirect investment, or roughly $100 per kilowatt of power capacity, is required in railroads.

Under what circumstances would the investment in railroads be this large? We may accept the round figure of $100,000 per mile as the capital investment in a railroad. *If a railroad were built solely for the purpose of supplying fuel to a 100,000 kilowatt generating plant,* the railroad investment per kilowatt would be $1 for each mile of length or $100 for a 100-mile road.

---

[36] *Electric Power Development in the U.S.S.R., op. cit.* Derived from figures given in various places in that report, in particular, p. 486 for the range of investment in thermal stations allowing for the mining and transportation of coal, p. 107 for the average hours of operation per kilowatt of capacity, and pp. 390–392 and 415–420 for some investment figures on coal transportation and mining.

Since coal haulage for a plant of this size would not use the railroad to capacity, the investment per kilowatt per mile of railroad would be proportionately smaller for larger generating stations.

Capital saving, based upon a reduction in the requirement for railroad construction, will be considerable under the following conditions: (a) that the railroads in question are required only for the transportation of coal to be used in power stations, and could otherwise be dispensed with or their construction at least postponed; (b) that the railroads are not already in existence; and (c) that a substantial mileage of railroads would be required for each 100,000 kilowatt capacity of the generating stations served by them. The last condition implies that the extent of the country in question is considerable, that the coal resources are highly localized, and the density of generating capacity not very great.

The first condition, the dispensability of railroads if coal haulage is eliminated, is debatable. To be sure, coal makes up a large fraction of rail tonnage in an industrial country, but there are other goods that must be hauled in bulk to manufacturing plants or to markets. Perhaps the best that could be hoped for would be to restrict the railroad system to a coarse network of trunk lines, depending upon roads, trucks, and water courses to connect the system with all cities but the largest and with rural areas. Even here, a net saving of only about $50,000 per mile could be anticipated, since the investment in highways would probably be in the neighborhood of 50% of the investment in railroads.

The second condition is satisfied to a greater or lesser degree in most backward countries. India is quite well supplied with railroads, except for the southern peninsula, but China is quite deficient in the interior, while Brazil and Mexico have large areas without railroads.

The third condition is present only in countries of considerable size, and it is easy to see why the great distances in Russia have made the problem particularly acute there. Even if the large part of the heavy industry were located near coal deposits (and the proximity of iron ore to coal in both India and China makes this the most sensible solution in those two countries) there would still be the problem of supplying power plants not so located. Hence, even if the capital saving could not be realized for the former, it might be advantageous to build atomic plants to serve the remainder of the country. Whether it would be generally advantageous to disperse industries other than iron and steel in this manner will be examined in the next section of this chapter.

The important backward areas of the world where distances are considerable, and where the indirect capital saving might be significant, include, in addition to Russia, western and northern India, the interior of China, the interior of Brazil, the interior of Mexico, and the southern part of Africa.

The smaller distances of Eastern Europe make the prospective savings there less considerable. Since, as we pointed out earlier, successful industrialization may involve a race between capital accumulation and population increase, even moderate reductions or postponements in capital requirements cannot be ignored.

When we consider atomic energy in competition with hydroelectric energy, the possibilities of capital savings appear to be somewhat more favorable. The investment per kilowatt of generating capacity in hydroelectric plants in the United States ranges from $95 to $479 with an average of $175.[37] In countries like India, with highly seasonal rainfall, the capital requirement may approach the higher figure. A 300-mile high-voltage transmission line (about the longest distance presently attainable) for a 300,000 kilowatt station might cost $21 million, or $70 per kilowatt.[38] Hence the total capital cost might range from $95 to $550 per kilowatt. Comparing this with the intermediate estimate of $215 for the atomic plant, we find a maximum capital saving of $335 per kilowatt, $218 per $1,000 of capital in manufacturing, or about 18% of the total investment. While this is not a negligible saving, it must be remembered that the estimate given is a maximum, and that practically all doubts have been resolved in favor of the atomic plant.

We should also note that to plan and build railroads and mines requires not only capital but also time before the production and delivery of energy can begin. This may not be true to the same extent of an atomic plant, once the technology of atomic power is well established. But it is difficult to set a value on this particular advantage of atomic over conventional power resources.

2. Regional Development and Atomic Energy

Thus far the discussion has been concerned largely with the industrialization of whole nations. There remains an important question of where *within* these nations the industrialization is to take place. Historically, the seaports have been the foci of industrialization, both because these were the points of easiest access to western trade, and also because industrialization often depended upon the importation of coal for power. If private investment were to be the principal source of capital for postwar industrialization, there is little doubt that this condition would continue. The hinterlands of many backward countries, and none more than China, have as yet only very tenuous relations with the world economy. The new industrial populations of the coastal cities have been largely furnished by migration from the agricultural hinterland.

[37] Barrows, *op. cit.,* p. 690.
[38] Bauer and Gold, *op. cit.,* p. 50.

However far the industrialization of the coastal sectors of backward countries proceeds, this industrialization will probably have only a moderate effect upon the subsistence economy of the interior.   It is entirely possible, of course, for an agricultural area to become a part of a modern industrial economy without itself being industrialized, a prime example being the western plains of the United States.   On the other hand, a cheap interchange of products between the industrial areas and the hinterland must at least be possible.   Once this condition is met, through the development of an adequate transportation network, there arises the further question of whether some relocation of industry to the agricultural regions would not be advantageous.

In countries which propose to industrialize by plan—and many backward countries do—the question must be fairly met whether it is preferable to concentrate the industry in a few areas, and let the effects of industrialization filter into the rest of the economy, or whether a deliberate attempt should be made to distribute the new industry widely in agricultural areas.   Rather than bring the whole nation into the orbit of the world economy, a more modest aim for the larger backward countries might be to create regional economies with local industries to process local agricultural products for local consumption, and to supply the simpler capital requirements—tools, machinery, and fertilizer—of agriculture.   This type of regional approach has entered into many of the plans for industrialization and for "Valley Authorities."

It has often been noted that much manufacturing activity is, under any circumstances, market oriented.   The development of industry in close geographical proximity to agriculture might (a) simplify the migration of the "surplus" population, (b) provide seasonal factory work for farmers and farm labor, (c) ease the transition of the displaced handicraft worker to the new economy, (d) facilitate efforts to increase agricultural productivity by producing necessary equipment, and (e) accelerate the acculturation of the agricultural population to an industrialized society.   When overall social costs, rather than costs to the individual firm, are considered, there may very well be major advantages in favor of a policy of this kind.[39]

[39] We cannot do better at this point than to quote the comments of two Chinese sociologists:
   "Historically, it is true that the industrial revolution was achieved through the concentration of machine equipment and of population.   The improvement in technology has been, so far, largely parallel with the development of urban centers.   However, this was mainly due to the employment of steam power at the first stage of industrial development.   When electric power was introduced, the trend toward concentration of industry changed. . . .
   "A decentralized pattern for China's new industry is suggested, mainly in order to improve the people's livelihood. . . . such industry must be confined to the manufacturing of consumers' goods.   For heavy industry a concentrated plant is necessary.   Therefore, another problem arises.   What kind of industry, heavy or light, should be

272 INDUSTRIALIZATION OF BACKWARD AREAS [XIV.

What would be the role of atomic power in facilitating the decentralization of industry in countries that determine upon this course? Possible savings in capital requirements have already been examined. In addition, atomic power might be brought into areas that would otherwise not have available power from any source at a reasonable cost: i.e. areas that (a) do not have local supplies of coal, oil, or gas; (b) are not within easy range of hydroelectric developments; and (c) do not have access (by reason of distance, cost, or foreign exchange difficulties) to imported supplies of cheap coal or oil.

There are not many areas of the world that meet all these criteria, but among them are several of real importance: Pakistan, central and southern India, and the east coast of South America, particularly Argentina. (China's coal deposits are widely dispersed throughout that country.)

In both of the areas mentioned there are considerable possibilities for hydroelectric development, but the seasonality of river flows (in India) and considerable transmission distances will make the costs high. In the absence of atomic power, it is likely that industrialization in India will be highly concentrated in the provinces around Calcutta because of the concentration there of coal and iron. Madras and Bombay are connected by rail with Calcutta, but the distance is a thousand miles, and even ocean transport of fuel would be costly. The interior of the peninsula is poorly served by railroads, and is an area whose agriculture could greatly benefit from irrigation. There are some possibilities for hydroelectric schemes in this area, but several of the most favorable power sites have already been developed to supply power to Bombay.

In the past, industrialization in Argentina and Brazil has been concentrated almost entirely in a few coastal cities. There are a number of potential waterpower sites within a few hundred miles of Rio de Janeiro, but almost none within transmission distance of Buenos Aires. These coastal cities have depended largely on imported coal and are sorely pressed by occurrences like the British coal shortage. Large-scale hydroelectric projects have been proposed for the western parts of both Argentina and Brazil, but any considerable industrialization in the west would involve large population shifts and probably extensions of the railroad networks. Even if these projects are undertaken, atomic power at moderate rates would be very welcome in the port cities and in the plains east of the mountains. But the reduction within Brazil of the rich Minas Geraes iron ore might be undertaken also with local hydroelectric power.

---

emphasized in China in the immediate years after the war? . . .

"[This] depends upon . . . the international order. . . . if the postwar world is again to be governed by power alone, China will have no alternative but to give first place to heavy industry and armaments." Fei and Chang, *op. cit.*, pp. 308–312.

We have attempted to illustrate the extent to which atomic power might facilitate the industrialization of backward areas. We have seen that a technological scarcity of energy resources is an important limitation upon industrialization in Japan and South America, and in large parts of India, but not in China or the islands of Indonesia. The latter countries appear to have ample fuel supplies, and even in South America and India hydro-electric developments may supply part of the necessary power.

On the contrary, the chief limit upon the possible rate of industrialization in most backward areas appears to be the scarcity of capital. This is particularly significant since it is only by means of a rather rapid industrialization that these countries can hope to escape the pressure of a growing population against limited agricultural resources. The use of atomic power might reduce the capital required for industrialization by an amount which tentative estimates show might be significant.

If atomic power were available, industrialization might be less closely tied than otherwise to the principal coal deposits and waterpower. Whether this would be advantageous in the long run depends upon a number of intangible economic and sociological factors.

REFERENCES

AKERS, SIR WALLACE, "Metallurgical Problems Involved in the Generation of Useful Power from Atomic Energy," *Journal of the Institute of Metals*, Vol. 73, July 1947, pp. 667–680.

"Aluminum Plants and Facilities," *Report of the Surplus Property Board to Congress, September 21, 1945*, Washington, Government Printing Office, 1945, 131 pp.

American Iron and Steel Institute, *Annual Statistical Report, 1940*, New York, 112 pp.; *1945*, 130 pp.; *1947*, 192 pp.

————, "Steel Facts," New York, April 1947, 8 pp.

"Atomic Energy, Its Future in Power Production," *Chemical Engineering*, Vol. 53, October 1946, pp. 125–133.

"Atomic Energy 1949," *Business Week*, No. 1026, April 30, 1949, pp. 67–74.

AYRES, EUGENE, "The Fuel Problem," *Scientific American*, Vol. 181, December 1949, pp. 32–39.

BACHER, R. F., "The Development of Nuclear Reactors," Speech before the American Academy of Arts and Sciences, Boston, Mass., February 9, 1949, AEC Press Release, 12 pp.

BACHER, R. F., and R. P. FEYNMAN, "Introduction to Atomic Energy," in *The International Control of Atomic Energy*, U. S. Department of State, Publication 2661, Washington, Government Printing Office, 1946, pp. 9–23.

BAIN, H. FOSTER, and T. T. READ, *Ores and Industry in South America,* Council on Foreign Relations, New York, Harper and Bros., 1934, 381 pp.

BARROWS, H. K., "Hydro-Generated Energy," *Proceedings of the American Society of Civil Engineers*, Vol. 64, April 1938, Part I, pp. 675–701.

BAUER, JOHN, and NATHANIEL GOLD, *The Electric Power Industry*, New York, Harper and Bros., 1939, 347 pp.

BAYKOV, ALEXANDER, *The Development of the Soviet Economic System*, Cambridge, England, Cambridge University Press, 1946, 514 pp.

BELL, ROSCOE E., "Transportation Costs of Sulphur," Pacific Northwest Coordination Committee, April 1948, U. S. Department of the Interior (mimeographed release), 4 pp.

BELL, ROSCOE E., and DONALD T. GRIFFITH, "Transportation Costs as They Affect New Phosphorus Industries in the West," Western Phosphate Fertilizer Program, June 1947, U. S. Department of the Interior (mimeographed release), 25 pp.

BLISS, CHARLES A., *The Structure of Manufacturing Production*, New York, National Bureau of Economic Research, 1939, 231 pp.

BOGOLEPOV, M., "On Sales Prices in Industry," *Planned Economy*, 1936, No. 5, pp. 76–77.

BORST, LYLE B., "Industrial Application of Nuclear Energy," *The Commercial and Financial Chronicle*, Vol. 167, March 4, 1948, pp. 4, 32.

"Britain's Atomic Pile," *Discovery*, Vol. 9, August 1948, p. 233.

British Information Services, *Labor and Industry in Britain*, Vol. 5, September–October 1947.

Business Week, No. 958, January 10, 1948, pp. 97–98; No. 1062, January 7, 1950, p. 26.

CAMP, J. M., and C. B. FRANCIS, *Making, Shaping and Treating of Steel*, fifth edition, Pittsburgh, Carnegie-Illinois Steel Corporation, 1940, 1440 pp.

Chemical Engineering, Vol. 55, February 1948, p. 107.

CHEYNEY, EDWARD P., *An Introduction to the Industrial and Social History of England*, New York, The Macmillan Co., 1910, 317 pp.

Chicago Plan Commission, *Master Plan of Residential Land Use of Chicago*, Chicago, 1943, 134 pp.

China Handbook, 1937–1943, Compiled by the Chinese Ministry of Information, New York, The Macmillan Co., 1943, 876 pp.

CLARK, COLIN, *The Conditions of Economic Progress,* London, Macmillan and Co., 1940, 504 pp.
————, *The Economics of 1960,* London, Macmillan and Co., 1943, 118 pp.
"Coal," *Fortune,* Vol. 35, March 1947, pp. 85–99, 202–206.
"Coal: The Fuel Revolution," *Fortune,* Vol. 35, April 1947, pp. 99–105, 238–254.
"Coal: The 'Pitt Consol' Adventure," *Fortune,* Vol. 35, July 1947, pp. 99–105, 134–147.
Coal Resources of the World, Study sponsored by the Executive Committee of the Twelfth International Geological Congress, Toronto, Canada, 1913, W. McInnes *et al.* (eds.), Toronto, Morang and Co., 1913, 3 vols., 1266 pp.
COCKCROFT, J. D., Essay on "Nuclear Reactors," in *An International Bibliography on Atomic Energy,* United Nations Atomic Energy Commission, Vol. II, preliminary edition, February 14, 1949 (not paged).
Columbia River and Minor Tributaries, Prepared by the U. S. Army Corps of Engineers, Published by the 73rd Congress, 1st Session, as House Document No. 103, Washington, Government Printing Office, 1933–1934, Vol. I, 564 pp.
"Cost of Energy Generation, Second Symposium on Power Costs," *Proceedings of the American Society of Civil Engineers,* Vol. 64, April 1938, Part I, pp. 638–735.
CURTIS, H. A., A. M. MILLER, and J. N. JUNKINS, "TVA Estimates Favorable Costs for Concentrated Superphosphate," *Chemical and Metallurgical Engineering,* Vol. 43, November 1936, pp. 583–587; December 1936, pp. 647–650.
DALY, M. C., "An Approximation to a Geographical Multiplier," *Economic Journal,* Vol. 50, June–September 1940, pp. 248–258.
DAVIDSON, W. F., "Atomic Power and Fuel Supply," *Atomics,* Vol. 5, October 1949, pp. 4–14.
————, "Nuclear Energy for Power Production," Paper delivered at the Fuel Economy Conference of the World Power Conference, The Hague, 1947 (conference proof), 12 pp.
Demographic Studies of Selected Areas of Rapid Growth, New York, Milbank Memorial Fund, 1944, 158 pp.
DEWHURST, J. FREDERIC, "Foreign Trade and Investment," in *America's Needs and Resources,* J. F. Dewhurst *et al.* (eds.), New York, Twentieth Century Fund, 1947, pp. 511–535.
"Diesel Locomotives," *The Lamp,* Vol. 29, June 1947, pp. 2–7.
The Economist, Vol. 157, September 10, 1949, p. 561.
Edison Electric Institute Bulletin, Vol. 14, July 1946, p. 233.
Edison Electric Institute, *Statistical Bulletin, 1942,* New York, 40 pp.; *1946,* 36 pp.; *1947,* 39 pp.; *1948,* 38 pp.
Electric Power Development in the U.S.S.R., Prepared for the Third World Power Conference, Washington, D.C., 1936, by the scientific staff of the Krzizhanovsky Power Institute of the Academy of Sciences of the U.S.S.R., Moscow, INRA Publishing Society, 1936, 496 pp.
Electrical Merchandising, Vol. 77, January 1947.
Engineering and Mining Journal, Vol. 149, May 1948, p. 106; Vol. 149, June 1948, p. 68.
ENGLE, N. H., H. E. GREGORY, and R. MOSSÉ, *Aluminum,* Chicago, Richard D. Irwin, Inc., 1945, 404 pp.
EZEKIEL, MORDECAI, "Industrial Possibilities Ahead," in *Towards World Prosperity,* M. Ezekiel (ed.), New York, Harper and Bros., 1947, pp. 14–29.
FEI, HSIAO-TUNG, and CHIH-I CHANG, *Earthbound China,* Chicago, The University of Chicago Press, 1945, 319 pp.
FEILER, A., and J. MARSCHAK (eds.), *Management in Russian Industry and Agriculture,* by G. Bienstock, S. Schwarz, and A. Yugow, New York, Oxford University Press, 1944, 198 pp.
FIES, M. H., and JAMES L. ELDER, "Laboratory and Field-Scale Experimentation in the Underground Gasification of Coal," Paper prepared for the United Nations Scientific Conference on the Conservation and Utilization of Resources, Lake Success, N.Y., May 25, 1949, 19 pp.
FURNIVALL, J. S., *Netherlands India,* Cambridge, England, Cambridge University Press, 1944, 502 pp.

GATZKE, P., "Electric Heating Systems in Furnaces for Ceramic Firing," *Ceramic Industry*, Vol. 27, October 1936, pp. 267–269.

GILLILAND, E. R., "Heat Transfer," in *The Science and Engineering of Nuclear Power*," Clark Goodman (ed.), Cambridge, Mass., Addison-Wesley Press, 1947, Vol. I, Chapter X.

"The Glass Industry," Prepared by *The American Glass Review* for *The Development of American Industries,* J. G. Glover and W. B. Cornell (eds.), New York, Prentice-Hall, Inc., 1946, pp. 469–485.

GOODMAN, CLARK, "Distribution of Uranium, Thorium and Beryllium Ores and Estimated Costs," Lecture notes (No. 26) at Clinton National Laboratories, 1947 (unclassified), 7 pp.

——, "Future Developments in Nuclear Energy," *Nucleonics*, Vol. 4, February 1949, pp. 2–16.

GORDON, LINCOLN, "Power and Fuels," in *Industrial Location and National Resources,* National Resources Planning Board, Washington, Government Printing Office, 1943, pp. 156–180.

GOULD, J. M., *Output and Productivity in the Electric and Gas Utilities, 1899–1942,* New York, National Bureau of Economic Research, 1946, 195 pp.

GREGORY, J. S., and D. W. SHAVE, *The U.S.S.R., A Geographical Survey,* London, George G. Horeb, 1944, 636 pp.

GUSTAFSON, J. K., "Uranium Resources," Speech delivered before a Metallurgic Colloquium, Massachusetts Institute of Technology, Cambridge, Mass., March 9, 1949, AEC Press Release, 14 pp.

HAMMOND, J. L., and BARBARA HAMMOND, *The Rise of Modern Industry,* New York, Harcourt, Brace and Co., 1926, 281 pp.

HARTSHORNE, RICHARD, "Location Factors in the Iron and Steel Industry," *Economic Geography*, Vol. 4, July 1928, pp. 241–252.

"The Heat Pump," *Electrical Engineering*, Vol. 67, April 1948, pp. 338–348.

HIRSHFELD, C. F., and R. M. VAN DUZER, JR., "Heat-Generated Energy," *Proceedings of the American Society of Civil Engineers*, Vol. 64, April 1938, Part I, pp. 647–674.

HOBSON, HAROLD, "Integration of Electric Utilities in Great Britain," *Transactions of the Third World Power Conference,* Washington, Government Printing Office, 1938, Vol. VII, pp. 615–642.

"How to Heat a House," *Fortune*, Vol. 38, September 1948, pp. 109–113.

HUBBARD, G. E., *Eastern Industrialization and Its Effect on the West,* London, Oxford University Press, 1935, 395 pp.

International Bank for Reconstruction and Development, Releases No. 126 and 162, January 27 and December 14, 1949.

IRVINE, JOHN W., JR., "Heavy Elements and Nuclear Fuels," in *The Science and Engineering of Nuclear Power,* Clark Goodman (ed.), Cambridge, Mass., Addison-Wesley Press, 1947, Vol. I, Chapter XI.

ISARD, WALTER, and JOHN B. LANSING, "Comparisons of Power Cost for Atomic and Conventional Steam Stations," *The Review of Economics and Statistics,* Vol. 31, August 1949, pp. 217–228.

IYENGAR, S. KESAVA, "Industrialisation and Agriculture in India Post-War Planning," *Economic Journal,* Vol. 54, June–September 1944, pp. 189–205.

JATHAR, G. B., and S. G. BERI, *Indian Economics,* third edition, London, Cambridge University Press, 1931, 2 vols.

JEFFRIES, ZAY, "Metals and Alloys of the Future," *American Metal Market,* Vol. 54, March 28, 1947, pp. 7, 12.

KNOWLTON, A. E., "Fourth Steam Station Cost Survey," *Electrical World,* Vol. 112, December 2, 1939, pp. 51–66.

KOWARSKI, L., "Atomic Energy Developments in France," *Bulletin of the Atomic Scientists,* Vol. 4, May 1948, pp. 139–140, 154–155.

KRIEG, EDWIN H., "Progressive Engineering Offsets Increased Costs of Steam-Electric Power Generation," *Civil Engineering,* Vol. 18, April 1948, pp. 14–18.

KUZNETS, SIMON, *Seasonal Variations in Industry and Trade,* New York, National Bureau of Economic Research, 1933, 455 pp.

LACHMANN, KURT, "The Balkan Countries," in *Towards World Prosperity*, Mordecai Ezekiel (ed.), New York, Harper and Bros., 1947, pp. 173–185.

League of Nations, *Statistical Yearbook, 1942–44*, Geneva, 1945, 315 pp.

LEVERETT, M. C., "Some Engineering and Economic Aspects of Nuclear Energy," U. S. Atomic Energy Commission, Declassified Document, MDDC-1304, Oak Ridge, Tenn., December 3, 1948, 10 pp.

LILIENTHAL, DAVID, "Atomic Energy and American Industry," Speech before the Economic Club of Detroit, October 6, 1947, AEC Press Release, 20 pp.

———, Press Conference on "Uranium Supplies," Denver, Colorado, December 17, 1948, AEC Press Release, 6 pp.

———, Testimony in Hearings before the Subcommittee of the Committee on Appropriations, on the Independent Offices Appropriation Bill for 1950, House of Representatives, 81st Congress, 1st Session, Part 1, Washington, Government Printing Office, 1949, pp. 1068–1279.

LOVERING, T. S., *Minerals in World Affairs*, New York, Prentice-Hall, Inc., 1944, 394 pp.

MANDELBAUM, K., *The Industrialisation of Backward Areas*, Monograph No. 2 of the Oxford Institute of Statistics, Oxford, Basil Blackwell, 1945, 111 pp.

MARTIN, T. M. C., "Railway Motive Power for the Pacific Northwest," U. S. Department of the Interior, Bonneville Power Administration, August 1946, 36 pp.

MENKE, J. R., "Nuclear Fission as a Source of Power," *Econometrica*, Vol. 15, October 1947, pp. 314–334.

MERCIER, ERNEST, "National Power and Resources Policies," *Transactions of the Third World Power Conference*, Washington, Government Printing Office, 1938, Vol. IX, pp. 93–118.

METZLER, LLOYD A., "Exchange Rates and the International Monetary Fund," in *International Monetary Policies*, Board of Governors of the Federal Reserve System, Postwar Economic Studies No. 7, Washington, September 1947, pp. 1–45.

MIKAMI, H. M., "World Iron-Ore Map," *Economic Geology*, Vol. 39, January-February 1944, pp. 1–24.

MIKHAILOV, N., *Land of the Soviets*, New York, Lee Furman and Co., 1939, 351 pp.

MILES, FRED C., "Locomotives Ordered and Built in 1946," *Railway Age*, Vol. 122, January 4, 1947, pp. 110–113.

MOMENT, SAMUEL, Unpublished manuscript on the aluminum industry, prepared for the Columbia Basin Study of estimated industrial development and power requirements in the Pacific Northwest, Bonneville Power Administration, August 15, 1947, 42 pp.

National District Heating Association, *Handbooks*, New York (published yearly).

———, *Proceedings*, Vol. 19, 1928, p. 60; Vol. 30, 1939, p. 100; Vol. 34, 1943, p. 76.

National Electric Light Association, *Statistical Bulletin, 1931*, New York, July 1932, 17 pp.

National Industrial Conference Board, *The Economic Almanac for 1949*, New York, 1948, 560 pp.

New York Times, September 18, 1947, p. 27, col. 1; June 15, 1948, p. 23, col. 1; September 11, 1949, Section III, p. 1, col. 6; November 29, 1949, p. 9, col. 3; December 1, 1949, p. 17, col. 2; March 8, 1950, p. 24, col. 2.

NORDBERG, BROR, "Sweden's Modern Cement Plant," *Rock Products*, Vol. 49, January 1946, pp. 78–83.

OHLIN, BERTIL, *Interregional and International Trade*, Cambridge, Mass., Harvard University Press, 1933, 617 pp.

ORCHARD, JOHN E., *Japan's Economic Position*, New York, Whittlesey House, McGraw-Hill Book Co., 1930, 504 pp.

ORROK, GEORGE A., *Proceedings of the American Society of Civil Engineers*, Vol. 63, December 1937, pp. 1884–1892.

PENROD, E. B., "A Review of Some Heat-Pump Installations," *Mechanical Engineering*, Vol. 69, August 1947, pp. 639–647.

PEP (Political and Economic Planning), *The British Fuel and Power Industries*, London, October 1947, 406 pp.

PIKE, SUMNER, "The Work of the United States Atomic Energy Commission," Speech before Cooper Union Forum, New York City, January 13, 1948, AEC Press Release, 13 pp.

————, "Atomic Energy in Relation to Geology," Speech at the Annual Meeting of the American Association of Petroleum Geologists and other organizations, St. Louis, Missouri, March 16, 1949, AEC Press Release, 8 pp.

PRICHARD, LEWIS G., HENRY O. PARSONS, ROBERT R. STEWART, and ROSCOE E. BELL, Unpublished manuscript on the phosphate fertilizer industry, prepared for the Columbia Basin Study of estimated industrial development and power requirements in the Pacific Northwest, Bonneville Power Administration, June 1947, 36 pp.

PRYCE, M. H. L., "Atomic Power: What Are the Prospects?" Bulletin of the Atomic Scientists, Vol. 4, August 1948, pp. 245–248.

RAMSEYER, C. F., "Sponge Iron—Its Possibilities and Limitations," Iron and Steel Engineer, Vol. 21, July 1944, pp. 35–44, 72.

ROPER, E. C., "The Union of Soviet Socialist Republics," in Towards World Prosperity, Mordecai Ezekiel (ed.), New York, Harper and Bros., 1947, pp. 186–214.

RYAN, WILLIAM F., "Power Plant Construction Costs and Implications," Power Plant Engineering, Vol. 51, August 1947, pp. 116–120.

SALZMAN, M. G., "Design, Construction and Operation Control Rising Costs of Hydro Power," Civil Engineering, Vol. 18, April 1948, pp. 23–27.

SCHURR, SAM H., "Atomic Power in Selected Industries," Harvard Business Review, Vol. 27, July 1949, pp. 459–479.

————, Discussion on "The Social and Economic Significance of Atomic Energy," with comments by Philip Sporn and Jacob Marschak, American Economic Review, Papers and Proceedings, Vol. 37, May 1947, pp. 110–117.

SIMON, HERBERT A., "Effects of Increased Productivity Upon the Ratio of Urban to Rural Population," Econometrica, Vol. 15, January 1947, pp. 31–42.

SMYTH, H. D., Atomic Energy for Military Purposes, Princeton, Princeton University Press, 1946, 308 pp.

SPEDDING, F. H., "Chemical Aspects of the Atomic Energy Problem," Bulletin of the Atomic Scientists, Vol. 5, February 1949, pp. 48–50.

SPORN, PHILIP, "Cost of Generation of Electric Energy," Proceedings of the American Society of Civil Engineers, Vol. 63, December 1937, pp. 1925–1935.

————, Discussion on "The Social and Economic Significance of Atomic Energy," American Economic Review, Papers and Proceedings, Vol. 37, May 1947, pp. 110–115.

————, "The Sixth Ingredient—The Heat Pump," Edison Electric Institute Bulletin, Vol. 12, August 1944, pp. 240–247, 272.

STANSFIELD, ALFRED, The Electric Furnace for Iron and Steel, New York, McGraw-Hill Book Co., 1923, 453 pp.

STROCK, CLIFFORD, Heating and Ventilating's Engineering Data Book, New York, The Industrial Press, 1948.

SVIRSKY, LEON, "The Atomic Energy Commission," Scientific American, Vol. 181, July 1949, pp. 30–43.

SZILARD, LEO, Testimony in Hearings before the Special Committee on Atomic Energy, U. S. Senate, 79th Congress, 2nd Session, Part 2, Washington, Government Printing Office, 1946, pp. 267–300.

SZILARD, LEO, and W. H. ZINN, "Emission of Neutrons by Uranium," Physical Review, Vol. 56, October 1, 1939, pp. 619–624.

"Technological Control of Atomic Energy Activities," in The International Control of Atomic Energy, U. S. Department of State, Publication 2661, Washington, Government Printing Office, 1946, pp. 143–195.

Tennessee Valley Authority, Annual Report, 1944, Washington, Government Printing Office, 180 pp.

THAKURDAS, P., et al., A Plan of Economic Development for India, New York, Penguin Books, 1945, 105 pp.

THELANDER, T., "Some Results of Electric Traction on the Swedish State Railways," Paper presented at the Fuel Economy Conference of the World Power Conference, The Hague, 1947 (conference proof), 12 pp.

THOMAS, C. A., "Nuclear Power," in The International Control of Atomic Energy, U. S. Department of State, Publication 2661, Washington, Government Printing Office, 1946, pp. 121–127.

TROXEL, EMERY, *Economics of Public Utilities*, New York, Rinehart and Co., 1947, 892 pp.

United Nations, Atomic Energy Commission Group, *An International Bibliography on Atomic Energy*, Vol. I, Lake Success, New York, 1949, 45 pp.; Vol. II, preliminary edition, February 14, 1949 (not paged).

———, Department of Economic Affairs, *A Survey of the Economic Situation and Prospects of Europe*, Geneva, 1948, 206 pp.

———, *Statistical Yearbook, 1949*, Lake Success, New York, 1949, 482 pp.

U. S. Atomic Energy Commission, "AEC Reactor Safeguard Committee to Visit United Kingdom," AEC Press Release, August 31, 1949, 3 pp.

———, "AEC Selects Contractor to Build First Nuclear Reactor at Testing Station in Idaho," AEC Press Release, November 28, 1949, 5 pp.

———, *Atomic Energy Development, 1947–1948*, Fifth Semiannual Report, Washington, Government Printing Office, 1949, 213 pp.

———, "Major Classes of Uranium Deposits," AEC Press Release, December 9, 1948, 2 pp.

———, Press Release, November 10, 1949, p. 1.

———, *Third Semiannual Report*, Washington, Government Printing Office, 1948, 49 pp.

———, *Fourth Semiannual Report*, Washington, Government Printing Office, 1948, 192 pp.

———, Seminar on the Disposal of Radioactive Wastes, January 24–25, 1949, "Digest of Proceedings," AEC Press Release, January 30, 1949, 56 pp.

U. S. Bureau of the Census, *Census of 1940, Population, Number of Inhabitants*, Vol. I, Washington, Government Printing Office, 1942, 1236 pp.

———, *Census of 1940, Mineral Industries, 1939*, Vol. I, Washington, Government Printing Office, 1944, 876 pp.

———, *Census of Manufactures, 1939*, Washington, Government Printing Office, 1942, 3 vols.

———, *Census of Manufactures, 1947*, Preliminary report on "Fuels and Electric Energy Consumed," Washington, November 18, 1949, 5 pp.; Section on "Industrial Inorganic Chemicals," Washington, Government Printing Office, 1949, 8 pp.

———, *Statistical Abstract of the United States, 1947*, Washington, Government Printing Office, 1038 pp.; *1948*, 1054 pp.

U. S. Bureau of Labor Statistics, "Full Employment Patterns, 1950," Washington, May 1946, Appendix A.

———, *Handbook of Labor Statistics*, 1947 edition, Washington, Government Printing Office, 1948, 220 pp.

———, *Monthly Labor Review*, Vol. 69, November 1949.

U. S. Bureau of Mines, *Coal Mining in Europe*, by G. S. Rice and I. Hartmann, Bulletin 414, Washington, Government Printing Office, 1939, 369 pp.

———, *Foreign Minerals Surveys* (published irregularly).

———, *Minerals Yearbook, 1940*, Washington, Government Printing Office, 1514 pp.; *1945*, 1689 pp.; *1946*, 1629 pp.; *1947*, Section on "Uranium, Radium and Thorium" (preprint), pp. 1199–1216; *1948*, Section on "Coal—Bituminous and Lignite" (preprint).

———, "Production of Sponge Iron," by C. E. Williams, E. P. Barrett, and B. M. Larsen, Bulletin 270, Washington, Government Printing Office, 1927, 175 pp.

———, "Recovery of Aluminum from Kaolin by the Lime-Soda Sinter Process," by Frank J. Cservenyak, R. I. 4069, Washington, May 1947, 59 pp.

———, "Use of Sponge Iron in Steel Production," by R. C. Buehl, M. B. Royer, and J. P. Riott, R. I. 4096, Washington, July 1947, 74 pp.

U. S. Department of Agriculture, "Double Superphosphate," by A. L. Mehring, Circular No. 718, Washington, Government Printing Office, December 1944, 24 pp.

U. S. Department of Commerce, *Foreign Commerce Weekly*, Vol. 19, May 12, 1945, pp. 10–13, 47–50.

———, *Foreign Commerce Yearbook, 1937*, Washington, Government Printing Office, 1938, 413 pp.

———, *Survey of Current Business*, Vol. 29, 1949.

U. S. Department of State, *Energy Resources of the World;* preliminary edition, 1945; Publication 3428, Washington, Government Printing Office, 1949, 128 pp.

————, *The International Control of Atomic Energy, Scientific Information Transmitted to the United Nations Atomic Energy Commission,* Publication 2661, Washington, Government Printing Office, 1946, 195 pp.

————, *A Report on the International Control of Atomic Energy,* Publication 2498, Published as House Document No. 709, 79th Congress, 2nd Session, Washington, Government Printing Office, March 16, 1946, 61 pp.

U. S. Federal Coordinator of Transportation, *Freight Traffic Report,* Washington, 1935, Vol. III, 287 pp.

U. S. Federal Power Commission, "Consumption of Fuel for Production of Electric Energy, 1946," Washington, 17 pp.

————, Natural Gas Investigation, "Problems of Long-Distance Transportation of Natural Gas," Docket No. G-580, Staff Report, Washington, November 1947, 128 pp.

————, Natural Gas Investigation, *Testimony of Commissioners Nelson Lee Smith and Harrington Wimberly,* Docket No. G-580, Washington, Government Printing Office, 1948, 498 pp.

————, *Power Requirements in Electrochemical, Electrometallurgical, and Allied Industries,* Washington, Government Printing Office, 1938, 125 pp.

————, *Steam-Electric Plant Construction Cost and Annual Production Expenses, 1938–1947,* Washington, 406 pp.

————, *The Use of Electric Power in Transportation,* Power Series No. 4, Washington, Government Printing Office, 1936, 61 pp.

U. S. Interstate Commerce Commission, *Comparative Statement of Railway Operating Statistics, 1946 and 1945,* Washington, Government Printing Office, 1947, 63 pp.

————, Ex Parte No. 162, No. 148, "Increased Railway Rates, Fares, and Charges," December 5, 1946, in *I.C.C. Reports,* Washington, Government Printing Office, Vol. 266, pp. 537–625.

————, "Monthly Comment on Transportation Statistics" (mimeographed); March 11, 1948, 19 pp.; August 15, 1949, 19 pp.; October 11, 1949, 17 pp.

————, "Postwar Capital Expenditures of the Railroads," Washington, Government Printing Office, March 1947, 75 pp.

————, *Statistics of Railways in the United States, 1939,* Washington, Government Printing Office, 583 pp.; *1945,* 608 pp.

U. S. National Resources Committee, *Energy Resources and National Policy,* Washington, Government Printing Office, 1939, 435 pp.

————, *The Structure of the American Economy,* Washington, Government Printing Office, 1939, Part I, 396 pp.

U. S. Tariff Commission, *Flat Glass and Related Glass Products,* Report No. 123, Second Series, Washington, Government Printing Office, 1937, 277 pp.

————, *Report on Cement,* Report No. 38, Second Series, Washington, Government Printing Office, 1932, 26 pp.

————, *Window Glass,* Washington, Government Printing Office, 1929, 44 pp.

WAGNER, C. F., and J. A. HUTCHESON, "Nuclear-Energy Potentialities," *Westinghouse Engineer,* Vol. 6, July 1946, pp. 125–127.

Wall Street Journal, October 20, 1948, p. 1, col. 6.

WEST, M. E., *Brick and Tile,* W.P.A. National Research Project, Report No. N-2, Philadelphia, 1939, 212 pp.

"The Western Steel Industry," *Utah Economic and Business Review,* Vol. 3, June 1944, p. 19.

WHEELER, J. A., "The Future of Nuclear Power," *Mechanical Engineering,* Vol. 68, May 1946, pp. 401–405.

————, "Inspection of Manufacturing Processes: Possibility of Detection," in *Reports and Discussion on Problems of War and Peace in the Atomic Age,* The Committee on Atomic Energy of the Carnegie Endowment for International Peace, New York, 1946, pp. 52–58.

WILSON, CARROLL, Testimony in Hearings before the Subcommittee of the Committee on Appropriations, on the Independent Offices Appropriation Bill for 1950, House of Representatives, 81st Congress, 1st Session, Part 1, Washington, Government Printing Office, 1949, pp. 1068–1279.

WILSON, ROBERT E., "Oil from Coal and Shale," in *Our Oil Resources,* Leonard M. Fanning (ed.), New York, McGraw-Hill Book Co., 1945, pp. 210–229.

World Oil Atlas, 1946, Section 2 of *The Oil Weekly,* Vol. 121, May 20, 1946, p. 29.

World Power Conference, *Power Resources of the World,* London, 1929, 170 pp.

———, *Transactions of the First World Power Conference,* London, 1924, 5 vols.

———, *Transactions of the Third World Power Conference,* Washington, D.C., 1936, 10 vols.

YAWORSKI, N., V. SPENCER, *et al., Fuel Efficiency in Cement Manufacture, 1909–1935,* W.P.A. National Research Project, Report No. E-5, Philadelphia, 1938, 92 pp.

YELLOTT, JOHN I., "The Coal-Burning Gas Turbine Locomotive," *The Analysts Journal,* Vol. 3, No. 1, 1947, pp. 18–31.

INDEX OF NAMES

SUBJECT INDEX

Agriculture, 89

Aluminum; bauxite requirements for production of, 105; cost of production of, 107; cost reductions through locational change in production of, 109, 110, 111, 115, 116; cost of transportation to market of, 110, 115; cost of transporting raw materials for production of, 107, 108, 109, 115; energy cost reductions in production of through atomic power, 108, 111, 115, 116, 117, 118; future power costs in production of, 112–113, 116, 117–118; growth in production of, 111–112; importance of energy cost in production of, 87, 106, 107, 108, 116; influence of energy in location of production of, 86, 105, 106; location of bauxite for production of, 105, 108; locational change of production site of through atomic power, 108–110, 115, 116; locational flexibility of production site of through atomic power, 114; material requirements for production of, 106; natural gas as energy source for production of, 113, 114; power rates needed for locational change of production site of, 109, 110, 113, 115; power required for growth in production of, 112, 115, 116; production of in different countries, 117; production and market locations of, compared, 109–110; production of from ores other than bauxite, 114, 115; production processes of, 105–106; production of in the U.S.S.R., 117, 118; significance of in industrialization, 265–266

Atomic energy control; Acheson-Lilienthal plan for, 36–39; economic implications of, 35–39; effect of on breeding, 38; effect of on cost, 35, 38, 39; effect of on location of atomic power plants, 36; effect of on size of atomic power industry, 35–36, 37; effect of on time scale of development, 36

Atomic power cost, 9, 23–33, 89–92; as affected by rate of capacity use, 32, 91; as affected by size of plant, 33; California Report on, 25, 26, 27, 28, 29, 30, 31; changes of over time, 27–28, 29; components of, 31–32, 90–91; conceptual basis of estimates of, 23, 26, 30–33; constancy in, 91–92, 114; depreciation allowance in, 33, 34–35; economies of scale of, 27, 33, 198; estimated minimum of, 14, 24–25, 90; foreign exchange requirements for, 75–77, 85; geographic variations in, 4, 82, 114; importance of fixed charges in, 32, 91–92, 114; and investment in plant, 26–29; maintenance charges in, 35; obsolescence allowance in, 33, 34; rate of fixed charges for, 29, 33–35, 71–72; realizable borderline of, 92; significance of for industry analysis, 90; significance of minimum cost of, 90; Thomas Report on, 25, 26, 27, 28, 29, 30, 31, 34

Atomic power plants; centralization of chemical processing in, 16, 27; cost of chemical processing facilities in, 27; description of, 14; economic implications of chemical processing in, 16, 27; economic significance of required plant and equipment of, 13–18; shielding for, 5; similarity of to conventional power plant, 14, 23; see also Reactors

Atomic power uses, 5–7; for airplanes, 6; for combined heat-power production, 201–202; for direct heat, 7, 139, 159, 199–215; industrial applicability of, 86, 87, 89; for locomotives, 6, 183, 196–198; for mobile units, 6; for naval vessels, 6; in stationary power plants, 6

Atomic powered district heating; approach to economic analysis of, 210–211; cost reductions through, 215; exclusion of commercial and industrial demand for, 214; feasible locations for, 212, 213–214, 215; location of reactor for, 214; necessary population density for, 211–212; in urban redevelopment, 214fn.; weakness of population density figures in selecting feasible locations for, 212, 213, 214

"Backward" areas; industrialization process in, 248–253; potential importance of atomic power in, 220, 261–264, 266–273

Breeding of fissionable substance, 8–10, 12, 13, 21, 37, 38, 39, 197

Brick; feasibility of combined production of with cement, 148; feasibility of electrothermal production of, 144; feasibility of production of in kilns using nuclear heat, 144, 145; fuel consumption in production of, 88; importance of energy

286

SUBJECT INDEX

cost in production of, 87, 144; mechanization of production of, 145fn.; process change in production of, 88; production location of, 144; size of plants for production of, 144–145; size of reactors for use in production of, 145

Brookhaven National Laboratory, 10

Capital accumulation; effects of atomic power on requirements for, 239–246, 270–272; effects of on economic development, 237–238; requirements of for economic development, 256–257

Cement; costs needed for using nuclear heat in production of, 139–141; electrothermal process of production of through atomic power, 138–139, 141, 143; energy requirements for production of, 88, 135, 136, 137, 139, 143; feasibility of combined production of with brick, 145; feasibility of production of in kilns using nuclear heat, 139, 140, 141, 142, 143, 145; importance of energy cost in production of, 87, 135, 136, 137, 142; material requirements for production of, 135–136; methods of using atomic power in production of, 136, 142; power rates needed for electrothermal production of, 137–138; process change in production of, 88; production of in electric kilns, 136–137; production location of, 136, 142; production process for, 135–136; rate of capacity utilization of plants for production of, 140; size of nuclear reactors for use in production of, 141–142; size of plants for production of, 142

Chlorine and caustic soda; cost of transportation of, 120; cost of transporting salt for production of, 120; energy cost reductions in production of through atomic power, 122; energy requirements for production of, 119, 120, 121, 123; importance of energy cost in production of, 87, 119, 121, 122, 123; location of salt for production of, 120; locational change in site of production of through atomic power, 123; material requirements for production of, 120; production location for, 120, 122, 123; production processes for, 119–120; weight loss in production of, 119, 120

Coal; cost of, 21, 43–44, 60–61, 64; cost of transportation of, 23, 60–61, 67; difficulty of recruiting miners of, 67; increasing mechanization of mining of, 68, 69; miners' wages, 66–67; production of, by countries, 61, 62; requirements of for industrialization, 263–264; shifting geographic pattern of production of, 73; subsidization of in the U.S.S.R., 63; use of in electricity generation, 55, 57, 60; see also Fuel, Gasification of coal, and Synthetic liquid fuels

Coal resources; by countries, 262; world distribution of, 61

Copper, 89

Cost comparisons between atomic and conventional power; as affected by capital requirements, 268–269; as affected by difference in cost composition, 65–77; conceptual basis of, 23–24, 26, 30–33, 41–50; see also International cost comparisons between atomic and conventional power

Cost components of atomic and conventional power, 32, 65; trends of, compared, 65–71

Demand for energy; as affected by locational shifts, 224; as affected by process changes, 223; estimation of for U. S., 226–231; in industrialization, 261–263; "trigger effects" of, 232–234

District heating; cost of heat distribution in, 202–208; cost of heat generation in nuclear reactor in, 208–209; cost of service connection in, 211; design temperature range in, 203–204; economies of scale in, 204–205; population density as measure of volume to heat in, 203, 204, 206; rate of capacity utilization in, 205; rate of fixed charges on investment in, 206; relation of annual degree days to, 206; relative costliness of heat distribution in, 208; significance of heat losses in with atomic power, 209–210; size of service area for, 204; variables determining heat requirements of, 206, 207; variables determining required capacity of, 202–205; see also Atomic powered district heating and Residential heating

Economic development; as influenced by industrial effects of atomic power, 84–85, 143, 178–181, 195–196, 264; as result of atomic power, 84–86, 89; capital requirements for, 256–257; effect on real income of, 253–255; importance of rate of, 254–255; limiting factors for, 255–256; supply of capital for, 257–260

Electric power; cost of from conventional energy sources, 55–62; demand for, 227–232; energy sources of, 54–55, 57; percentage of total energy use of, 40; production of, by countries, 55, 56; transmission cost of, 30fn., 92; see also Hydroelectric power and Thermal power costs